RADICAL
AGRICULTURE

*the text of this book is printed
on 100% recycled paper*

# Contributors

George Baker
6513 Orono Court
Springfield, Virginia 22152
Peter Barnes
432 28th Street
San Francisco, California
Wendell Berry
Port Royal, Kentucky 40058
Murray Bookchin
61 Bon Aire Circle, Apt 7156
Suffern, New York, 10901
John Elter
130 Laburnum Crescent
Rochester, New York 14620
Jerry Goldstein
c/o Rodale Press
Emmaus, Pennsylvania 18049
Sheldon Greene
c/o Greene, Kelly, Halloran
345 Franklin
San Francisco, California 94102
Jim Hightower
no current address, currently
campaign director for Fred Harris for President
Last address: c/o Agribusiness
Accountability Project
Washington, D.C.
Nick Kotz
c/o Washington *Post*
1150 15th Avenue N.W.
Washington, D.C. 20005
Don Marier
Route 2, Box 90A
Milaca, Minnesota 56353

William McLarney
c/o New Alchemy Institute
Box 432
Woods Hole, Massachusetts
02543
Darryl McLeod
Dept of Agricultural Economics
University of California
Berkeley, California
Robin Myers
2075 Second Avenue 12E
New York, New York 10029
Bill and Helga Olkowski
1307 Acton
Berkeley, California 94702
Michael Perelman
Dept of Agricultural Economics
California State College
Chico, California
Warren Pierce
c/o Community Environmental
Council
109 E De La Guerra
Santa Barbara, California 93101
Paul Relis
2664 Puesta del Sol
Santa Barbara, California 93105
John Todd
c/o New Alchemy Institute
Box 432
Woods Hole, Massachusetts
02543
Ron Weintraub
3360 W 4th Street
Williamsport, Pennsylvania 17701

# RADICAL
# AGRICULTURE

EDITED BY
## RICHARD MERRILL

**HARPER COLOPHON BOOKS**
Harper & Row, Publishers
NEW YORK  HAGERSTOWN  SAN FRANCISCO  LONDON

A hardcover edition is published by New York University Press.

RADICAL AGRICULTURE Copyright © 1976 by Richard Merrill.
All rights reserved. Printed in the United States of America. No part of
this book may be used or reproduced in any manner without written permission except in the case of brief quotations embodied in critical articles
and reviews. For information address Harper & Row, Publishers, Inc., 10
East 53d Street, New York, N.Y. 10022. Published simultaneously in Canada
by Fitzhenry & Whiteside Limited, Toronto.

First HARPER COLOPHON edition published 1976

LIBRARY OF CONGRESS CATALOG CARD NUMBER: 73–5469

STANDARD BOOK NUMBER: 06–090437–6

77   78   79   80   5   4   3

"Where Cities and Farms Come Together" by Wendell Berry. Copyright ©
1972 by Wendell Berry, excerpted from the essay "Discipline and Hope" in
*A Continuous Harmony*. Reprinted by permission of Harcourt Brace Jovan-
ovich, Inc. and the author.

"Land Reform in America" by Peter Barnes. Copyright © 1971 by Harrison-
Blaine, Inc., of New Jersey. Reprinted by permission of *The New Republic*.

"Agribusiness" by Nick Kotz. Copyright © 1971 by the Washington *Post*.
Excerpted from "Agribusiness," October 3, 4, 5, 1971, by Nick Kotz.

"Corporate Accountability and the Family Farm" by Sheldon L. Greene. Presented in Subcommittee Hearings. *The Role of Giant Corporations in the
American and World Economies: Corporate Secrecy, Agribusiness*. Subcommittee on Monopoly. Gaylord Nelson, Chairman. Washington, D.C.
2 March 1972.

"Hard Tomatoes, Hard Times: The Failure of the Land Grant College
Complex" by Jim Hightower. Copyright © 1972. Reprinted by permission
of Schenkman Publishing Company, Inc., and the Land Grant College Task
Force, part of the Agribusiness Accountability Project, an independent nonprofit research organization, 1000 Wisconsin Avenue, Washington, D.C.

"Organic Force" by Jerome Goldstein. Copyright © 1973 by Rodale Press.
Reprinted from *The New Food Chain: An Organic Link Between Farm and
City*, edited by Jerome Goldstein with the permission of Rodale Press, Inc.

as the Ancients
Say wisely, have a care o' th'main chance,
And Look before you ere you leap;
For As you sow, y' are like to reap.
—SAMUEL BUTLER
*Hudibras*, PART 2, CANTO 2

*For Yedida*
*and all those who seek freedom*
*within themselves*
*and on some land*

# Contents

# Preface

*Radical*: 1: of, relating to, or preceding from a root.

—*Webster's Dictionary*

There is a growing feeling in our culture today that the era of cheap abundant food is over, and that the cornucopia has been a short-term marvel with long-term costs to society. These costs include the loss of food quality,[1] the destruction of our rural culture and environment, the rise of centralized food monopolies, and the consequences of a vast migration of people from farms to cities. Since 1948 over 25 million people have been relocated to urban centers by high technology and agribusiness economy. In less than two generations there has been a revolutionary change in the means of food production and in the patterns of human settlement and food distribution in this country.

The abandonment of farmlands and the separation of people from their land and food resources have become symbols of our social "progress." According to this view the success of our society can be measured by the degree to which our rural culture becomes a labor force for the urban machine and ceases to be a steward of the rural environment. But it is by no means obvious that the emigration of rural communities and the industrialization of agriculture have produced a just, stable, and fulfilling society. In fact, as this book suggests[2] there is much to indicate that just the opposite has been the case, and that we have become affluent at the expense of agriculture, not because of it.

This is a radical notion. But then more and more people seem to be arguing in radical terms for new approaches to the way food is grown and distributed. In the cities the community food cooperatives of the sixties have evolved into a second wave of co-op organizing that has been spurred by rising food prices and consumer reform groups. Other older co-ops have formed regional federations linked by co-op truckers, local farmers, and newsletters. The recent growth of buying clubs, food cooperatives, farmers markets, consumer warehouses, and co-op federations[3] emphasizes the current shift of grass-roots food economies from inner-city radicals to broad-based support.

The growth of interest in community gardens and urban agriculture seems to be more than just a returning trend in our war-prosperity-depression cycle. Victory gardens have become inflation gardens, only this time local food production will probably have more lasting survival value, especially in a society whose food resources are controlled more and more by big business and a fossil-fuels technology.

Others in the cities, depressed by pollution, overcrowding, and an unbearable complexity have taken a harder look at the advantages of a simpler and healthier environment. For the first time since the Depression there has been a net movement of people away from cities into more rural areas. It is hard to say whether this represents a temporary trend or permanent transition in the patterns of human settlement. It is very likely that speculation and rising land prices will continue to make rural opportunity possible only for the rich. Hence the importance of grass-roots land trusts and the activities of land reform groups to break up land monopolies and provide farmlands for small growers and postindustrial communities.[4]

Recently some states have enacted legislation that would protect prime farm lands from development and in effect place them where they belong—in the public trust. Other states have proposed initiatives, antitrust legislation, and "Family Farm" acts that would prevent nonfarm investors from seeking tax shelters and quick returns from capital-intensive agriculture and capital gains from inflated farmland values. Although these actions will help proprietary farmers to realize their full potential, there is still no wide acceptance of the fact that big farms are less efficient and more costly to society than small farms or that big farms tend to mine the land and that small farmers are in a much better position to conserve energy, use fewer petrochemicals, and husband the land in a rational way.

But there are signs that even this may be changing. It is now recognized that farm corporations that began operation after 1966 or so were too late to capitalize on rising efficiencies. High land prices, pollution control, poor farm management, and diseconomies of scale have forced many of the vertically integrated corporations (like Tenneco) out of farm production, at least for the time being.

Another recent note of optimism was the 1975 California Agricultural Labor Relations Act, which was passed as a compromise for regulating labor relations in California agriculture. The act was designed to protect farm workers from unfair labor practices and to ensure collective bargaining in good faith. The effectiveness and consequences of the law are still unclear, but it may be an important precedent.

In California and a few other states, a small but increasing number of farm labor groups have also organized to become successful growers themselves. These efforts demonstrate that the true power of rural democracy today resides in the formation of ECONOMI-CALLY viable farm communities that can EXIST in spite of agribusiness. The next step will hopefully be the formation of ECOLOGICALLY stable rural communities that can PERSIST.

There continues to be a great deal written about the limits and hazards of modern farm technology and of the pollution and health problems caused by salt fertilizers, livestock chemicals, pesticides, etc. There is even growing evidence that the production potential of agriculture has already been reached and that higher yields will not warrant the costs of new technologies.[5] But the recent "energy crisis" has produced deeper insights into the major weakness of modern agriculture. It is fueled, fertilized, and fed by a dwindling energy resource (fossil fuels). In other words, agriculture has become vulnerable BECAUSE of its energy-intensive technologies. It remains for the agricultural establishment to squarely face one of the basic dilemmas of our time. What happens to agriculture when its traditional energy base is exhausted?

It is not surprising, then, that rising energy prices have forced many farmers to reexamine some of the older methods of soil husbandry (manure spreading, green manuring, and strip cropping) and some of the newer methods of composting, integrated pest control, mixed cropping systems, renewable energy resources (solar-wind power and organic fuels), and the reintegration of livestock with agronomy. Some of these alternatives may already be cost effective.[6]

Real advances in the TECHNIQUE of Radical Agriculture will only come when there are new priorities in agricultural research. The problem is that there seems to be little consensus as to what the fundamental questions should be for a postindustrial system of agriculture. For one thing, most researchers are still operating from a basic set of assumptions that prevent new questions from being asked: for example, that the sole purpose of agriculture is to produce high yields; that fossil fuels and other cheap energy will continue to supply agriculture with all of its energy needs; and that agriculture can only operate "efficiently" in an industrial milieu. Instead, if we assume that fossil fuels are exhaustible, that an equally important role of agriculture is to sustain farmlands with ecologically wise methods, and that the single strategy of farm chemicals, mechanized cultivation, and genetic engineering is open to question and alternatives, then we are led into new areas of agricultural research.

Another reason for the lack of research direction in Radical Agriculture is that most of the work remains in specialized disciplines like agricultural economics, soil microbiology, humus biochemistry, organic waste recycling, integrated pest control, solar-energy technology, etc. Very few workers seem to be integrating results from different fields and approaching an economic and ecological alternative to conventional agriculture from a systems point of view . . . from the practical experiences of a new-age farm community.

Also, many people see alternatives to conventional high-energy agriculture as simply a return to some prechemical farming ethic. We will surely have to use some of the old techniques once again. But there are also many new and ecologically sophisticated techniques of food production that have proven themselves during the last few years. Typically, they are written off as economically inefficient. One can only wonder when the consciousness will change.

There is a final barrier to the alternatives suggested in this and in other books and that is the isolation that now exists between urban and rural cultures. This segregation has profoundly affected our world view. Today most problems are "urban" problems. We seem to miss the point that the decaying urban condition has its origins in the decaying rural condition. A radical agriculture, geared to the needs of a postindustrial society, must begin in the cities as well as the farms. Then we can finally come together.

Needless to say this book would not have been possible without the ideas and contributions of the individual authors. I would especially like to thank Murray Bookchin, Darryl McLeod, Michael Perelman, and John Todd for their advice and encouragement. My sincerest appreciation also goes out to the following people for their conversations, suggestions, and inspirations: Fay Bennett, Wilson Clark, Peter Corbetta, Ken Hagen, Stuart Hill, John Jeavons, Steve Kaffka, David Katz, Donald Landenberger, Phil LaVeen, E. F. Schumacher, Lee Swenson, Dr. Paul Taylor, Charles Walters, Harvey Wasserman, and to my mentor, Joseph P. Connell.

—RICHARD MERRILL

### Notes

1. R. H. Hall, *Food for Naught* (New York: Harper and Row, 1974).

2. See also: P. Barnes ed., *The People's Land* (Emmaus, Pa.,: Rodale Press, 1975); M. Allaby et al., *Losing Ground* (London: Friends of the Earth, 1975); J. Goldstein, *The New Food Chain* (Emmaus, Pa.: Rodale Press, 1973); J. Hightower, *Eat Your Heart Out: Food Profiteering in America* (New York: Quadrangle, 1975); D. Mitchell, *The Politics of Food* (Toronto: James Lorimer & Co., 1975).

3. Listings of the Chicago Food Project include over 1,000 major food co-ops. There are perhaps 2 to 4 times that many.

4. In California, where land monopolies are a tradition, major land reform activities have shifted from the cities (National Coalition for Land Reform) to the rural areas (National Land for People, Fresno).

5. J. G. Horsfall, and C. R. Frink, *Perspective on Agriculture's Future: Rising Costs . . . Rising Doubts.* Limits to Growth Symposium, Dallas, Texas, October 21, 1975.

6. W. Lockeretz et al., *A Comparison of Organic and Conventional Farms in the Corn Belt.* Center for the Biology of Natural Systems (St. Louis, Mo.: Washington University; 1975).

# PART I

# Land and Culture

# RADICAL AGRICULTURE

## *Murray Bookchin*

Agriculture is a form of culture. The cultivation of food is a social and cultural phenomenon unique to man. Among animals, anything that could remotely be described as food cultivation appears ephemerally, if at all; and even among humans, agriculture developed little more than ten thousand years ago. Yet in an epoch when food cultivation is reduced to a mere industrial technique, it becomes especially important to dwell on the cultural implications of "modern" agriculture—to indicate their impact not only on public health but also on humanity's relationship to nature and the relationship of human to human.

The contrast between early and modern agricultural practices is dramatic. Indeed, it would be very difficult to understand the one through the vision of the other, to recognize that they are united by any kind of cultural continuity. Nor can we ascribe this contrast merely to differences in technology. Our agricultural epoch—a distinctly capitalistic one—envisions food cultivation as a business enterprise to be operated strictly for the purpose of

Murray Bookchin teaches in the Social Ecology Studies Program at Goddard College and at the Center for New Studies at Ramapo College. He is the author of *Our Synthetic Environment, Post-Scarcity Anarchism,* and *The Limits of the City.*

generating profit in a market economy. From this standpoint, land is an alienable commodity called "real estate," soil a "natural resource," and food an exchange value that is bought and sold impersonally through a medium called "money." Agriculture, in effect, differs no more from any branch of industry than does steelmaking or automobile production. In fact, to the degree that food cultivation is affected by nonindustrial factors such as climatic and seasonal changes, it lacks the exactness that marks a truly "rational" and scientifically managed operation. And lest these natural factors elude bourgeois manipulation, they too are the objects of speculation in future markets and between middlemen in the circuit from farm to retail outlet.

In this impersonal domain of food production, it is not surprising to find that a "farmer" often turns out to be an airplane pilot who dusts crops with pesticides, a chemist who treats soil as a lifeless repository for inorganic compounds, an operator of immense agricultural machines who is more familiar with engines than botany, and, perhaps most decisively, a financier whose knowledge of land may beggar that of an urban cab driver. Food, in turn, reaches the consumer in containers and in forms so highly modified and denatured as to bear scant resemblance to the original. In the modern, glistening supermarket, the buyer walks dreamily through a spectacle of packaged materials in which the pictures of plants, meat, and dairy foods replace the life forms from which they are derived. The fetish assumes the form of the real phenomenon. Here, the individual's relationship to one of the most intimate of natural experiences—the nutriments indispensable to life—is divorced from its roots in the totality of nature. Vegetables, fruit, cereals, dairy foods, and meat lose their identity as organic realities and often acquire the name of the corporate enterprise that produces them. The "Big Mac" and the "Swift sausage" no longer convey even the faintest notion that a living creature was painfully butchered to provide the consumer with that food.

This denatured outlook stands sharply at odds with an earlier animistic sensibility that viewed land as an inalienable, almost sacred domain, food cultivation as a spiritual activity, and food consumption as a hallowed social ritual. The Cayuses of the Northwest were not unique in listening to the ground, for the

"Great Spirit," in the words of a Cayuse chief, "appointed the roots to feed the Indians on."[1] The ground lived, and its voice had to be heeded. Indeed, this vision may have been a cultural obstacle to the spread of food cultivation; there are few statements of the hunter against agriculture that are more moving than Smohalla's memorable remarks: "You ask me to plow the ground. Shall I take a knife and tear my mother's breast? Then when I die she will not take me to her bosom to rest."[2]

When agriculture did emerge, it clearly perpetuated the hunter's animistic sensibility. The wealth of mythic narrative that surrounds food cultivation is testimony to an enchanted world brimming with life, purpose, and spirituality. Ludwig Feuerbach's notion of God as the projection of man omits the extent to which early man is stamped by the imprint of the natural world and, in this sense, is an extension or projection of it. To say that early humanity lived in "partnership" with this world tends to understate the case; humanity lived as *part* of this world—not beside it or above it.

Because the soil was alive, indeed the mother of life, to cultivate it was a sacred act that required invocatory and appeasing rituals. Virtually every aspect of the agricultural procedure had its sanctifying dimension, from preparing a tilth to harvesting a crop. The harvest itself was blessed, and to "break bread" was at once a domestic ritual that daily affirmed the solidarity of kinfolk as well as an act of hospitable pacification between the stranger and the community. We still seal a bargain with a drink or celebrate an important event with a feast. To fell a tree or kill an animal required appeasing rites, which acknowledged that life inhered in these beings and that this life partook of a sacred constellation of phenomena.

Naive as the myths and many of these practices may seem to the modern mind, they reflect a truth about the agricultural situation. After having lost contact with this "prescientific" sensibility— at great cost to the fertility of the land and to its ecological balance— we now know that soil is very much alive; that it has its health, its dynamic equilibrium, and a complexity comparable to that of any living community. Not that the details that enter into this knowledge are new; rather, we are *aware* of them in a new and holistic way. As recently as the early 1960s, American agronomy generally viewed soil as a medium in which living organisms were largely extraneous to the chemical management of food cultivation. Having saturated the soil with nitrates, insecticides, herbicides, and an appalling variety of toxic compounds, we have become the victims of a new type of pollution that could well be called "soil pollution." These toxins are the hidden additives to the dinner table, the unseen specters that return to us as the residual products of our exploitative attitude toward the natural world. No less significantly, we have gravely damaged soil in vast areas of the earth and reduced it to the simplified image of the modern scientistic viewpoint. The animal and plant life so essential to the development of a nutritive, friable soil is diminished and in many places approaches the sterility of impoverished, desertlike sand.

By contrast, early agriculture, despite its imaginary aspects, defined humanity's relationship to nature within sound ecological parameters. As Edward Hyams observes, the attitude of people and their culture is as much a part of their technical equipment as are the implements they employ. If the "axe was only the physical tool which ancient man used to cut down trees" and the "intellectual tool enabled him to swing his axe" effectively, "what of the spiritual tool?" This "tool" is the "member of the trinity of tools which enables men to control and check their actions by reference to the 'feeling' which they possess for the consequences of the changes they make in their environments." Accordingly, tree-felling would have been limited by their state of mind as early men "believed that trees had souls and were worshipful, and they associated certain gods with certain trees. Osiris with acacia; Apollo with oak and apple. The temples of many primitive peoples were groves. . . ." If the mythical aspects of this mentality are evident enough, the fact remains that the mentality as such "was immensely valuable to the soil community and therefore, in the long run, to man. It meant that no trees would be wantonly felled, but only when it was absolutely necessary, and then to the accompaniment of propitiatory rites which, if they did nothing else, served constantly to remind tree-fellers that they were doing dangerous and important work."[3] One may add that if culture can be regarded as a "tool," a mere shift in emphasis would easily make it possible to regard tools as part of culture. This different emphasis comes closer to what Hyams is trying to say than does his own formulation. In fact, what uniquely marks the bourgeois mentality is the debasement of art, values, and rationality to mere tools—a mentality that has even infiltrated the radical critique of capitalism if one is to judge from the tenor of the Marxian literature that abounds today.

A radical approach to agriculture seeks to transcend the prevailing instrumentalist approach that views food cultivation merely as a "human technique" opposed to "natural resources." This radical approach is literally ecological in the strict sense that the land is viewed as an *oikos*—a *home*. Land is neither a "resource" nor a "tool," but the *oikos* of myriad kinds of bacteria, fungi, insects, earthworms, and small mammals. If hunting leaves this *oikos* essentially undisturbed, agriculture by contrast affects it profoundly and makes humanity an integral part of it. Human beings no longer indirectly affect the soil; they intervene into its food webs and biogeochemical cycles directly and immediately.

Conversely, it becomes very difficult to understand human social institutions without referring to the prevailing agricultural practices of a historical period and, ultimately, to the soil situation to which they apply. Hyams's description of every human community as a "soil community" is unerring; historically, soil types and agrarian technological changes played a major, often decisive, role in determining whether the land would be worked cooperatively or individualistically—whether in a conciliatory manner or an exploitative one—and this, in turn, profoundly affected the prevailing system of social relations. The highly centralized empires of the ancient world were clearly fostered by the irrigation works required for arid regions of the Near East; the cooperative medieval village, by the open-field strip system and the moldboard plow. Lynn White, Jr., in fact, roots the Western coercive attitude toward nature as far back as Carolingian times, with the ascendancy of the heavy European plow and the consequent tendency to allot land to peasants not according to their family subsistence needs but "in proportion to their contribution to the ploughteam."[4] He finds this changing attitude reflected in Charlemagne's efforts to rename the months according to labor responsibilities, thereby revealing an emphasis on work rather than on nature or deities. "The old Roman calendars had occasionally shown genre scenes in human activity, but the dominant tradition (which continued in Byzantium) was to depict the months as passive personifications bearing symbols of attributes. The new Carolingian calendars, which set the pattern for the Middle Ages, are very different: they show a coercive attitude toward natural resources. They are definitely northern in origin; for the olive, which loomed so large in the Roman cycles, has now vanished. The pictures change to scenes of ploughing, harvesting, wood-chopping, people knocking down acorns for the pigs, pig-slaughtering. Man and nature are now two things, and man is master."[5]

Yet not until we come to the modern capitalist era do man and nature separate as almost complete foes, and the "mastery" by man over the natural world assumes the form of harsh domination, not merely hierarchical classification. The rupture of the most vestigial corporate ties that once united clansfolk, guildsmen, and the fraternity of the *polis* into a nexus of mutual aid; the reduction of everyone to an antagonistic buyer or seller; the rule of competition and egotism in every arena of economic and social life—all of this completely dissolves any sense of community, whether with nature or in society. The traditional assumption that community is the authentic locus of life fades so completely from human consciousness that it ceases to exercise any relevance to the human condition. The new starting point for forming a conception of society or of the psyche is the isolated, atomized man fending for himself in a competitive jungle. The disastrous consequences of this outlook toward nature and society are evident enough in a world burdened by explosive social antagonisms, ecological simplification, and widespread pollution.

*Radical agriculture seeks to restore humanity's sense of community: first, by giving full recognition to the soil as an ecosystem, a biotic community; and second, by viewing agriculture as the activity of a natural human community, a rural society and culture.* Indeed, agriculture becomes the practical, day-to-day interface of soil and human communities, the means by which both meet and blend. Such a meeting and blending involves several key presuppositions. The most obvious of these is that humanity is part of the natural world, not above it as "master" or "lord." Undeniably, human consciousness is unique in its scope and insight, but uniqueness is no warrant for domination and exploitation. Radical agriculture, in this respect, accepts the ecological precept that variety does not have to be structured along hierarchical lines as we tend to do under the influence of hierarchical society. Things and relations that patently benefit the biosphere must be valued for their own sake, each unique in its own way and contributory to the whole— not one above or below the other and fair game for domination.

Variety, in both society and agriculture, far from being constrained, must be promoted as a positive value. We are now only too familiar with the fact that the more simplified an ecosystem— and, in agriculture, the more limited the variety of domesticated stocks involved—the more likely is the ecosystem to break down. The more complex the food webs, the more stable the biotic structure. This insight, which we have gained at so costly an expense to the biosphere and to ourselves, merely reflects the age-old thrust of evolution. The advance of the biotic world consists primarily of the differentiation, colonization, and growing web of interdependence of life-forms on an inorganic planet—a long process that has remade the atmosphere and landscape along lines that are hospitable for complex and increasingly intelligent organisms. The most disastrous aspect of prevailing agricultural methodologies, with their emphasis on monoculture, crop hybrids, and chemicals, has been the simplification they have introduced into food cultivation—a simplification that occurs on such a global scale that it may well throw back the planet to an evolutionary stage when it could support only simpler forms of life.

Radical agriculture's respect for variety implies a respect for the complexity of a balanced agricultural situation: the innumerable factors that influence plant nutrition and well-being; the diversified soil relations that exist from area to area; the complex interplay between climatic, geological, and biotic factors that make for the differences between one tract of land and another; and the variety of ways in which human cultures react to these differences. Accordingly, the radical agriculturist sees agriculture not only as science but also as art. The food cultivator must live on intimate terms with a given area of land and develop a sensitivity for its special needs—needs that no textbook approach can possibly encompass. The food cultivator must be part of a "soil community" in the very meaningful sense that she or he belongs to a unique biotic system as well as to a given social system.

Yet to deal with these issues merely in terms of technique would be a scant improvement over the approach that prevails today in agriculture. To be a technical connoisseur of an "organic" approach to agriculture is no better than to be a mere practitioner of a chemical approach. We do not become "organic farmers" merely by culling the latest magazines and manuals in this area, any more than we become healthy by consuming "organic" foods acquired from the newest suburban supermarket. What basically separates the organic approach from the synthetic is the *overall* attitude and praxis the food cultivator brings to the natural world as a whole. At a time when organic foods and environmentalism have become highly fashionable, it may be well to distinguish the ecological outlook of radical agriculture from the crude "environmentalism" that is currently so widespread. Environmentalism sees the natural world merely as a habitat that must be engineered with minimal pollution to suit society's "needs," however irrational or synthetic these needs may be. A truly ecological outlook, by contrast, sees the biotic world as a holistic unity of which humanity is a part. Accordingly, in this world, human needs must be integrated with those of the biosphere if the human species is to survive. This integration, as we have already seen, involves a profound respect for natural variety, for the complexity of natural processes and relations, and for the cultivation of a mutualistic attitude toward the biosphere. *Radical agriculture, in short, implies not merely new techniques in food cultivation, but a new non-Promethean sensibility toward land and society as a whole.*

Can we hope to achieve fully this new sensibility solely as individuals, without regard to the larger social world around us?

Radical agriculture, I think, would be obliged to reject an isolated approach of this kind. Although individual practice doubtless plays an invaluable role in initiating a broad movement for social reconstruction, ultimately we will not achieve an ecologically viable relationship with the natural world without an ecological society. Modern capitalism is inherently antiecological: The nuclear relationship from which it is constituted—the buyer-seller relationship —pits individual against individual and, on the larger scale, humanity against nature. Capital's law of life of infinite expansion, of "production for the sake of production" and "consumption for the sake of consumption," turns the domination and exploitation of nature into the "highest good" of social life and human self-realization. Even Marx succumbs to this inherently bourgeois mentality when he accords to capitalism a "great civilizing influence" for reducing nature "for the first time simply [to] an object for mankind, purely a matter of utility. . . ." Nature "ceases to be recognized as a power in its own right; and the theoretical knowledge of its independent laws appears only as a strategem designed to subdue it to human requirements . . ."[6]

In contrast to this tradition, radical agriculture is essentially libertarian in its emphasis on community and mutualism, rather than on competition, an emphasis that derives from the writings of Peter Kropotkin[7] and William Morris. This emphasis could justly be called ecological before the word "ecology" became fashionable, indeed, before it was coined by Ernst Haeckel a century ago. The notion of blending town with country, of rotating specifically urban with agricultural tasks, had been raised by so-called utopian socialists such as Charles Fourier during the Industrial Revolution. Variety and diversity in one's workaday activities—the Hellenic ideal of the rounded individual in a rounded society—found its physical counterpart in varied surroundings that were neither strictly urban nor rural, but a synthesis of both. Ecology validated this ideal by revealing that it formed the precondition not only for humanity's psychic and social well-being but for the well-being of the natural world as well.

Our own era has gone further than this visionary approach. A century ago it was still possible to reach the countryside without difficulty even from the largest cities and, if one so desired, to leave the city permanently for a rural way of life. Capitalism had not so completely effaced humanity's legacy that one lacked evidence of neighborhood enclaves, quaint life-styles and personalities, architectural diversity, and even village society. Predatory as the new industrial system was, it had not so completely eliminated the human scale as to leave the individual totally faceless and estranged. By contrast, we are compelled to occupy even quasi-rural areas that have become essentially urbanized, and we are reduced to anonymous digits in a staggering bureaucratic apparatus that lacks personality, human relevance, or individual understanding. In population, if not in physical size, our cities compare to the nation-states of the last century. The human scale has been replaced by the inhuman scale. We can hardly comprehend our own lives, much less manage society or our immediate environment. Our very self-integrity, today, is implicated in achieving the vision that utopians and radical libertarians held forth a century ago. In this matter, we are struggling not only for a better way of life *but for our very survival.*

*Radical agriculture offers a meaningful response to this desperate situation in terms not of a fanciful flight to a remote agrarian refuge but of a systematic recolonization of the land along ecological lines.* Cities are to be decentralized—and this is no longer a utopistic fantasy but a visible necessity which even conventional city planning is beginning to recognize—and new ecocommunities are to be established, tailored artistically to the ecosystems in which they are located. These ecocommunities are to be scaled to human dimensions, both to afford the greatest degree of self-management possible and personal comprehension of the social situation. No bureaucratic, manipulative, centralized administration here, but a voluntaristic system in which the economy, society, and ecology of an area are administered by the community as a whole, and the distribution of the means of life is determined by need, rather than by labor, profit, or accumulation.

But radical agriculture carries this tradition further—into technology itself. In contemporary social thought, technology tends to be polarized into highly centralized labor-extensive forms on the one hand and decentralized, craft-scale labor-intensive forms on the other. Radical agriculture steers the middle ground established by an ecotechnology: It avails itself of the tendency toward miniaturization and versatility, quality production, and a balanced combination of mass manufacture and crafts. For side by side with the massive, highly specialized fossil-fuel technology in use today, we are beginning to see the emergence of a new technology—one that lends itself to the local deployment of many energy resources on a small scale (wind, solar, and geothermal)—that provides a wider latitude in the use of small, multipurpose machinery, and that can easily provide us with the high-quality semifinished goods that we, as individuals, may choose to finish according to our proclivities and tastes. The rounded ecocommunities of the future would thereby be sustained by rounded ecotechnologies.[8] The people of these communities, living in a highly diversified agricultural and industrial society, would be free to avail themselves of the most sophisticated technologies without suffering the social distortions that have pitted town against country, mind against work, and humanity against itself and the natural world.

Radical agriculture brings all of these possibilities into focus, for we must begin with the land if only because the basic materials for life are acquired from the land. This is not only an ecological truth but a social one as well. The kind of agricultural practice we adopt at once reflects and reinforces the approach we will utilize in all spheres of industrial and social life. Capitalism began historically by undermining and overcoming the resistance of the traditional agrarian world to a market economy; it will never be fully transcended unless a new society is created on the land that liberates humanity in the fullest sense and restores the balance between society and nature.

## Notes

1. T. C. McLuhan, ed., *Touch the Earth* (New York: Outerbridge & Lazard, 1971), p. 8.

2. Ibid., p. 56.

3. Edward Hyams, *Soil and Civilization* (London: Thames & Hudson, 1952), pp. 274, 276.

4. Lynn White, Jr., *Medieval Technology and Social Change* (New York: Oxford Univ. Press, 1962), p. 56.

5. Ibid., p. 57.

6. Karl Marx, *Grundrisse*, ed. and trans. David McLellan (New York: Harper & Row, 1971), p. 94.

7. See especially P. Kropotkin, *Fields, Factories and Workshops Tomorrow* (New York: Harper & Row, 1974); *Mutual Aid* (Boston: Sargent Publishers, 1955), and also: *Conquest of Bread* (New York: New York University Press, 1972).

8. See Murray Bookchin, *Post-Scarcity Anarchism* (Berkeley: Ramparts Press, 1972).

# 2

## WHERE CITIES AND
## FARMS COME TOGETHER

*Wendell Berry*

The mentality of organic agriculture is not a technological mentality—though it does concern itself with technology. It does not merely ask what is the easiest and cheapest and quickest way to reach an immediate aim. It is, rather, a complex and radical attitude toward the problem of our relationship to the earth. It is concerned with the long-term questions of what humans need from the earth and what duties and devotions humans owe the earth in return for the satisfaction of their needs. It understands that the terms of a lasting agriculture are not human terms, that the final terms are nature's, that an agriculture—and for that matter, a culture—that holds in ignorance or contempt the truths and the mysteries of nature is doomed to failure, for it is out of control.

At least since the time of Henry Adams, numerous critics and historians have been concerned with the disintegration of the syntheses of disciplines that made the medieval cathedral one of the supreme articulations of humanity's relation to God. Only recently have we begun to be aware of the disintegration of an even more ancient and fundamental synthesis—that of the old peasant and yeoman agriculture, which still stands as the best articulation of

Wendell Berry, a poet-farmer, teaches English at the University of Kentucky in Port Royal, Kentucky.

humanity's relation to the world. This was not simply an agriculture; at best, it was also a *culture* of such deep-rooted and complex wisdom that it preserved the fertility of the earth under the most intensive human use. It was a culture that made men the preservers rather than the parasites of the sources of their life. The organic movement has its roots in this ancient agriculture that was so wise and profound a bond between human beings and their fields. And it is the rise of the organic movement that affords us a perspective from which we can understand the consequences of the disintegration of that bond—a distintegration that now palpably threatens the destruction not merely of human culture, but of human life as well.

Nearly all the old standards, which required rigorous discipline, have now been replaced by a new standard of efficiency, which requires not discipline, not a mastery of means, but rather a carelessness of means, a relentless subjection of means to immediate ends. The standard of efficiency displaces and destroys the standards of quality because, by definition, it cannot even consider them. Instead of asking a man what he can do well, it asks him what he can do fast and cheap. Instead of asking the farmer to practice the best husbandry, to be a good steward and trustee of his land and his art, it puts irresistible pressures on him to produce more and more food and fiber more and more cheaply, thereby destroying the health of the land, the best traditions of husbandry, and the farm population itself. And so when we examine the principle of efficiency as we now practice it, we see that it is not really efficient at all. As we use the word, efficiency means no such thing, or it means short-term or temporary efficiency, which is a contradiction in terms. It means cheapness at any price. It means hurrying to nowhere. It means the profligate waste of humanity and of nature. It means the greatest profit to the greatest liar. What we have called efficiency has produced among us, and to our incalculable cost, such unprecedented monuments of destructiveness and waste as the strip-mining industry, the Pentagon, the federal bureaucracy, and the family car.

Real efficiency is something entirely different. It is neither cheap (in terms of skill and labor) nor fast. Real efficiency is long-term efficiency. It is to be found in means that are in keeping with and preserving of their ends, in methods of production that preserve the sources of production, in workmanship that is durable and of high quality. In this age of consumerism, planned obsolescence, frivolous horsepower, and surplus manpower, those salesmen and politicians who talk about efficiency are talking, in reality, about spiritual and biological death.

Specialization, a result of our nearly exclusive concern with the form of exploitation that we call efficiency, has in its turn become a destructive force. Carried to the extent to which we have carried it, it is both socially and ecologically destructive. That specialization has vastly increased our knowledge, as its defenders claim, cannot be disputed. But I think that one might reasonably dispute the underlying assumption that knowledge per se, undisciplined knowledge, is good. For, although specialization has increased knowledge, it has also fragmented it. And this fragmentation of knowledge has been accompanied by a fragmentation of discipline. That is, specialization has tended to draw the specialist toward the discipline that will lead to the discovery of new facts or processes within a narrowly defined area, and it has tended to lead him away from or distract him from those disciplines which might enable him to see the *effects* of his discovery upon human society or upon the world. It has tended to value the disciplines that pertain to the gathering of knowledge and to its immediate use and to devalue those that pertain to its ultimate effects.

Nowhere are these tendencies more apparent than in agriculture. For years the agricultural specialists have tended to think and work in terms of piecemeal solutions and annual production, rather than in terms of a whole and coherent system that would maintain the fertility and the ecological health of the land over a period of centuries. Focused nearly exclusively upon so-called "efficiency" with respect to production, as if the only discipline pertinent to agriculture were that of economics, they have eagerly abetted a rapid industrialization of agriculture, which is potentially catastrophic, both in the ecological deterioration of farm areas and in the diminishment, the dispossession, and the displacement of the rural population.

Ignoring the ample evidence that a healthy agriculture is a highly diversified one, using the greatest possible variety of animals and plants, and that it returns all organic wastes to the soil, the specialists of the laboratories have promoted the specialization of farms, encouraging one-crop agriculture and the replacement of humus by chemicals. And as the pressures of urban populations upon the land have grown, the specialists have turned, more and more, not to the land but to the laboratory. Ignoring the considerable historical evidence that to have a productive agriculture over a long period of time it is necessary to have a stable and prosperous rural population closely bound in sympathy and association to the land, the specialists have either connived in the dispossession of small farmers by machinery and technology or have actively encouraged their migration into the cities.

The result of the short-term vision of these experts is a whole series of difficulties that together amount to a rapidly building ecological and social disaster, which at present there is little disposition to regret, much less to correct. The organic wastes of our society, for which our land is starved and which in a sound agricultural economy would be returned to the land, are instead flushed out through the sewers to pollute the streams and rivers and, finally, the oceans; or they are burned, the smoke polluting the air; or they are wasted in other ways. Similarly, the small farmers who in a healthy society would be the mainstay of the country—whose allegiance to their land, continuing and deepening in association from one generation to another, would be the motive and guarantee of good care—are forced out by the economics of efficiency to become immigrants and dependents in the already overcrowded cities. In both instances, by the abuse of knowledge in the name of efficiency, assets have been converted into problems.

Modern agricultural practice concentrates almost exclusively on the productive phase of the natural cycle. The means of production become more elaborate all the time, but the means of return—the building of health and fertility in the soil—are reduced more and more to the shorthand of chemicals. According to the industrial vision of it, the life of the farm does not rise and fall in the turning cycle of the year; it goes on in a straight line from one harvest to another. In the long run, this may well be more productive of waste than of anything else. It wastes the soil. It wastes the animal manures and other organic residues that industrialized agriculture fails to return to the soil. And what may be our largest agricultural waste is not usually recognized as such, but is thought to be both an urban product and an urban problem: the tons of garbage and sewage that are burned or buried or flushed into our rivers. This, like all waste, is the abuse of a resource. It was ecological stupidity of exactly this kind that destroyed Rome. The chemist Justus Liebig wrote that "the sewers of the immense metropolis engulfed in the course of centuries the prosperity of Roman peasants. The Roman Campagna would no longer yield the means of feeding her population; these same sewers devoured the wealth of Sicily, Sardinia and the fertile lands of the coast of Africa."

To recognize the extent and the destructiveness of our "urban waste" is to recognize the shallowness of the notion that agriculture is only another form of technology to be turned over to a few specialists. The sewage and garbage problem of our cities suggests, rather, that a healthy agriculture is a cultural organism, not merely a universal necessity but a universal obligation as well. It suggests that, just as the cities exist within the ecology, they also exist within agriculture. It suggests that, like farmers, city dwellers have agricultural responsibilities: to use no more than necessary, to waste nothing, to return organic residues to the soil.

We are being virtually buried by the evidence that those disciplines by which we manipulate *things* are inadequate disciplines. Our cities have become almost unlivable because they have been built to be factories and vending machines rather than communities. They are conceptions of the desires for wealth, excitement, and ease—all illegitimate motives from the standpoint of community, as is proved by the fact that without the community disciplines that make for a stable, neighborly population, the cities have become scenes of poverty, boredom, and disease.

The rural community—that is, the land and the people—is being degraded in complementary fashion by the specialists' premise that the exclusive function of the farmer is production and that his major discipline is economics. On the contrary, both the function and the discipline of the farmer concern provision: He must provide; he must look ahead. He must look ahead, however, not in the economic mechanistic sense of anticipating a need and fulfilling it, but in the sense of using methods that preserve the source. In his work, sound economics becomes identical with sound ecology. The farmer is not a factory worker; he is the trustee of the life of the topsoil, the keeper of the rural community in precisely the same way the dweller in a healthy city is not an office or a factory worker, but part and preserver of the urban community. It is in thinking of the whole citizenry as factory workers—as readily interchangeable parts of an entirely mechanistic and economic order—that we have reduced people to the most abject and aimless of nomads and have displaced and fragmented our communities.

An index of the health of a rural community—and, of course, of the urban community, its blood kin—might be found in the relative acreages of field crops and tree crops. By tree crops I mean not just those orchard trees of comparatively early bearing and short life, but also the fruit and nut and timber trees that bear late and live long. It is characteristic of an unsettled and anxious farm population—a population that feels itself, because of economic threat or the degradation of cultural value, to be ephemeral—that it farm almost exclusively with field crops within economic and biological cycles that are complete in one year. This has been the dominant pattern of American agriculture. Stable, settled populations, assured both of an economic sufficiency in return for their work and of the cultural value of their work, tend to have methods and attitudes of a much longer range. Though they also have generally farmed with field crops, established farm populations have always been planters of trees. In parts of Europe, according to J. Russell Smith's important book *Tree Crops*, the steep hillsides were covered with orchards of chestnut trees that were kept and maintained with great care by the farmers.[1] Many of the trees were ancient, and when one began to show signs of dying, a seedling would be planted beside it to replace it. Here is an agricultural discipline that could only develop among farmers who felt secure—as individuals, and also as families and communities—in their connection to their land. Such a discipline depends not just on the younger men in the prime of their workdays but also on the older men, the keepers of tradition. The model figure of this agriculture is an old man planting a young tree that will live longer than a man and that he himself may not live to see in its first bearing. And he is planting, moreover, a tree whose worth lies beyond any conceivable market prediction. He is planting it because the good sense of doing so has been clear to men of his place and kind for generations. The practice has been continued because it is ecologically and agriculturally sound; the economic soundness of it must be assumed. While the planting of a field crop, then, may be looked upon as a "short-term investment," the planting of a chestnut tree is a covenant of faith.

The metaphor governing the distortions of efficiency and specialization has been that of the laboratory. The working assumption has been that nature and society, like laboratory experiments, can be manipulated by processes that are, for the most part, comprehensible toward ends that are, for the most part, foreseeable. But the analogy, as any farmer would know instantly, is too simple, for both nature and humanity are vast in possibility, unpredictable, and ultimately mysterious. Sir Albert Howard has spoken to this problem: "Instead of breaking up the subject into fragments and studying agriculture in piecemeal fashion by the analytical methods of science, appropriate only to the discovery of new facts, we must adopt a synthetic approach and look at the wheel of life as one great subject and not as if it were a patchwork of unrelated things."[2] A much more appropriate model for the agriculturist, scientist, or farmer is the forest, for the forest, as Howard pointed out, "manures itself" and is, therefore, self-renewing; it has achieved that "correct relation between the processes of growth and the processes of decay that is the first principle of successful agriculture." A healthy agriculture can take place only within nature and in cooperation with its process, not in spite of it and not by "conquering" it. Nature, Howard points out, in elaboration of his metaphor, "never attempts to farm livestock; she always raises mixed crops; great pains are taken to preserve the soil and to prevent erosion; the mixed vegetable and animal wastes are converted into humus; *there is no waste* [emphasis mine]; the process of growth and the processes of decay balance one another; ample provision is made to maintain large reserves of fertility; the greatest care is taken to store the rainfall; both plants and animals are left to protect themselves against disease."

The fact is that farming is not a laboratory science but a science of practice. It would be, I think, a good deal more accurate to call it an art, for it grows not only out of factual knowledge but out of cultural tradition; it is learned not only by precept but by example, by apprenticeship; and it requires not merely a competent knowledge of its facts and processes but also a complex set of attitudes, a certain culturally evolved stance, in the face of the unexpected and the unknown. That is to say that it requires *style* in the highest and richest sense of that term.

One of the most often repeated tenets of contemporary optimism asserts that "a nation that can put men on the moon certainly should be able to solve the problem of hunger." This proposition seems to me to have three important flaws, which I think may be taken as typical of our official view of ourselves:

1. It construes the flight to the moon as a historical event of a complete and coherent significance, when in fact, it is a fragmentary event of very uncertain significance. Americans have gone to the moon as they came to the frontiers of the New World: with their minds very much on getting there, very little on what might be involved in staying there. I mean that because of our history of waste and destruction here, we have no assurance that we can survive in America, much less on the moon. And until we can bring into balance the processes of growth and decay, the white man's settlement of this continent will remain an incomplete event. When a Japanese peasant went to the fields of his tiny farm in the preindustrial age, he worked in the governance of an agricultural tradition thousands of years old, which had sustained the land in prime fertility during all that time, in spite of the pressures of a population that in 1907 had reached a density of "more than three people to each acre."[3] Such a farmer might look upon his crop year as a complete and coherent historical event, suffused and illuminated with a meaning and mystery that were both its own and the world's because in his mind and work, agricultural process had come into an enduring and preserving harmony with natural process. To him, the past confidently promised a future. What are we to say, by contrast, of a society that places no value at all upon such a tradition or such a man, that instead works the destruction of such imperfect agricultural traditions as it has, that replaces farm people with machines, that values the techniques of production far above the techniques of land maintenance, and that has espoused as an ideal a depopulated countryside farmed by a few technicians for the supposedly greater benefit of hundreds of millions crowded into cities and helpless to produce food or any other essentials for themselves?

2. The agricultural optimism that bases itself upon the moon landings assumes that there is an equation between agriculture and technology or that agriculture is a kind of technology. This grows out

of the much-popularized false assumptions of the agricultural specialists, who have gone about their work as if agriculture were answerable only to the demands of economics, not to those of ecology or human culture, just as most urban consumers conceive eating to be an activity associated with economics but not with agriculture. The discipline proper to eating, of course, is not economics but agriculture. The discipline proper to agriculture, which survives not just by production but also by the return of wastes to the ground, is not economics but ecology. And ecology may well find its proper disciplines in the arts, whose function is to refine and enliven *perception*, for ecological principle, however publicly approved, can be enacted only upon the basis of each man's perception of his relation to the world.

Under the governance of the laboratory analogy, the *device*, which is simple and apparently simplifying, becomes the focal point and the standard rather than the human need, which is complex. Thus, an agricultural specialist, prescribing the best conditions for the use of a harvesting machine, thinks only of the machine, not its cultural or ecological effects. And because of the fixation on optimum conditions, big-farm technology has come to be highly developed, whereas the technology of the family farm, which must still involve methods and economics that are "old-fashioned," has been neglected. For this reason, and others perhaps more pressing, small-farm technology, along with the small farmers, is rapidly passing from sight. As a result, we have an increasing acreage of supposedly "marginal," but potentially productive, land for the use of which we have neither methods nor people—an alarming condition in view of the likelihood that someday we will desperately need to farm these lands again.

The drastic and incalculably dangerous assumption is that farming can be considered apart from farmers, that the land may be conceptually divided in its use from human need and human care. The assumption is that moving a farmer into a factory is as simple a cultural act as moving a worker from one factory to another. It is inconceivably more complicated and final. American agricultural tradition has been for the most part inadequate from the beginning, and we have an abundance of diminished land to show for it. But American farmers are nevertheless an agricultural population of long standing. Most settlers who farmed in America farmed in Europe. The farm population in this country therefore embodies a knowledge and a set of attitudes and interests that have been literally thousands of years in the making. This mentality is, or was, a great resource upon which we might have built a truly indigenous agriculture, fully adequate to the needs and demands of American regions. Ancient as it is, it is destroyed in a generation in every family that is forced off the farm into the city—or in less than a generation, for the farm mentality can survive only in sustained vital contact with the land.

A truer agricultural vision would look upon farming not as a function of the economy or even of the society but as a function of the land, and it would look upon the farm population as an indispensable and inalienable part of the ecological system. Among the Incas, according to John Collier, the basic social and economic unit was the tribe, or *ayllu*, but he says "the *ayllu* was not merely its people, and not merely the land, but people and land wedded through a mystical bond."[4] The union of the land and the people was indissoluble, like marriage or grace. Chief Rekayi of the Tangwena tribe of Rhodesia, in refusing to leave his ancestral home which had been claimed by the whites, is reported in a recent newspaper account to have said: "I am married to this land. I was put here by God . . . and if I am to leave, I must be removed by God who put me here." This altogether natural and noble sentiment was said by the Internal Affairs Minister to have been "Communist inspired."

3. The notion that the moon voyages provide us assurance of enough to eat exposes the shallowness of our intellectual confidence, for it is based upon our growing inability to distinguish between training and education. The fact is that a man can be made an astronaut much more quickly than he can be made a good farmer, for the astronaut is produced by training and the farmer by education. Training is a process of conditioning, an orderly and highly efficient procedure by which a man learns a prescribed pattern of facts and functions. Education, on the other hand, is an obscure process by which a person's experience is brought into contact with his place and his history. A college can train a person in four years; it can barely begin his education in that time. A person's education begins before his birth in the making of the disciplines, traditions, and attitudes of mind that he will inherit, and it continues until his death under the slow, expensive, uneasy tutelage of his experience. The process that produces astronauts may produce good soldiers and factory workers and clerks; it will never produce good farmers or good artists or good citizens or good parents.

White American tradition, as far as I know, contains only one coherent social vision that takes such matters into consideration, and that is Thomas Jefferson's. Jefferson's public reputation seems to have dwindled to that of Founding Father and advocate of liberty, author of several documents and actions that have been enshrined and forgotten. But in his thinking, democracy was not an ideal that stood alone. He saw that it would have to be secured by vigorous disciplines, or its public offices would become merely the hunting grounds of mediocrity and venality. And so those who associate his name only with his political utterances miss both the breadth and depth of his wisdom. As Jefferson saw it, two disciplines were indispensable to democracy: on the one hand, education, which was to produce a class of qualified leaders, an aristocracy of "virtue and talents" drawn from all economic classes; and on the other hand, land, the widespread possession of which would assure stable communities, a tangible connection to the country, and a permanent interest in its welfare. In language which recalls Collier's description of the *ayllu* of the Incas and the language of Chief Rekayi of the Tangwenans, he wrote that farmers "are tied to their country, and wedded to its liberty and interests, by the most lasting bonds." And: ". . . legislators cannot invent too many devices for subdividing property. . . ." And: ". . . It is not too soon to provide by every possible means that as few as possible shall be without a little portion of land. The small landholders are the most precious part of a state. . . ." For the discipline of education of the broad and humane sort that Jefferson had in mind, to produce a "natural aristocracy . . . for the instruction, the trusts, and government of society," we have tended more and more to substitute the specialized training that will most readily secure the careerist in his career. For the ownership of "a little portion of land" we have, and we apparently wish, to substitute the barbarous abstraction of nationalism, which puts our minds within the control of whatever demagogue can soonest rouse us to self-righteousness.

On September 10, 1814, Jefferson wrote to Dr. Thomas Cooper of the "condition of society" as he saw it at that time: ". . . We have no paupers, the old and crippled among us, who possess nothing and have no families to take care of them, being too few to merit notice as a separate section of society. . . . The great mass of our population is of laborers; our rich . . . being few, and of moderate wealth. Most of the laboring class possess property, cultivate their own lands . . . and from the demand for their labor are enabled . . . to be fed abundantly, clothed above mere decency, to labor moderately. . . . The wealthy . . . know nothing of what the Europeans call luxury." This has an obvious kinship with the Confucian formula "that the producers be many and that the mere consumers be few; that the artisan mass be energetic and the consumers temperate. . . ."

In the loss of that vision, or of such a vision, and in the abandonment of that possibility, we have created a society characterized by degrading urban poverty and an equally degrading affluence—a society of undisciplined abundance, which is to say a society of waste.

## Notes

1. J. Russell Smith, *Tree Crops: A Permanent Agriculture* (Old Greenwich, Conn.: Devin-Adair Co., 1953).

2. Albert Howard, *An Agricultural Testament* (New York: Oxford University Press, 1940).

3. F. H. King, *Farmers of Forty Centuries: Permanent Agriculture in China, Korea and Japan* (London: Jonathan Cape, 1926; reprint ed., Emmaus, Pa.: Rodale Press).

4. John Collier, *Indians of the Americas* (New York: W. W. Norton, 1947).

# 3

# LAND REFORM IN AMERICA

## *Peter Barnes*

Before the year 2000 the concept of private property in land as we know it today will no longer exist in the United States. The idea that a single man or a single corporation can do anything they want with a large chunk of land is an idea that's on its way out. The process of reform has already begun in this country, and we are going to arrive at a system that takes land out of private ownership and puts it into public trust.

—*Maine Times*, 7 JULY 1972

The great cause of the inequality in the distribution of wealth is inequality in the ownership of land.

—HENRY GEORGE

### The Case for Redistribution

It's hard for people in cities to appreciate the need for land reform in the United States. Most of us have been so cut off from the land that, through ignorance, we accept present landholding patterns as desirable or inevitable. They are neither.

---

Peter Barnes is the West Coast editor of *The New Republic*.

What are the advantages of giving land to the few instead of to the many? Efficiency is supposed to be the main one: Big farms, we're told by agribusiness spokesmen, can produce more food at less cost and thus save the consumer money. That same thinking underlies Soviet collectives. What's overlooked is that in societies in which tractors are relatively inexpensive to own or rent, economies of scale contribute to agricultural abundance only marginally. Beyond a certain point, there's nothing gained by having one vast farm in place of several smaller ones. In fact, small farms are often more productive per acre because their owners work harder and take better care of the soil.

Large farms in America are efficient at some things—they excel at tapping the federal treasury and exploiting hired labor. Take away these privileges and the small farmer looks extremely good. As for saving the consumer money, the chief reason food prices have remained relatively low is not large-scale efficiency—it is intense competition. Allow a handful of agribusiness giants to gain control of the market, and prices will assuredly rise a lot more than they have. [Clearly the thrust of this statement has come true. Due to a variety of events, including corporate food monopolies, food prices of the mid 1970s can no longer be considered "relatively low."—Ed.]

There is, furthermore, the question of how much and what kind of efficiency is desirable. American agriculture is, if anything, too efficient; its chronic problem is not underproduction but surplus. It is the only industry in which people are paid *not* to produce. The argument that ever-increasing agricultural efficiency is a desirable national goal is, therefore, unsound. Moreover, what kind of efficiency are we talking about? When a large grower increases his profit margin by replacing farmworkers with a fancy new machine, he's not doing anybody but himself a favor. The farmworkers, now unemployed, drift to already overcrowded cities, where no jobs await them either. Welfare rolls and social tensions rise, transferring to society at large the ultimate cost of "efficiency" on the large farm.

If the advantages of large landholdings (except to those who own them) are scant, the harmful effects are legion. Several have already been noted: the impoverishment of millions of rural families and the migration to cities of millions more, with little education or hope for improvement. We expect poor Americans to lift themselves up the economic ladder, and yet, by cutting them off from productive landownership, we knock out the bottom rungs.

The vitality of community life in rural America has also suffered because of maldistributed land. Main Street businesses are not appreciably aided by large absentee landowners who purchase their supplies in distant cities, or by underpaid migrants who buy nothing, or by sharecroppers forced to shop at the company store. A study in the 1940s by Walter Goldschmidt, a California sociologist, found that communities in small-farm areas have a more sizable middle class, more stable income patterns, better schools, and more active civic groups than do communities where large landholdings predominate. A recent incident in Mendota, California—a town surrounded by large farms—helps explain why. A group of citizens wanted to establish a special taxing district for construction of a hospital, the nearest one being forty miles away. Three agribusiness giants that owned more than half the land in the proposed district opposed the plan and killed it. Two of the companies were based in other California cities, and the third, Anderson Clayton, was headquartered in Houston.

Protection of the environment also tends to be less of a concern to large corporations, who've been despoiling the American landscape for the better part of two centuries, than to small farmers who live on their land. Companies farming for tax or speculative reasons, for example, seek to maximize earnings over the short run. They can milk the soil, deplete the underground water supply, or poison the land with pesticides, knowing full well that they will eventually sell. Resident farmers who hope to pass on their land to their offspring cannot be so careless with nature's gifts. Moreover, small-scale farming lends itself much more readily than does large-scale monoculture to biological pest control—a technique that must increasingly be adopted if we are to avoid ecological disaster.

If there's little to be said for large landholdings on social or environmental grounds, neither can it be said that they are inevitable. Land concentration in America, particularly in the South and West, is not the result of inscrutable historical forces, but of a long train of government policies, sometimes in the form of action, often of inaction. English grants to large landholders in the colonial

South and Mexican grants in the West could have been broken up at several convenient historical moments but were allowed to remain intact. Vast expanses of public lands were given away in large chunks to speculators rather than in small parcels to settlers. Tax and labor laws, reclamation projects, and government-financed research have encouraged large-scale corporate agriculture to the detriment of independent small farmers and landless farmworkers. On top of all this have come the government's ultimate reward to big landholders—cash subsidies—mainly for being big.

Why, then, do we need land reform in America? About the only thing that can be said for large landholdings is that they exist and, in the spirit of free enterprise, ought to be left untouched. This is the strongest argument in favor of leaving things as they are. Land, however, is not like other forms of wealth in our economy, which we allow to be accumulated without limit: It is a public resource; it is finite; and it is where people live and work. Free enterprise does not merely imply the right to be big. It also implies the right to start. As corporate farms become increasingly integrated with processors and distributors, as they advance toward the technological millennium in which ten-mile-long fields are sowed and harvested by computer-controlled machines, the right to get a start in agriculture will be obliterated—as it almost already has been today. Americans must decide whether they want the rich to get richer or the poor to have a chance. Agriculture is one of the few places where the poor *can* have a chance. If it is closed off, if the profits of the few are given precedence over the needs of the many, the consequences can be only unpleasant.

There are additional reasons why it's time to reform landholding patterns in the United States. Frederick Jackson Turner talked seventy years ago of the frontier as a "safety valve" for urban discontent. If ever the cities needed a safety valve, it is now. Urban problems are virtually insoluble; city residents seem on the verge of a mass psychic breakdown. The exodus from the countryside must not only be stopped, it must be dramatically reversed.

One approach to the problem of population dispersal is to build new communities on rural lands now owned by speculators. This will undoubtedly happen, but it's far from enough. It is much more important to revive existing rural communities and to do so by enabling greater numbers of people to live decently off the land. There is no shortage of people who would remain on the land, or return to it, if they could do so at a higher than subsistence level. Many Mexican-Americans, blacks, and Indians would be among them. So would many whites who have become drained, physically and spiritually, by city living. The difficulty is that the frontier is long gone. That's why reform, as opposed to the giving away of unsettled land, is essential.

*Land reform is also needed to increase the number of people in the United States who are free.* This may sound silly in a country that presumes to be a breeder of free men. Yet ever-increasing numbers of Americans are not really free to assume responsibilities or to make major decisions affecting their lives. They work for large corporations or government bureaucracies or on assembly lines. They are not their own bosses, not proud of their work, and are not motivated to exercise their full rights as citizens. Farming has traditionally been a bastion of the independent small businessman who won't take guff from anybody and who prides himself on the quality of his work. But now farming, too, is becoming computerized and corporatized. Its executives wear silk ties and share the attitudes of other wealthy executives; its workers are powerless, dispensable hirelings. If agriculture goes the way of the auto industry, where will our independent citizens come from?

American land policy should have as its highest priority the building of a society in which human beings can achieve dignity. This includes the easing of present social ills, both rural and urban, and the creation of a lasting economic base for democracy. A second priority should be to preserve the beauty and fertility of the land. Production of abundant food should be a third goal, but it need not be paramount and is not, in any case, a problem.

To achieve these goals, a multitude of reforms should be carried out. First and most important, small-scale farming must be made economically viable so that present small farmers can survive and new ones get started. Unless it is done, there is no point in changing landholding patterns to favor smaller units.

There's no secret to making small-scale farming viable; it can be accomplished by eliminating the favors bestowed upon large farms. Federal tax laws that encourage corporate farming for tax-loss and speculative purposes should be changed, even if this means closing the capital gains loophole. Labor laws should guarantee to farmworkers a minimum wage equal to that of other workers and should make the knowing employment of illegal aliens a crime punishable by imprisonment. This would put an end to one of the large landholders' major competitive advantages—their ability to exploit great numbers of poor people—and would allow self-employed farmers to derive more values from their own labor.

Also essential to the future viability of small-scale farming is some protection against conglomerates. There is no way a small farmer can compete against an oil company or against a vertically integrated giant like Tenneco, which not only farms tens of thousands of acres but also makes its own farm machinery and chemicals, and processes, packages, and distributes its own foods. Such conglomerates aren't hurt by a low price for crops; what they lose in farming they can pick up in processing or distributing or, for that matter, in oil. The small farmer, on the other hand, has no outside income and no tolerance for soft spots. What he needs is legislation that would prohibit corporations or individuals with more than fifty thousand dollars, say, in nonfarming income from engaging in farming—in effect, a forceful antitrust policy for agriculture.

Once small-scale farming is made viable, the second major area for change involves the redistribution of land. The guiding principles behind redistribution are that land should belong to those who work and live on it and that holdings should be of reasonable, not feudal proportions. These are not revolutionary concepts; America recognized them in the Pre-emption, Homestead, and Reclamation acts and is merely being asked to renew that recognition.

A convenient place to start is with enforcement of the Reclamation Act of 1902, which provides that large landholders in the West who accept subsidized water must agree to sell their federally irrigated holdings in excess of 160 acres at prewater prices within ten years. The Reclamation Act has never been properly enforced for a variety of reasons. One is that, through one stratagem or another, large landholders have escaped having to sell their excess lands. Another is that even in the few cases in which large landowners *have* agreed to sell, their prices have been so high and terms so stiff that only the wealthy could afford to buy. Occasionally, as in parts of the San Joaquin Valley at the moment, prewater prices as approved by the Bureau of Reclamation are so out of line—higher, in fact than prevailing market prices—that even wealthy persons have not seen fit to purchase excess lands put up for sale under the law.

To assure not only the sale of excess landholdings but also their availability at prices that persons of limited means can afford, Rep. Robert Kastenmeir (D-Wisc.), Jerome Waldie (D-Calif.), and others have introduced legislation that would authorize the federal government itself to buy up all properties in reclamation areas that are either too big or owned by absentees. The government would then resell some of these lands, at reasonable prices and on liberal terms, to small resident farmers and retain others as sites for new cities or as undeveloped open space. The plan would actually earn money for the government because the lands would be purchased at true prewater prices and resold at a slight markup. The money thus earned could be used for education, conservation, or other purposes.

Other plans for enforcing the Reclamation Act are worth study. For example, the federal government could purchase irrigated lands in excess of 160 acres and lease them back to individual small farmers or to cooperatives. Or it could buy large landholdings in reclamation areas with long-term "land bonds," which it then would redeem over forty years with low interest payments made by the small farmers to whom the land was resold. This would amount to a subsidy for the small farmers who bought the land, but it would be no more generous than the current subsidy to large landholders who buy federal water.

Of course, land redistribution should go beyond the western areas served by federal reclamation projects; in particular, it should reach into the South. Thaddeus Stevens's old proposal for dividing up the large plantations into forty-acre parcels is unrealistic today, but an updated plan, with due compensation to present owners, could be devised and implemented.

Another objective toward which new policies should be directed is preserving the beauty of the land. Reforms in this area are fully consistent with a restructuring of landholding patterns. Thus, a change in local tax laws so that land is assessed in accordance with its use would benefit small farmers and penalize developers. Zoning rural land for specific uses, such as agriculture or new towns, would similarly help contain suburban sprawl and ease the pressure on small farmers to sell to developers or speculators. If as a result of new zoning laws the value of a farmer's land were decreased, he would be compensated for that loss.

An indefinite moratorium should also be placed on further reclamation projects, at least until the 160-acre and residency requirements are enforced, and even then, they ought to be closely examined for environmental impact. Schemes are kicking about to bring more water to southern California and the Southwest from northern California, the Columbia River, and even Alaska. These plans ought to be shelved. Federal revenues that would be spent on damming America's last wild rivers could, in most cases, be more fruitfully devoted to such purposes as redistributing croplands.

Policy changes in other areas should complement the major reforms outlined above. Existing farm-loan programs, for example, should be greatly expanded so that new farmers can get started in agriculture. Farming cooperatives, which can be a starting point for workers unable to afford an entire farm, should be encouraged through tax laws and credit programs. Research funds spent on developing machinery for large-scale farming should be rechanneled into extension programs for small farmers and co-ops.

It won't be easy to enact any of these reforms. Friends of large-scale agribusiness are strategically scattered throughout the Agriculture, Interior, and Appropriations committees of Congress and are equally well ensconced within the present administration. Small-farmer associations like the Grange, the National Farmers Union, and the National Farmers Organization don't have nearly the clout of the American Farm Bureau Federation, the big grower associations, and the giant corporations themselves. The pro-industry land policy "experts" who formed the Public Land Law Review Commission, which reported its findings last year, were no friendlier to small-scale farming: They recommended repeal of the Reclamation Act's 160-acre limitation and residency requirement and adoption of policies favorable to large-scale mechanized agriculture.

Nevertheless, there are some grounds for optimism. Many citizens and public officials are coming to realize that rural America ought to be revived, cities salvaged, welfare rolls reduced, and they see that present policies aimed at achieving these objectives are not working. Environmentalists who for years have pointed to the dangers of intensive agriculture and the need for prudent rural land use, are finally getting an audience. The list of organizations that have recently urged vigorous enforcement of the 160-acre limitation includes the AFL-CIO, the Sierra Club, Common Cause, the National Education Association, the Grange, and the National Farmers Union. That's not enough to sweep Congress off its feet, but it's a good start.

*The ultimate political appeal of land reform is that it places both the burden and opportunity of self-improvement upon the people themselves. It can give hundreds of thousands of Americans a place to plant roots and a chance to work with dignity. Can we deny them that chance?*

## Addendum

Since the previous article was written in early 1971, there has been a blossoming of interest in various aspects of land reform. The Senate Migratory Labor Subcommittee, chaired by Senator Adlai E. Stevenson III, conducted three days of hearings in California that focused attention on the problems of small farmers and landless farmworkers. A family farm act, which would ban large conglomerates from agriculture, was cosponsored by dozens of senators and congressmen. The National Coalition for Land Reform, headed by former Senator Fred Harris was launched. And a national conference on land reform, held in San Francisco in April 1973, agreed that radically new approaches to land tenure were needed in America. At that meeting, the following "Declaration of Participants" was drafted:

Land is a precious and finite resource and the birthright of the people. Its ownership and control, and the associated economic and political power, must be widely distributed.

A sound land policy should regulate the use of the land in the public interest; keep the land in the hands of those who live and work on it; put the land in trust for the public good; and prevent it from falling into the hands of large corporations and wealthy individuals who are absentee owners. It should preserve and strengthen the family farm, make it possible for people on the land to earn a decent living, and provide conditions that revitalize rural communities. Government policies that encourage absentee owner-ship, the corporate takeover of agriculture, and the exodus of people from rural areas to the cities must be reversed.

A broad range of Americans have a vital interest in regaining control of land from absentee corporations in which it is unduly concentrated. We urge environmental organizations, labor unions, independent bankers, small business groups, farmers and farm organizations, cooperative members, consumers, and low-income and minority groups to join forces in shaping and implementing policies that preserve land and jobs, create parks and housing, provide recreational and economic opportunities, and protect legiti-mate land ownership and use while discouraging the abuses of concentrated absentee ownership and irresponsible economic and political power.

Energy resources on public lands should be developed by public entities in the public interest, and the give-away of public resources to large corporations must cease.

We urge national officials responsible for law observance to en-force national policy as expressed in acreage limitations and resi-dency requirements for receivers of federal irrigation water. We also favor government purchase of excess lands at the prewar price specified by law.

Timber on public lands should be made available on a preferen-tial basis to independent woodcutters and cooperatives.

Absentee corporations with nonfarm interests should be barred from agriculture.

Tax laws that encourage ownership of land by speculators, cor-porations, and absentee landlords should be repealed. "Tax loss" farming, preferential treatment for capital gains, depletion allow-ances, and underassessment of corporate landholdings must be eliminated. Large landholdings should be discouraged through the use of progressive property taxes, taxes on unearned increment in land value, and increases in severance taxes. The proceeds of taxes on large landholdings should be used to provide human social services such as schools and health care for rural people, to preserve wilderness land, and to encourage small-scale farming, timber, and other rural enterprises through the financing of land banks and trusts.

Indians have been stripped of their historic claims to land. Shame-ful violations now occurring should be halted and the Indian land base preserved. In any land reform policy the Indian trust relation-ship with the federal government and Indian water rights should remain undisturbed. Treaties, executive orders, and hunting and fishing rights should be upheld. To provide for economic growth the tax-free status of Indian lands should be maintained. The policy of individual allotment was designed to alienate the land and break up the reservations; a positive government program should be initiated to consolidate these allotments under tribal control where the tribe so requests.

Land granted to railroads by federal and state governments other than rights-of-way presently retained by railroads should revert to original Indian tribes now on those lands. Mineral rights claimed or appropriated should also revert to said tribes. Remaining land not needed for rights-of-way should revert to the public domain. Additional mineral rights claimed or appropriated by railroads should also revert to the public domain.

Hispanic-Americans have been robbed of common and private lands on a massive scale throughout the Southwest in gross violation of the Treaty of Guadalupe Hidalgo of 1848. These legitimate land claims must be resolved.

In any distribution or disposal of surplus lands, Indians and Hispanic-Americans with historic claims, along with low-income people generally, should receive priority.

The technical and financial assistance programs of the U.S. Department of Agriculture and the land grant colleges should be completely restructured. Their services should be offered exclusively to family farmers, farmworkers, and consumers. Moreover, the USDA should promote the development of new technologies that enable people to earn a decent living on family-size farms. The crop-supply–management program should be restored to its original objective of helping family farmers remain on the land by assuring that they receive a fair return for their labor and investment. New efforts, such as providing long-term, low-interest loans, should be given high legislative priority to allow able and willing young people to enter family farming.

The exploitation of rural labor must be halted immediately. We support the efforts of farmworkers, woodcutters, and other rural workers to organize effective unions and associations. Decent minimum wages, unemployment compensation, and union protection similar to the provisions of the original Wagner Act should be extended to all rural workers and the exploitation of child labor must be abolished. We support the nonviolent struggle of the United Farm Workers and the boycott of nonunion lettuce and grapes.

Abuse of the land must also be halted at once. Strip mining should be prohibited. Severe penalties should be applied to corporate offenders in all cases of excessive water and air pollution and irresponsible timber and mining operations. A shift away from monoculture which depletes the natural fertility of the soil should be a long-range goal of national farm policy.

We condemn the monopolistic market power of big corporations like General Mills, Del Monte, International Paper, Tenneco, Consolidated Coal, and ITT in food, timber, coal, and other land-based industries. We call for the vigorous enforcement of antitrust laws in these industries.

To achieve these policies aimed at reclaiming America's land for the benefit of her people, we conferees resolve to carry on the work begun at this First National Conference on Land Reform. We resolve to return to our regions and localities to help educate and organize a network of people committed to a redirection of the nation's land policies at the local, state, and federal levels.

We urge the creation of a national land reform coalition to perform the following functions: (1) actively educate opinionmakers and public officials about the need for new policies toward land and people; (2) develop concrete programs to implement the principles contained in these resolutions; (3) stimulate and support local organizing efforts; and (4) publish and circulate a newspaper among all those interested in building a land reform movement.

One approach to land tenure that has aroused growing interest is the nonprofit land trust. Essentially, a land trust is a corporation designed to hold land in perpetuity for humane and ecological purposes. Lessors of trust-held land may use it, farm it, and build houses on it, but they can't own or sell it. The most notable example of a land trust is the Jewish National Fund which holds about 60 percent of the arable land in Israel in permanent trust for small farmers and cooperatives. In Britain, the National Trust holds many scenic and historical properties in trust for future generations, as does the Nature Conservancy in the United States. Small agricultural trusts have recently been created in Georgia and in Maine. For a detailed explanation of the theory and practice of land trusts, the interested reader is referred to *The Community Land Trust Guide*, available for three dollars from the International Independence Institute, West Road, Box 183, Ashby, Massachusetts 01431.

The most formidable problem confronting land trusts, rural cooperatives, and alternative agricultural ventures generally is financing. Banks usually won't lend money to land trusts or small co-ops, and government sources have all but dried up. The need for substantial long-term funding has stimulated exploration of some novel political initiatives. One idea that has emerged in California is that of a state trust fund, financed by a severance tax on oil and minerals and a capital gains tax on the rise in land value. Both taxes would fall almost entirely on the biggest land

and resource profiteers. The revenues from the taxes would be earmarked for open-space land acquisition, grants to nonprofit land trusts, and low-income cooperatives. Further information on the trust-fund approach is available from the Center for Rural Studies, 345 Franklin Street, San Francisco, California 94102.

### References

Advisory Commission on Intergovernmental Relations. *Urban and Rural America*. Washington, D.C.: Government Printing Office, 1968.

Baker, George L. "Land is Power: The Kingdom of Railroads." *Nation*, 12 March 1973, pp. 334–339.

Barnes, Peter, and Casalino, L. *Who Owns the Land?: A Primer on Land Reform in the USA*. Berkeley, Calif.: Center for Rural Studies, 1972.

Clawson, Marion. *America's Land and Its Uses*. Published for Resources for the Future, Inc. Baltimore: Johns Hopkins Press, 1972. A compact, nontechnical account of the nation's land and its many uses; a consideration of past history, present trends, and future possibilities, with chapters on recreation, agriculture, and urbanization.

*Community Economics*. An occasional bulletin. Issue on land. Cambridge, Mass.: Center for Community Economic Development, May 1972. $1.00.

Contant, Florence. "Land Reform: A Bibliography." Cambridge, Mass.: Center for Community Economic Development, 1972. 40¢. A nine-page bibliography on land reform in the United States and abroad; land banking, land problems, land use, land as a corporate investment, land ownership.

Rose, John Kerr. *Survey of National Policies on Federal Land Ownership*, Senate Report, 85th Congress, Doc. No. 338. Washington, D.C.: Government Printing Office, 1957. Surveys historic policies on federal lands, as well as studies, proposals, and bills to divest certain public lands to the states and private economy.

*Stewardship*. New York: Open Space Institute, 145 E. 42 St., 1965.

U.S. Dept. of Agriculture. *Land Tenure in the U.S.: Development and Status*. Economic Research Bulletin No. 338. Washington, D.C.: Government Printing Office, 1969.

# PART II

# Agriculture and Agribusiness

# 4

---

## AGRIBUSINESS

### Nick Kotz

The farmhouse lights are going out all over
America. And every time a light goes out, this
country is losing something. It is losing the precious skills of a family farm system that has given
this country unbounded wealth. And it is losing
free men.
>—Oren Lee Stanley
>National Farmers Organization

### Conglomerates Reshape American Food Supply

The name "Tenneco" is not yet a household word to American
consumers, but it weighs heavily on the minds of the nation's
embattled farmers and of government officials who worry about
the cost of food and the fate of rural America, for Tenneco, Inc.,
the thirty-fourth largest U.S. corporation and fastest-growing conglomerate, has become a farmer. Its new activities symbolize an
agricultural revolution that may reshape beyond recognition the
nation's food supply system. Dozens of the largest corporations,
with such unfarmlike names as Standard Oil, Kaiser Aluminum,
and Southern Pacific have diversified into agriculture.

---

Nick Kotz is a reporter for the Washington *Post*.

What concerns farmers, processors, and wholesalers is that the new breed of conglomerate farmers does not just grow crops or raise cattle. The corporate executives think in terms of "food supply systems" in which they own or control production, processing, and marketing of food. "Tenneco's goal in agriculture is integration from seedling to supermarket," the conglomerate reported to its stockholders. Its resources to achieve that goal include 1970 sales of $2.5 billion, profits of $234 million, and assets of $4.3 billion in such fields as oil production, shipbuilding, and manufacturing.

The conglomerate invasion of agriculture comes at a time when millions of farmers and farmworkers have already been displaced, contributing to the problems of rural wastelands and congested cities. More than 100,000 farmers a year are quitting the land, and more than 1.5 million of those who remain are earning less than poverty-level farm incomes. Their plight is severe. Although the U.S. census still counts 2.9 million farmers, 50,000 grow one-third of the country's food supply and 200,000 produce more than one-half of all food. The concentration of production is especially pronounced in such crops as fruit, vegetables, and cotton. In 1965, 3,400 cotton growers accounted for 34 percent of sales, 2,500 fruit growers had 46 percent of sales, and 1,600 vegetable growers had 61 percent of the market.

The medium to large-size "family farms"—annual sales of $20,000 to $500,000—survived earlier industrial and scientific revolutions in agriculture. They now face a financial revolution in which the traditional functions of the food supply system are being reshuffled, combined, and coordinated by corporate giants. "Farming is moving with full speed toward becoming part of an integrated market-production system," says Eric Thor, an outspoken farm economist and director of the Agriculture Department's Farmer Cooperative Service. "This system, once it is developed, will be the same as industrialized systems in other U.S. industries."

Efforts to bar large corporations from farming have come too late, says Thor: "The battle for bigness in the food industry was fought and settled thirty-five years ago—chain stores versus 'Ma and Pa stores.'" Corporate takeover of the poultry industry did result in lower consumer prices. But for numerous food products, corporate farming has not lowered grocery costs because the price of raw food materials is not a significant factor in determining final retail prices. For example, the cost of a food container is sometimes more than the farmer receives for the food packaged in it.

The new corporate farmers account for only 7 percent of total food production, but they have made significant inroads in certain areas. Twenty large corporations now control poultry production. A dozen oil companies have invested in cattle feeding, helping shift the balance of production from small Midwestern feed lots to 100,000-head lots in the high plains of Texas. Just three corporations—United Brands, Purex, and Bud Antle, a company partly owned by Dow Chemical—dominate California lettuce production. The family farmer still rules supreme only in growing corn, wheat, and other grains, and even here constantly larger acreage, machinery, credit, and higher prices are needed for the family farmer to stay profitably in business.

Even the largest independent California farmers question how they can compete with a corporation which can, at least in theory, own or control virtually every phase of a food supply system. Tenneco can plant its own vast acreage. (Tenneco announced in 1972 that it was stopping growing on its own land but continuing its other food production functions.) It can plow those fields with its own tractors which can be fueled with its own oil. It can spray its crops with its own pesticides and utilize its own food additives. It then can process its food products in its own plants, package them in its own containers, and distribute them to grocery stores through its own marketing system.

Financing the entire operation are the resources of a conglomerate with billions in assets, hundreds of millions in tax-free oil income, and interests in banking and insurance companies. Tenneco, according to reports filed with the Securities and Exchange Commission, had in 1969 gross oil income of $464 million and taxable oil income of $88.7 million. Yet due to federal tax breaks, Tenneco not only paid no taxes on that income, but had a tax credit of $13.3 million.

The type of food system being put together by Tenneco and other conglomerates frustrates and frightens independent farmers. They see every element of the food business acquiring market power but themselves. On one side, they confront the buying power of giant food chains. Now they must also compete with conglomerates that can take profits either from production, processing, or marketing. The individual farmer usually does not have such options. The giant competitors also benefit from a variety of government subsidies on water, crops, and income taxes.

Contrary to popular notion and most galling to the efficient large independent farmer, the corporate giants generally do not grow food cheaper than they do. Numerous USDA and university studies show that enormous acreage is not needed to farm efficiently. For example, maximum cost-saving production efficiency is generally reached at about 1,500 acres for cotton, less than 1,000 acres for corn and wheat, and 110 acres for peaches. Thousands of independent family farmers possess such needed acreage and farm it with the same machinery and techniques used by their rivals. In fact, studies show that the largest growers incur higher farm production costs as they employ more workers and administrators.

The nation's fruit and vegetable growers are not strangers to the tough competition of agribusiness. For many years, they have wrestled with the market power of chain stores and major food processors. They sell to canners such as Del Monte, Libby-McNeil & Libby, Green Giant Co., H. J. Heinz Co., and Minute Maid Corp. (a subsidiary of Coca-Cola). Each of these canners also competes with the independent farmer by growing large amounts of its own food supply. But the new conglomerate represents a different kind of competition. The older agribusiness corporations are primarily food companies and must make money somewhere in the food distribution system. Such is not necessarily the case with the new conglomerate farmers, for whom millions of dollars of agribusiness investment may represent only a fraction of their total holdings. Only 4 percent of Tenneco's sales are from agriculture. The new conglomerates utilize a variety of federal tax provisions that permit them to benefit from tax-loss farming and then profit again by taking capital gains from land sales. In fact, the conglomerates may find their food investments profitable even without earning anything from them. The profits may come from land speculation, federal crop subsidies, or generous federal tax laws.

## Agribusiness Threatens Family Farm

Although admitting the increasing concentration of corporate power in fruit and vegetable production and the corporate take-over of poultry farming, USDA officials generally contend that this phenomenon will not spread to other farm products. Many mid-western cattle, hog, and grain farmers disagree. They fear that the cattle and hog feeding business, their best source of income, may follow the pattern in which independent poultry growers were wiped out. About twenty corporations including Allied Mills, Ralston Purina, and Pillsbury Co., originally went into poultry production as a means of developing markets for their feed. Farmers were signed up to grow the corporation's poultry using their feed. According to USDA studies, the poor but once inde-pendent poultry farmers are still as poor as contract workers, earning about fifty-four cents an hour. The corporations, however, contend that they have benefited small farmers with a steady, if small, source of income. And, they say, they have given consumers lower priced chicken and turkey.

However, the contest between the family farmer and the con-glomerates is incredibly unequal. There is Tenneco with its $4.3 billion in assets and its ability to employ its own land, tractors, pesticides, oil, processing plants, and marketing system. On the other side, there are the small farmers trying to hold on in the face of a dilemma. For example, if the processors and conglom-erates gain control of hog and cattle feeding, then midwestern family farmers will have to get all their income from growing corn, wheat, and soybeans. Farmers fear they cannot survive if their only function is to provide grain for an integrated food system in which most profits are taken further up the food chain of animal feeding, processing, marketing, and retail sales.

The threat to the "family farm," and to the way of life it repre-sents, is so strong that even the American Farm Bureau Federation, the nation's largest and most conservative farm organization, shows symptoms of upheaval. In the past, the AFB has consistently and vigorously opposed federal intervention in the farm economy. But today it is swallowing its ideology and asking for federal laws to strengthen individual farmers in dealing with the new corporate forces in agriculture.

The stakes in this struggle between independent farmers and the giant new farm corporations are immense: Food is the nation's largest business with $114 billion in annual retail sales. More than $8 billion in annual farm exports keep the U.S. balance of trade from becoming an economic disaster. The question of who in agriculture is to share in this bounty, and on what terms, is at the root of the National Farmers Organization's militance and the Farm Bureau's philosophical turnaround.

Will the family farm survive in the years ahead? Or will agriculture become—like steel, autos, and chemicals—an industry dominated by giant conglomerate corporations such as Tenneco? In that case, the nation will have lost its prized Jeffersonian ideal, praised in myth and song, of the yeoman farmer and independent landowner as the backbone of America. What will become of rural America if the greatest migration in history—forty million to the cities in fifty years—is further accelerated? Farmers have provided the economic base of the small towns, and that base is becoming perilously small. What will be the effect of a rural wasteland on the American political system? The power of the farm lobby and the small towns, already in sharp decline, has traditionally provided a counterbalancing force to the politics of the big cities, and a buffer area in times of social crises. On all these questions, the symptoms are not encouraging for the family farm system. A million farms are eliminated every ten years and only 2.9 million remain.

The NFO plan for saving the family farmer includes legislation prohibiting farming by large conglomerate corporations, closing loopholes that promote tax-loss farming by nonfarmers, and providing easier financial credit for young farmers. But the NFO has little confidence in getting help from a Congress in which the farm vote has shrunk into political insignificance. Its basic strategy is to organize farmers into bargaining blocks of sufficient power to raise prices for their beef, hogs, grain, and other commodities from the marketplace.

The National Farmers Organization's ultimate goal is to protect the "family farmers" of the world from forces over which they have minimal control—giant food chains, food manufacturers, and conglomerates that are attempting to bring to agriculture the industrial bigness, "efficiency," and control that characterizes much of the American economy.

### Alternatives

Several years ago the Farm Bureau organized voluntary bargaining associations but learned, to its surprise, that its old friends and philosophical allies in agribusiness were not cooperative. Agribusiness corporations such as Campbell Soup Co., Green Giant Co., Del Monte Corp., and Pillsbury Co. flatly refused to sit down at the bargaining table. Many Farm Bureau members suddenly looked at their prestigious organization in a different light. The Farm Bureau had built a $4-billion empire selling life insurance and supplies to farmers. But what, asked farmers, had the Farm Bureau done for them?

Still another approach to increased farmer power is taken by advocates of giant farmer cooperatives, which already are powerful in the dairy industry and in California citrus. The co-ops believe that farmers must compete by creating their own vertically integrated systems of production, processing, and marketing. The giant dairy co-ops also seek to win higher prices under government-approved marketing orders by exercising political muscle in campaign financing. The dairy co-ops poured $432,000 into a 1972 Republican campaign chest for President Nixon's reelection. [The ethics and legality of this contribution have, of course, been seriously questioned and litigated. —Ed.]

Some farmers complain, however, that the "super co-ops" have become just another kind of conglomerate giant from which they get few benefits. For example, Sunkist Growers, Inc., which dominates 80 percent of California citrus, is a many-layered, pyramid-shaped corporation. Small growers are at the bottom. Contrary to general knowledge, the processors at the top of the "super co-op" include major private corporations as well as farmer-owner processors. Critics contend that decisions are made and profits are taken at the top of the pyramid with too little consideration paid to the economic interests of the small grower.

In general, the various plans of farm groups to save the family farm face an uncertain future. Their legislative and organizational prospects are seriously weakened by traditional divisions in their own ranks. The NFO is suspicious of the Farm Bureau and is itself distrusted as too "radical" by other farmers. The National Farmers Union, which represents midwestern grain producers, has its own legislative goals. Other farmers, including cattlemen, fear that mandatory bargaining—a Farm Bureau proposal—will merely stimulate further vertical integration by the conglomerates. Faced with the prospects of collective bargaining, giant meat packers, canners, and sugar refiners may respond by growing even more of their own raw food materials.

Indeed, it is difficult to design legislation to meet the differing problems of Iowa corn producers and California fruit growers. Furthermore, the agriculture committees of Congress are confronted with new conflicts of interest. In the past, these committees had little trouble satisfying both big farmers and corporate food processors. The big farmer and conglomerate both benefited from farm subsidy payments, a cheap labor supply, and foreign aid food programs. But Senate and House agriculture committees are increasingly faced with difficult choices—resolving new conflicts between independent farmers and the corporations. Agribusiness, led by the National Canners Association, National Broiler Council, and the American Meat Institute, strongly opposes bargaining legislation. These committees give considerable weight—as do many economists—to the agribusiness argument that farm commodity prices are determined on a day-to-day basis in a highly competitive world market and that rigid bargaining legislation might well weaken the ability of American agriculture to compete in world trade. They are concerned, too, about maintaining the vigorous competition that now exists among food processors who fight for position in retail stores and who seek to satisfy shifting consumer preferences that often are geared to price. Processors want to retain this pricing flexibility and fear the rigidities that could come from enforced bargaining.

The political disputes and maneuverings are still largely regarded by consumers, urban politicians, and the news media as intramural issues involving "the farm problem." But the broadest issue involves the future shape of America and of its rural communities. There is the strong, compelling desire in rural America to maintain the family farm and the small town. Independent farmers question whether a way of life will be replaced by another industrialized system, administered by the forces of big labor and big industry. And migrant farmworkers, struggling to organize, question whether society does not have some obligation to help the lowest-paid worker who is being replaced by machines.

The problems created in "rural America" by various government policies have prompted politicians and presidents to come up with new programs and new rhetoric to "save" the small towns and the small farms of the country. There have been, in recent years, "wars on poverty," "rural development" schemes, and efforts at "balanced national growth." But thus far, the powerful and impersonal forces of corporate agriculture have been the dominant factors in the changes sweeping the farm economy. The measures that might reverse the trend—strong farmworker labor unions, generous subsidies to small cooperatives, the redistribution of land from corporate farmers to individual farm entrepreneurs— have not been undertaken.

What is happening in American agriculture—bigness, concentration, and the resulting efficiency—may be good or bad for the country in the long run. But the implications of these tendencies are very serious. These implications include the following:

1. The future shape of the American landscape. Already in this country, 74 percent of the population lives on only 1 percent of the land. If present trends continue, only 12 percent of the American people will live in communities of less than 100,000 by the twenty-first century; 60 percent will be living in four huge megalopoli; and 28 percent will be in other large cities.
2. The further erosion of rural life, already seriously undermined by the urban migration. Today, 800,000 people a year are migrating from the countryside to the cities. Between 1960 and 1970 more than half of our rural counties suffered population declines. One result is the aggravation of urban pathology—congestion, pollution, welfare problems, crime, and the whole catalogue of central city ills.
3. The domination of what is left of rural America by agribusiness corporations. This is not only increasing the amount of productive land in the hands of the few but is also accelerating the migration patterns of recent decades and raising the specter of a kind of twentieth-century agricultural feudalism in the culture that remains.

In response to this vision of the future, the federal government in the 1960s undertook limited measures to stimulate the survival of the small farm and the small towns of America. The anti-poverty programs administered by the Office of Economic Opportunity (OEO) touched the problem in certain ways. [The dismantling of OEO by the Nixon administration was but one example of the government's support of big business in agriculture. —Ed.]

### Solution?

Some reformers argue that the small farmer can still be given a place in America if the government brings about "land reform," including enforcement of the 1902 Reclamation Law. This law originally was designed to protect the small farmer. It provided that government-irrigated land could not be owned by absentee landlords and that no individual could own more than 160 acres of government-irrigated land. The law has never been enforced. In California alone, corporate landholders continue to occupy and benefit from more than one million acres subject to the 160-acre limitation.

Unless present trends are reversed, the ultimate cost of the new conglomerate revolution in agriculture will be paid by the small towns of the Midwest and of California. Jack Molsbergen, a real estate man in Mendota, California, describes as "disastrous" the effects of conglomerate farming on his town in the western San Joaquin Valley. Conglomerate farmers such as Anderson Clayton & Co., the country's 185th largest corporation with 1970 sales of $639 million, contribute little to the local enonomy, says Molsbergen. The conglomerates buy their farm machinery and supplies directly from the factory and their oil directly from the refinery. When Mendota tried to build a hospital several years ago, says Molsbergen, Anderson Clayton and two other large corporate landowners blocked the project because it would increase their property taxes. "The guy who made the decision for Anderson Clayton lives in Phoenix," explains Molsbergen, "and if you live in Phoenix, you don't need a hospital in Mendota. These corporate guys don't go around with a Simon Legree mustache. They are nice men. It's just the way things are."

Agriculture Department economists do not see any future for

the Mendota, Californias of the country. "These towns represent the unfulfilled dreams of the people who went there," says USDA economist Warren Bailey. "They are going the same way as the neighborhood grocery. People want to shop where they have a choice. With air-conditioned cars and good roads, they choose to do their shopping in the cities. Iowa really doesn't have room for more than twelve regional centers. The small town will remain only as a pleasant place to live."

As matters now stand, the small towns will die and the small farmer and farmworker will be replaced without any of the attention and national debate that has focused on other economic disruptions. There is a marked contrast between national concern shown over the economic problems of a Lockheed and over the problems of 150,000 small North Carolina tobacco farmers, who soon will be displaced by a new tobacco harvester. Woodrow L. Ginsburg, research director of the Center for Community Change, contrasts that concern: "When tens of thousands of scientists and skilled technicians were threatened with loss of jobs in the aerospace industry, a host of industrialists, bankers, and others besieged Congress for large-scale loans and special legislation. But when even larger numbers of workers are threatened with loss of jobs in the tobacco industry, scarcely a voice is raised. What corporate executive speaks for such workers, what banker pleads for financial aid for them, what congressman or state official calls upon his colleagues to enact special legislation?"

Ginsburg believes no voice is heard because America lacks "a national rural policy that considers the needs and aspirations of the majority of rural Americans—farmworkers, small farmers, small independent businessmen, and the aged."

# CORPORATE ACCOUNTABILITY
# AND THE FAMILY FARM

## Sheldon L. Greene

> What is life and so-called liberty if the means of
> subsistence are monopolized? . . . The corporation
> has absorbed the (rural) community. The (rural)
> community must now absorb the corporation.
> —FARMERS' ALLIANCE

During the past few years, the world has been treated to a partial
aerial survey of the landscape of Mars. Similarly, the lunar surface
has been subjected to a limited but exhaustive geological survey.
These achievements of man represent the high points of a project
which has already cost the taxpayer in excess of $20 billion.

Mars and the moon, notwithstanding their mystery, have a
relatively small impact on the lives of the average American.
Much greater is the impact of General Motors, General Foods,
Purex, Tenneco, AT&T, and the one thousand corporations that
do 22 percent of the business of the United States. Equally significant to the average American is the nature of the land itself—its
resources, limitations, expendability, and the character of agriculture in the United States.

---

Sheldon Greene is cofounder of the National Coalition for Land Reform
and General Council for the Northern California Land Trust.

The interrelationship between these subjects is simply that, while the development of data by our society proceeds willy-nilly, the data that we most require in order to develop intelligent direction is either unavailable to us or deliberately kept from us. Initially, it must be said with respect to corporate data that business and public agencies do not hold in high regard the public's right to know.

Beyond the fact that businesses, public agencies, and organizations fail to disclose available information to the public, there is the problem of the government's failure to collect data important to the public and to public policy. One of the most important yet unheralded examples of this has been the government's failure to analyze the significant changes that have been occurring in rural America. We know that each year 100,000 farms are abandoned and that rural America has sustained a population loss of 40 million people in the last fifty years. Along with the abandonment of small farms and the migration to the cities of a heretofore rural population, has been the increasing entry into agriculture of multipurpose business interests, bringing with it an increase in farm size and absentee ownership of the land. Once-populous areas occupied by independent small landholders interspersed with small rural service communities are being transformed into feudalistic estates—possibly one of the most significant economic and social transformations to be experienced in our history. The phenomenon is infinitely more relevant than the recognition that the lines on Mars are not canals or that the moon is sorely deficient in cheese.

To my knowledge, neither federal nor state governments have undertaken a comprehensive study of something as basic as *who owns the land*—other than to maintain obscure county tax records. Nor has a comprehensive study been undertaken of land use.

On the corporate side, we are well aware of the deficiencies in data reporting, regarding not only the identification of profits and losses among subsidiaries of conglomerates but also the extent of ownership and interrelationship between corporations that are ostensibly unaffiliated.

Finally, because the data are unavailable, we are by no means certain of the impact of vertical integration on single-purpose businesses, whether agricultural, mercantile, or manufacturing.

## Agribigness

In agriculture, however, we do know that the entry of big business into agriculture has produced symptoms that smack of unfair trade practices, in many instances accelerating the demise of the small farm, drying up the farmer's credit, increasing his dependency on processors, decreasing his mobility and leverage on the market, increasing his debt burden without accompanying growth of return on investment, and so on. A brief analysis of this impact on agricultural production is useful.

As we well know, equating bigness with efficiency in agriculture is a misconception. Studies have demonstrated the family farm to be the most efficient unit of agricultural production. Summarizing the studies made on the subject of farm efficiency, G. P. Madden concluded, "All of the economies of size could be achieved by modern and fully mechanized one-man or two-man farms."[1] The study concluded that the major difference between the small and medium-sized farm and the large farm was simply that the latter had the potential to produce more profits for the farm owner.

*The issue for agriculture is less a question of farm size than it is the maintenance of market conditions that tend to assure a sufficient return on the farmer's investment and his labor.* Costs, the availability of credit, and market leverage are more critical factors; yet they are to a great extent unrelated to actual or potential efficiency.

A review of these factors unrelated to efficiency reveals that the family farmer is disadvantaged. Despite the fact that he bears all the risk of producing the food, must nurture the crop from year to year, often waiting for years before vines and trees reach maturity, he receives too often the least return of all components of the food delivery chain. For example, a fourteen-ounce bottle of ketchup, which costs the housewife about thirty cents, brings the farmer a little more than one penny. In contrast, the wholesaler, or middleman, who is a transient conduit between the farmer and the retailer, skims off as much as 40 percent of the price the consumer pays for market produce. Clerks in air-conditioned Safeway markets earn up to five dollars an hour in parts of California, providing them with a greater return than the farmer receives for his labor and three times the earnings of the farmworker. Consistent with the average farmer's deficient return on his investment, the farmworker's earnings average one-half the national industrial average.

There are those who would say that big business is the solution to the farm problem. In fact, the entry of big business into agriculture has caused much of the problem.

The poultry and egg industry, for example, has moved from production by small independent farmers into control by vertically integrated national poultry-feed suppliers such as Ralston Purina. In 1961, a California legislative committee completed a report on vertical integration of the egg and poultry field by nonagricultural corporate interests. The following extracts from the study reveal the negative impact of the transformation:

> The plight of the industry was traced to feed dealers and others moving into it. . . . They financed growers right and left, with the final result being an over-production which reduced grower profits to zero, while it still enabled the feed men to make money since, under their gross-profit-splitting contracts, they did not have to account for depreciation on the grower's plant or pay interest on his investment. While gigantic promotional efforts had more than doubled California consumption per capita, the growers were still not making money, apparently because the integrator had no real incentive . . . to raise wholesale prices to the level which would have brought his "hired hands" the profit from sales he had already taken on the feed he supplied to them.[2]

An official of the Department of Agriculture, in testimony before the House Agriculture Committee in April 1959, confirmed that farm losses in the poultry industry were attributable to "the rapid development of a specialized commercial production within the industry and the trend to contract farming and integration."

Paradoxically, overproduction led to further expansion. Reduced return meant that a farmer had to maintain a larger operation in order to obtain an adequate income for his family.

Business-hungry feed mills, equipment producers, investors anxious to find tax-saving devices—all contributed to the overproduction. Hatcheries burdened with overproduction contracted with farmers to simply raise the chickens, supplying both feed and birds and paying the farmer a fixed amount per dozen eggs[3]—an amount that was insufficiently related to his costs of production. Under vertical integration, the farmer claimed, "the margins are so low you need to maintain a volume in order to stay in."[4] The grower can't pay back his loans because of low prices, and the company, in order to make the investment bring in something, puts more chickens on the ranch, which depletes prices even further.[5]

Those farmers who resisted vertical integration became the victims of purchasers. Processors and wholesalers would keep the producers' price low, maximizing their profit on resale to retailers. Citing variations in the market unrelated to demand, they indicated that wholesalers "simply stated the price they wanted to pay . . . to force the poultryman out of business or into an integrated set-up." Wholesalers, they claimed, would stop buying when prices rose, forcing the prices down.[6]

Time has not corrected the problem experienced by egg and poultry producers. A recent University of California study of egg production in Riverside County (which produces almost 12 percent of the nation's eggs) shows that producers are "getting 8 to 10 cents less per dozen eggs than it costs to produce them."

The foregoing review of the transition from independent operator to external vertical integration in the poultry market foretells the future for field and tree crops and for meat production. Shrinking profits due to the manipulation of the market and costs of production, overproduction, and tax-loss farming are now being employed by conglomerates seeking to eliminate the family farmer or to make him a vassal of vertical integrators. *The greatest incentive that conglomerates and syndicates have to enter agriculture stems not from the profit motive but rather from our convoluted federal tax laws.* (See Chapters 4 and 6.)

### Unfair Competition

The result of conglomerate entry into agriculture is that the single-activity farmer must compete against producers who not only corner the market through vertical integration, but produce at a loss, deriving the benefit not from profits on the sale of agricultural production, but rather from tax gains and land speculation . . . or simply, tax-loss farming.

Overproduction, the extrinsic control of market and costs, and tax-loss farming continue to force efficient family farmers out of agriculture. Many of those who remain will be tied by contracts to vertically integrated conglomerates as mere vassals or, as one farmer put it, "hired hands." Enormous industrialized farms will run for miles, interspersed with labor camps. Merchants in rural communities will lose some of their markets; the body politic of freeholders will shrink; and agricultural areas will be controlled by dominant land-owning corporations whose board members reside in distant cities.[7]

Consumer interests could in the short run equate tax-loss farming with lower prices, but the conclusion would be premature. Seventy-five percent of the increase in food prices in recent years is attributable to nonfarm costs. Moreover, the recent disclosure that the monopoly conditions which prevail in the breakfast-food industry cost the consumer an estimated extra $200 million each year is an indication that the short-term gain is only a sugar-coated lemon.

Some conclusions can be reached regarding the trend toward vertical integration in agriculture, unwarranted as it is by economic considerations. There is a need to apply the principle of public trust to the affairs of our largest corporations in order to assure both that their conduct be known to those who will be affected by it and that it be not patently inimical to the general welfare of the society. Laws encouraging unfair competition in agriculture should be modified to assure that the federal government help those in need rather than those who not only don't need help but who should themselves be helping others.

## Public Trust Accountability

The notions of public trust (which has its heritage in English common law and in Anglo-American jurisprudence) and the duty to give an accounting are relevant to the conduct of the thousand largest corporations in the United States. A conglomerate whose aggregate sales represent one-tenth of one percent or more of the gross national product of the United States has a sufficiently substantial impact on the consumer, on commerce, on labor, on the cost of living—indeed, on the quality of life, the environment, the allocation of resources, and so on—to justify the application of the public trust doctrine to that enterprise. Accountability and the duty to disclose transcend the necessity of full and complete disclosure related to the periodic sale of stock. It is, or should be, an ongoing responsibility based upon the year-to-year affairs of the enterprise and its continuing effect on the society.

The concept is already manifested in the regulation of the banking industries in the United States. The insurance industry, for example, was the first industry subjected to affirmative regulation. It is unique because it is undertaken almost exclusively by the states, notwithstanding the interstate nature of the commercial activities of the industry. It is unique, also, in that the prospective contractual obligations of the insurance company, requiring continual solvency, have led the regulation of the industry to take on the quasi-public trust analogy. The essence of insurance regulation is to impose on the company the duty of full disclosure of its financial affairs, including an annual audit undertaken by officials of state insurance departments. Investments by insurance companies are strictly regulated, and rates are subject to disapproval in the interest of the consumer and the solvency of the company.

The public trust doctrine is readily applicable to the thousand largest corporations in the United States. Legislation could be enacted determining that any business entity whose sales in the aggregate exceeded one-tenth of one percent of the gross national product would occupy a quasi-public trust relationship to the United States. The duty to provide disclosure of its activities by way of accounting would be imposed upon the entity as a condition of its public trust relationship. As with the insurance, banking, and communications industries, a licensure provision could be included as well. Operations would be subject to periodic audit by way of verification of the information provided annually on a voluntary basis at the expense of the company.

### Equal Opportunity for Family Farmers

Beyond the corporate duty of public disclosure, a number of specific steps should be taken by the federal government to restore equal competition for the family farmer.

1. *Tax changes*: Current tax laws that provide conglomerates with unfair tax advantages should be reviewed and modified to reduce

the advantage deriving from land speculation and the competitive disadvantages experienced by persons earning the bulk of their income from agriculture alone.

a. Tax-loss farming could be minimized by prohibiting tax credits resulting from the setting off of losses in agriculture against profits earned by nonagricultural subsidiaries.

b. Speculation might be minimized by imposing a tax on increases in land values resulting from other than improvement of the land or increased economic value of the land attributable to increased earnings. The tax would be payable in the year in which the increase in value occurred. Owners who directly or indirectly derived their substantial earnings from agricultural production would be exempted.

c. To further reduce speculation, net profit from the sale of land could be taxed as ordinary income. An inordinate tax occurring in the year of the sale could be reduced by application of the income-averaging provisions.

2. *Acreage limit and residency requirement*: The existing laws establishing the small- and medium-size farmer as the basic agricultural unit of production in America might be enforced—specifically, the law limiting the supply of water from federal reclamation projects to resident farmers owning 160 acres or less. Many farmers who have contracted to divest themselves of excess acreage have not as yet done so. A measure is now pending in Congress, in both the House and the Senate, that would enable the federal government to purchase land in excess of the 160-acre limitation. If enacted into law, the bill could both reduce the acreage of some landowners and at the same time provide for the reapportionment of prime agricultural acreage among small farmers and farmworkers desirous of moving up to farm ownership.

3. *Family Farm Act*: Recognizing the unfair business advantages that conglomerates derive through tax-loss farming and land speculation, Congress should enact the Family Farm Act, which would altogether prohibit engagement in agricultural production by conglomerates or large, nonagriculturally based enterprises. The significance of this bill would be to place farmers on an equal competition footing.

4. *Encourage co-ops*: Small farmers can compete with large farmers efficiently if they are able to take advantage of economies of scale deriving from common purchasing, processing, and even marketing. A program of technical assistance should be initiated,

providing assistance to small farmers seeking to modernize plant and equipment who have combined in cooperatives that show a capability of reducing costs and maximizing gain from sale of produce.

5. *Farmworker farm ownership*: A related program should be established to provide seed money and ongoing technical assistance to farmworkers seeking to take an ownership position in agriculture. The program might be integrated with related government projects so that, for example, excess land purchased under the acreage limitation enforcement act would be leased to individual farmworkers who have formed agricultural cooperatives to take advantage of economies of scale resulting from cooperative purchasing, processing, and marketing. The seed money program would enable farmworker cooperatives to obtain loans from the Farmers Home Administration and commercial banking sources, providing for both capital development and operating funds. Technical assistance would carry the farmworkers over the transitional period, rounding out their skills and providing them with management training and experience.

6. *Land bank*: A federal land bank could provide low-interest loans and loan guarantees to enable Southern sharecroppers to purchase property, expand farms, or move to more advantageous arrangements with private owners or the federal government. Once again, the use of the cooperative would be tied to the provision of assistance and financing.

7. *Extension services*: A subordinate program would enable successful cooperatives to organize and finance ancillary services, such as rural health programs.

8. *Reasonable return*: Farmers who expend efforts to and, in fact, attain optimal efficiency in production and utilization of their resources should derive a reasonable return from the sale of their product, related to the return that industrial sales yield (known as "parity"). Similarly, farm laborers, providing an indispensable service in the food delivery chain, are entitled to parity with national industrial-wage averages. A farmworkers' bill of rights would correct the disparities between benefits accruing to industrial workers and to farmworkers under present laws and economic conditions. Farmworker minimum wages could be increased to close the gap between the average farmworker hourly wage and the average industrial wage in America. Similarly, benefits such as unemployment insurance could be extended to the farm labor force.

9. *Marketing leverage*: Since agriculture meets a national market—fruits and vegetables can be air-freighted from one end of the country to the other in a matter of hours—the question of over-production and resulting loss of income might be considered to be a national, rather than a regional, problem. Therefore, national marketing boards might be established to minimize unreasonable competition between farmers of competing regions. The marketing boards would function to restrict productivity to that which the market is likely to reasonably absorb, minimizing uneconomic surpluses that benefit neither farmers nor consumers and only maximize profits of middlemen.

   Although the national marketing boards would be voluntary, special privileges such as federal loan guarantees might be made available to farmers participating in the marketing boards, as an incentive to participation and to maximize their effectiveness.

10. *Collective bargaining with processors*: Legislation compelling processors and middlemen to bargain collectively with farm associations would assure farmers a better price for their produce.

11. *Zoning*: Attention should be given not to the solution of short-range problems but to the establishment of a system that will also preserve and maximize the utilization of our limited natural resources for the future. To this end, Congress and the states should institute a system of agricultural zoning, beginning with a national survey of land resources and present utilization. The second phase of the survey would be to establish, based on the climatological and soil conditions in each region, the most efficient uses to which the land might be put, in terms of specific agricultural, timber, or mineral productivity. Next, agricultural economists would ascertain the most efficient units of production for the various uses to which land in the sector might be put. Finally, variable acreage limitations would be established for all agricultural uses benefiting from some form of federal or state assistance, such as subsidies, loans, or services. These limitations would be nonrestrictive and would, rather, impose flexible guide-lines to assure the highest use of the land. If, for example, the optimum acreage for a farm best suited for midwestern grain crops was 400 acres, farms in excess of 440 acres engaged in grain production would either be ineligible for public assistance such as government loans or would pay a premium for such loans.

12. *Progressive real property tax*: A corollary to the variable acreage limitation and regional zoning program would be the imposition of a progressive real property tax based on value. The graduated

property tax would tend to reduce the advantage deriving from land held for speculative purposes and reduce the pressure for increased land values related purely to speculation rather than to increases in productivity-related income. It would also discourage the concentration of ownership of land and shift some of the burden for support of local and state functions from the homeowner to commercial interests.

## Appendix

NATIONAL COALITION FOR LAND REFORM
126 Hyde Street, Suite 101, San Francisco, Calif. 94102
1878 Massachusetts Avenue, Cambridge, Mass. 02140

STATEMENT OF AIMS: The National Coalition for Land Reform brings together citizens and organizations from all sections of the country who recognize the need for a more equitable distribution of land in rural America.

Coalition members believe that ownership of land by those who work and live on it is the key to alleviating rural poverty, easing urban overcrowding, reducing welfare costs and unemployment, protecting the rural environment, and building a stronger democracy.

Through educational, legal, and political action, the NCLR seeks to:

- assure existing small farmers a fair return and increase the number of self-employed farmers
- encourage agricultural cooperatives
- combat corporate feudalism
- make government and financial institutions more responsive to working farmers and the rural poor
- promote small-town businesses and rural economic development under local control
- preserve open spaces and diminish use of toxic chemicals

In particular, the NCLR seeks:
- enforcement of acreage limitations
- application of antitrust laws to agriculture
- restructuring of tax laws and subsidies to favor working farmers rather than large landowners and speculators

- new laws enabling rural Americans to acquire a proprietary interest in their local economies

## Notes

1. Size/efficiency relationships varied from crop to crop; however, with regard to the production of cling peaches, "average cost reached a minimum with an orchard size of 90 to 110 acres when mechanized practices were used." In the Imperial Valley, examination of vegetable farms having acreage that ranged higher than 2,400 acres disclosed that the farms under 640 acres "could produce almost as efficiently as any larger size." Producers of field crops such as cotton, alfalfa, milo, and barley "were found to achieve lowest average cost at about 640 acres." The report found, in fact, that in these areas, larger farms extending beyond 1,280 acres "were slightly less efficient." U.S. Dept. of Agriculture, *Economies of Size in Farming*, Agricultural Economic Report No. 107 (Washington, D.C.: Government Printing Office, 1969).

2. Report of Assembly Interim Committee on Agriculture, Vertical Integration, Family Farm, etc., January 1961, p. 13.

3. Ibid., p. 16.

4. Ibid., p. 23.

5. Ibid., p. 25.

6. Ibid., p. 17.

7. For an example of this trend, see "The Revolution in American Agriculture," *National Geographic* (February 1970).

## EFFICIENCY IN AGRICULTURE:
## THE ECONOMICS OF ENERGY

*Michael Perelman*

> "Would you tell me please, which way I
> ought to go from here?"
> "That depends a great deal on where you
> want to get to," said the Cat.
> "I don't much care where—" said Alice.
> "Then it doesn't matter which way you go,"
> said the Cat.
> "—so long as I get somewhere," Alice added. . . .
> —LEWIS CARROLL
> *Alice in Wonderland*

Agriculture in this country has changed greatly since the Second World War. Farm machinery and chemicals have greatly increased the quantity of food and fiber produced and decreased the number of farmers necessary for their production. Although the release of millions of small farmers from the land has often been cited as the *sine qua non* of progress, this agricultural revolution has had considerable disruptive impact on our society and on the rural environment in which it has occurred. The increased "efficiency" of our farms must therefore be assessed in terms of the costs of mechanized agriculture.

Michael Perelman is professor of economics at California State University, Chico.

Since 1940 the number of farms has been reduced from 6.3 million to 2.9 million, with 1 million of those disappearing since 1961. At the same time, the farm population has been reduced from 31.9 million (23.2 percent of the population in 1940) to 9.7 million (4.8 percent of the population in 1970). While these reductions have been taking place, there has been little change in the amount of farm acreage, the area remaining around the 1 billion acres first achieved in the mid-1930s (see Table I).

TABLE I
LAND AND PEOPLE IN U.S. AGRICULTURE: SOME RECENT TRENDS

| Year | Farm Pop. (1,000)[a] | Farm Pop. As % of Total[a] | No. of Farms (1,000)[a] | Aver. Farm Size (Acres)[a] | Land In Farms (1,000 acres)[b] |
|------|------|------|------|------|------|
| 1920 | 31,974 | 30.1 | 6,518 | 147 | 955,884* |
| 1930 | 30,529 | 24.9 | 6,546 | 151 | 990,112 |
| 1940 | 30,547 | 23.2 | 6,350 | 167 | 1,065,114 |
| 1950 | 23,046 | 15.3 | 5,648 | 213 | 1,161,420 |
| 1960 | 15,635 | 8.7 | 3,962 | 297 | 1,123,507 |
| 1970 | 9,712 | 4.8 | 2,924 | 383 | |

SOURCES: a. *Statistical Abstracts of the United States, 1971*, U.S. Dept. of Commerce, Washington, D.C.

b. *Agricultural Statistics, 1971*, U.S. Dept. of Agriculture, Washington, D.C.

* Excludes Alaska and Hawaii.

### The Basis of Modern Farm Efficiency

#### LABOR

According to one popular notion, modern agriculture is efficient because it has freed people from the need to be farmers . . . that it has developed as a labor-saving device. Former Secretary of Agriculture Clifford Hardin reflects this basic attitude when he writes:

Using a modern feeding system for broilers, one man can take care of 60,000 to 75,000 chickens. One man in a modern feedlot can now take care of 5,000 head of cattle. One man, with a mechanized system, can operate a dairy enterprise of 50 to 60 milk cows.

Agriculture, in short, does an amazingly efficient job of producing food.[1]

If we measure efficiency by output per man-hour, then we must agree with Secretary Hardin's analysis, and in that case, we should get on with the job of clearing the land of the inefficient small farmer to make way for the large modern farms that are capable of using the newest technology.

But, should we measure efficiency by output per man-hour? After all, no person alive can really feed 75,000 chickens alone. In reality, he is aided by many other people who have made the cages and grown the feed. But we don't see these other people at the broiler factory; in fact, some of them might have never set foot on a farm. Yet they are farmers nonetheless, for without these people producing the capital and other inputs, the modern farm would wither away. Today, although only 4.5 percent of the U.S. population works on farms, another 5 percent is engaged in producing supplies, and another 10 percent is employed in processing the food and bringing it to the consumer.[2]

Table II gives us some idea of the importance of nonfarm inputs to agriculture. The table lists total farm production expenses between 1954 and 1969. Notice that the cost of capital represents between one-third and one-fourth of all expenses. More than 10 percent of the total costs fall into the category of "miscellaneous" expenditures, which include pesticides, cotton ginning, and many similar items. These miscellaneous expenditures, as well as the capital on the farm, have diminished the farm labor force; as a result, hired labor represents less than 10 percent of total farm costs.

However, the important question we must ask is not how little labor we can use on the farm, as conventional wisdom seems to suggest, but whether society benefits from the replacement of farm labor by capital.

## PROFITABILITY

Because we cannot measure the total contribution of all people working in agriculture, we might use another measure of efficiency, namely profitability. Indeed, modern large-scale agriculture does appear profitable; otherwise major corporations would not be investing in these farms. However, this profitability owes a great deal to tax accountants and attorneys. Through their expertise, nonfarm businesses and wealthy individuals can "farm." They can raise cattle or develop an orchard. These operations will not turn a profit until the cattle or the trees reach maturity, and as long as they do not produce any profit, the owner can write off these expenses from his nonfarm income. And just when they are mature the owner can sell out at a profit and declare a capital gain so that he is taxed at a lower rate. The government has long been aware of the danger to the small farmer of these tax loopholes. In 1963, Secretary of the Treasury Douglas Dillon told the House Ways and Means Committee that the tax farmers "create unfair competition for farmers who may be competitors and who do not pay costs and expenses out of tax dollars but who must make an economic profit in order to carry on their farming activities."[3]

Second, until just recently farm subsidies favored the largest corporations. Payments were roughly proportional to farm sales, so that the large farms naturally got more than the small ones. Moreover, both price support and direct payment benefits of the farm commodity programs were more highly concentrated among the large farmers than was income itself.[4]

Moreover, the more successful the government is in maintaining high prices the more incentive the farmer has to raise his yields. High yields are good, except that the farmer of today takes shortcuts to high yields by using ecologically damaging farm chemicals; the small farmer uses less of these chemicals per acre than does the large farmer.[5]

Large farmers have other economic advantages over small farmers besides those conferred by government policy. In the first place, their buying power gives them leverage in the marketplace. They get cheaper inputs and lower interests. Table III shows the relationship between farm size and the costs of capital and other inputs. *These data suggest that profitability can have nothing to do with efficiency in agriculture.*

There is also the fact that many banks prefer not to lend any money to the "inefficient" small farmer, in spite of the fact that USDA officials admit, "We know from our studies in the Department of Agriculture that the rates of foreclosure and delinquency

TABLE II
FARM PRODUCTION EXPENSES, UNITED STATES,[a] 1955–1969
(IN MILLIONS OF DOLLARS)[d]

| Year | Feed, Livestock[b] and Seed | Fertilizer and Lime | Capital Equipment: Repairs, Operation,[c] Depreciation, and other Capital Consumption[d] | Hired Labor[e] | Taxes on Farm Property[f] | Interest on Farm Mortgage Debt[g] | Net Rent to Non-farm Landlords | Misc.[h] | Total Production Expenses |
|---|---|---|---|---|---|---|---|---|---|
| 1955 | 5,995 | 1,185 | 7,300 | 2,615 | 1,141 | 402 | 1,057 | 2,204 | 21,889 |
| 1960 | 7,935 | 1,315 | 8,210 | 2,923 | 1,502 | 628 | 1,010 | 2,829 | 26,352 |
| 1965 | 9,299 | 1,754 | 9,055 | 2,849 | 1,943 | 1,077 | 1,328 | 3,628 | 30,933 |
| 1966 | 10,448 | 1,952 | 9,508 | 2,889 | 2,108 | 1,205 | 1,442 | 3,854 | 33,406 |
| 1967 | 10,541 | 2,124 | 10,241 | 2,878 | 2,275 | 1,343 | 1,305 | 4,068 | 34,775 |
| 1968 | 10,338 | 2,125 | 10,851 | 3,045 | 2,526 | 1,477 | 1,308 | 4,342 | 36,012 |
| 1969 | 11,505 | 2,013 | 11,500 | 3,192 | 2,753 | 1,602 | 1,303 | 4,576 | 38,444 |

SOURCE: *Agricultural Statistics, 1970*, U.S. Dept. of of Agriculture, Washington, D.C.

[a] Includes Alaska and Hawaii, beginning 1960.

[b] Includes bulbs, plants, and trees.

[c] Includes expenditures for repairs and maintenance of farm buildings and other land improvements, petroleum fuel and oil, other motor vehicle operation, and repairs on other machinery.

[d] Estimated outlay necessary at current prices, for the replacement of capital equipment that has been used up during the year.

[e] Includes cash wages, perquisites, and Social Security taxes paid by employers.

[f] Includes taxes levied against farm real estate and farm personal property.

[g] Interest charges payable during the calendar year on oustanding farm-mortgage debt.

[h] Includes interest on non-real-estate debt, pesticides, ginning, electricity and telephone (business share), livestock marketing charges (excluding feed and transportation), containers, milk hauling, irrigation, grazing, binding materials, tolls for sirup, horses and mules, harness and saddlery, blacksmithing, hardware, veterinary services and medicines, net insurance premiums (crop, fire, wind, and hail), and miscellaneous dairy, nursery, greenhouse, apiary, and other supplies.

TABLE III

RELATIONSHIP BETWEEN FARM SIZE AND COST OF CAPITAL
AND OTHER PURCHASED INPUTS

| Farm Size (acres) | Interest on Operating Capital (6% Norm) | Volume Discounts | | | Total Difference from Base Cost per Acre |
| | | Fertilizers | Insecticides | Crop Dusting & Aerial Spraying | |
| --- | --- | --- | --- | --- | --- |
| 80 | 6.88% | 0% | 0 % | 0 % | $0.56 |
| 160 | 6.52% | 4% | 0 % | 0 % | —0.25 |
| 320 | 6.47% | 4% | 5 % | 0 % | —0.53 |
| 640 | 6.47% | 4% | 5 % | 12.5% | —1.27 |
| 1,280 | 6.15% | 10% | 8.5% | 17.5% | —3.96 |
| 3,200 | 5.90% | 10% | 14 % * | 25 % | —6.62 |

SOURCE: U.S. Senate, Committee on Agriculture and Forestry, *Farm Programs and Dynamic Forces in Agriculture*, Legislative Reference Service (Washington, D.C.: U.S. Gov. Printing Office, 1965).

* Only one observation.

are greater on big farm loans, for the large-scale farm units, than for smaller loans on family farms."[6]

So it turns out that the "inefficient" small farmer is a better risk than is his larger, more modern counterpart. But why should the large businesses go into farming if they are not more efficient? We have already touched on some of the reasons; to this list we can add two more: a desire for the economic integration of their industries, and land speculation (see Chapters 3–5).

The effect of favorable tax laws and cheap credit on large farms is that "high leverage [that is, the ability to use borrowed money] and capital gains . . . can convert a nominal rate of return on total investment of 1 percent or 2 percent into an effective rate of return on equity of 8 percent to 10 percent, or higher."[7] Land speculation and the opportunity for vertical integration make large farming even more profitable. Stock-market manipulations also play a role in making large-scale agriculture more attractive to nonfarm corporations.

In recent years, several of these corporations have spread their efforts into the countryside in search of profits.[8] Included are Dow Chemical, ITT, Coca-Cola (which owns Minute Maid), Gulf & Western, Kaiser Aluminum, Aetna Life Insurance, Goodyear, and Monsanto. In California forty-five corporations now own 3.7 million acres or nearly half of the farm land in the state, while nineteen corporations own about 21 percent of the timberland, and twenty-nine corporations own about 21 percent of the cropland.[9] On a national scale, two conglomerates, Purex and United Brands, now control about one-third of the green, leafy vegetable production in this country. Green Giant in Minnesota claims about 25 percent of the U.S. canned corn and peas market. Ralston Purina, which had 1971 sales of some $1.75 billion and which sells 14 percent of the U.S. livestock feed, is one of the nation's largest single producers of wheat, corn, and soybeans. Clifford Hardin is now vice-chairman of Ralston Purina, while the present Secretary of Agriculture, Earl Butz, is a former company director.

Because of this unequal competitive situation, many small farms have been unable to hold on. These trends have generally been accepted as "inevitable and desirable" by banks and companies with large holdings in agriculture. For example, Rudolph A.

Peterson of the Bank of America has stated that "what is needed is a program that will enable the small and uneconomic farmer —the one who is unwilling or unable to bring his farm to the commercial level by expansion or merger—to take his land out of production with dignity."[10]

A spokesman for Gates Rubber Corporation was a little more blunt: "The economists say that 40 percent of the people in agriculture are going to have to leave the farms eventually—we're just helping some of them to make the change."[11] Ironically, Gates Rubber has proved a failure in its venture in agriculture.

As more and more of these well-off farm interests go into farming, the price of land is bid up and the market for farm products becomes glutted. Thus, the price of farm products falls or fails to keep up with the prices of other goods. But the land speculators don't mind. In fact, it is to their benefit at tax time to be in an industry with a low rate of current earnings while equity rises with increasing land values.

On the other hand, the small farmer needs his income today to meet his current expenses. He can benefit from the rising land values only when he sells out and ceases to be a farmer. Thus, again it seems clear that *much of the profitability of large-scale farming has very little to do with economic efficiency.*

## MECHANIZATION

Probably the most impelling cause of the increase in farm efficiency has been the development of various kinds of machinery for planting, spraying, cultivating, harvesting, or any other farm activity that once required a large labor force. Without the modern machinery now available, large farms would be unworkable. Many farm operations must be accomplished in relatively short periods of time during the year, and the reliability of the machinery used to perform these chores on time and over large acreages is far greater than that of human labor. Most important, the use of machinery is far more economical than the use of labor.

The best-known form of agricultural mechanization is the tractor. As late as 1920, more than twenty million horsepower was provided by horses and mules.[12] These animals had to be fed from the land. With the adoption of the tractor, this land was freed to produce food-animals for humans. Not only was land freed by the tractor; labor was also freed because one man plowing with a tractor could do the work of several men plowing with a mule. Trends in the replacement of human energy by mechanical power are shown in Table IV.

TABLE IV
MECHANICAL POWER REPLACES HUMAN POWER ON
U.S. FARMS

| Year | Horsepower (millions) | Man-Hours on Crops (millions) | Cost of Operating and Maintaining Farm Capital (millions of dollars) |
|---|---|---|---|
| 1920 | 5 | 13,406 | (not available) |
| 1950 | 93 | 6,922 | 5,640 |
| 1960 | 154 | 4,590 | 8,310 |
| 1969 | 203 | 3,431 | 11,500 |

SOURCE: Austin Fox, *Demand for Farm Tractors in the United States,* U.S. Dept. of Agriculture, Economic Research Service, Report No. 103 (Washington, D.C.: U.S. Gov. Printing Office, November 1966).

Another advantage of mechanization is related to the division of labor in agriculture. Earlier economists were not very optimistic about improvements in agricultural productivity because they believed that it depended upon the division of labor and that there was not much scope for improvement in the division of labor on the farm.

Agriculture . . . is not susceptible of so great a division of occupations as many branches of manufacturers, because its different operations cannot possibly be simultaneous. One man cannot always be ploughing, another sowing, and another reaping. A workman who only practiced one agricultural operation would lie idle eleven months of the year. The same person may perform them all in succession, and have, in most climates, a considerable amount of unoccupied time.[13]

But in fact, agriculture did move quite fast in the direction of manufacturing because the mechanization of agriculture meant that many jobs were transferred from the farm to the factory. For instance, the growing of hay for horses was replaced by the refining of petroleum for tractors. Workers who were displaced by the new machines migrated to the city where many of them were employed in producing machines and other nonfarm inputs for their comrades who remained on the farm. Between 1919 and the present, U.S. industries have employed almost two million man-years per annum in the production of goods and services used in American agriculture.[14]

However, as we have seen, the *degree* to which labor is displaced is not necessarily an accurate estimate of agricultural efficiency. The development and utilization of expensive, large-scale machinery has a much more intensive effect on farm labor than does the use of small-scale machinery designed for small family-farm operations. As a result of grandiose mechanization, agriculture has taken on more and more of the aspects of industrialization. Fields are larger; rows are straighter; ditchbanks and fencerows are cleaner, and all together there is less diversity. All acreage is treated in a uniform, assembly-line way. The application of these techniques, that is, the forcing of biological systems into a highly artificial form, has caused numerous problems. Lack of diversity in the fields stimulates pest problems, for instance, by removing natural controls, while the regularity of plantings provides economic incentives for increased mechanization of fertilizer and herbicide application. These, in turn, require increasing applications of polluting farm chemicals and greater expenditures of energy resources. Mechanization per se is not the culprit; only the degree to which it has been applied.

## Energy Use In Agriculture

Until the age of industrialization, all societies had to work harder to feed themselves as their population grew; that is, a 1 percent increase in population meant a larger than 1 percent increase in the work required to feed everyone.[15] We have reversed this trend with industrialization only by means of harnessing the energy of fossil fuels. This stored-up energy has made it possible for the farmer to cut the soil with steel plows, to harvest with sophisticated machinery, and then to take his produce to cities hundreds or even thousands of miles away.

To show what high levels of energy consumption mean for agriculture, Fred Cottrell tried to compare the energy budgets of

Japanese and American farming. He found comparable statistics for two rice farms, one in Japan and the other in Arkansas. Each had approximately the same yield per acre. In Japan, an acre could be cultivated and harvested in about ninety man-days, which is equivalent to ninety horsepower hours. On the Arkansas farm, more than 1,000 horsepower hours of energy were used just to power the tractor and truck. Moreover, the nonresidential consumption of electrical energy exceeded 600 horsepower hours. Cottrell did not even include the energy required to produce the tractors and equipment.[16]

In more general terms, on a national level, U.S. farmers burned about 7 billion gallons of motor fuel in farm machinery, according to 1965 statistics.[17] In addition, a great deal of energy is used in the production of *farm equipment*. The farm implement industry alone uses the heat value of about four gallons of gasoline for every American, not counting the energy used by the supporting industries that supply that industry.[18]

The *fertilizer* industry also consumes large amounts of energy. Current technology requires about ten million calories for each kilogram of nitrogen fertilizer produced commercially.[19] In 1969, United States farms consumed about 7.5 million tons of nitrogen fertilizer which required about $2 \times 10^{14}$ BTUs, or about 1.5 billion gallons of gasoline. But then, nitrogen fertilizer makes up only one-fifth of our total commercial fertilizer supply.[20] A. B. Makhijani estimates that the overall average energy use in the fertilizer industry is a little less than 20 million BTUs per ton of fertilizer. Since the total 1969 fertilizer usage was about 400 million tons, Makhijani's figures represent a total of about 800 trillion BTUs, or a heat equivalent of more than thirty gallons of gasoline for every American.[21]

The energy cost of *petrochemicals* is obtained through the estimate of David Pimentel that the production and processing of one pound of pesticide or herbicide require about 11,000 kilocalories.[22] The use of these materials in 1970 was approximately 1.1 billion pounds,[23] thus implying the energy cost of about 48 trillion BTUs.

*Electricity* also contributes a great deal to the energy demands of farm production. In 1970, U.S. farms consumed more than 50 trillion BTUs of electrical energy.[24] However, the production of one BTU of electrical energy requires 3.07 BTUs of fuel input.[25] Thus, farm electrical production actually represents more than 150 trillion BTUs or an equivalent of about four gallons of gasoline for every American, not counting the energy used by the supporting industries which supply that industry.[26]

The energy cost of the *food processing* sector is also significant. Makhijani estimates that this sector consumes about $10^{15}$ BTUs or an amount comparable to the consumption of energy by farm machinery.[27]

The degree of waste in the food processing industry is illustrated in a study by Bruce M. Hannon, an engineer at the University of Illinois, who made a study of the McDonald's hamburger chain. Hannon calculated that this single chain of restaurants used up the energy equivalent of 12.7 million tons of coal in 1971, enough to supply the combined power needs of Pittsburgh, San Francisco, Washington, and Boston for an entire year. McDonald's consumes so much paper alone that it would require the sustained yield of a 315-square-mile forest to keep their restaurants supplied for a year.

Much of the energy used in the distribution and processing of food should be charged to the organization of agricultural production which has minimized production costs through regional specialization. This specialization requires that food be transported longer distances and also that much food be processed to avoid spoilage in the often circuitous road from farmer to consumer. In 1969, almost $5 billion was spent for transporting food by rail and inner-city trucks.[28]

One method of getting a handle on the energy cost of agriculture would be to combine the energy cost of operating and producing farm machinery with the energy cost of producing the farm chemicals and electricity plus the energy cost of the food processing industry. These activities require the equivalent of about 116 gallons of gasoline for every American (Table V) or over three and one-half times the number of calories consumed at the table,[29] in spite of the energy costs which are excluded from this calculation. For instance, farmers purchase products containing 360 million pounds of rubber (about 7 percent of the total United States rubber production) and 6½ million tons of steel in the form of trucks, farm machinery, and fences.[30]

TABLE V
GROSS ESTIMATES OF ANNUAL ENERGY USE IN AGRICULTURE*

| Source | Year | Equiv. Motor Fuel-Gallons† | Gal. Gas per U.S. Indiv. | BTU Equivalent‡ |
|---|---|---|---|---|
| FARM IMPLEMENTS | | | | |
| Operation | 1965 | 7 billion | 35.0 | 940 trillion |
| Production of | | 800 million | 4.0 | 108 trillion |
| FARM CHEMICALS | | | | |
| Production, Fertilizers | 1969 | 6 billion | 30.0 | 800 trillion |
| Production and Processing, Petrochemicals | | 360 million | 1.8 | 48 trillion |
| FARM ELECTRICITY | | | | |
| Consumed | 1970 | 370 million | 1.9 | 50 trillion |
| For Production of | | 1.1 billion | 5.6 | 150 trillion |
| FOOD PROCESSING | | 7.4 billion | 37.2 | 1,000 trillion |
| | | 23.03 billion | 115.5 | 3,096 trillion† |
| TOTAL U.S. ENERGY CONSUMPTION | 1970 | | | 64,000 trillion |

* See text.
† 135,000 BTU/gal.
‡ Total does not include energy used by food transportation or support industries of power plants and farm-equipment plants.

Some authors argue that agriculture is making strides forward in energy conservation. The most frequently cited example is the so-called no-till cultivation whereby herbicides are used instead of mechanical cultivation of the stubble. However, David Pimentel has calculated that no-till cultivation is actually a step backward because it increases the energy requirements of corn by about two gallons per acre.

Pimentel's work brings us directly back to our starting point of corn. According to his calculations, corn currently requires about eighty gallons of fuel per acre. More important, corn culture has declined in efficiency from the standpoint of energy. One calorie of energy input in corn produces 26 percent fewer calories of corn production than it did in 1945.

If efficiency is measured in terms of the conservation of energy, then American agriculture comes out very poorly. As noted by René Dubos:

> [Agricultural] efficiency . . . cannot be measured only in terms of agricultural yields. Another criterion is the amount of energy required for the production of a given amount of food. And when [modern] agriculture is judged on this basis, its efficiency is often found to be very low.

This conclusion has been stated in other terms:

> [In the future], the ability of the planet to support people will be extraordinarily sensitive to the energy input to agriculture in the form of equipment, fertilizers and other technology. This demonstrates that the central question for the future of man on this planet is energy availability.[31]

But modern agricultural technology is grounded on the faith that there is an abundance of energy. In fact, recently it has been stated quite often that controlled nuclear fission will be the next great breakthrough in agricultural technology. For example:

> As agriculturists, we must begin to assume active roles in earmarking nonfarm energy resources for future agricultural use because modern agriculture has become thoroughly dependent upon the availability of nonrenewable fossil fuels. We have reached a point in time when the food reaching the tables of American citizens costs as much or more in fossil-fuel energy than is represented in the food energy to be consumed.

I predict that the next major thrust of research will be in the application of nuclear power to agriculture, since nuclear power seems to be the cheapest . . . power source for the 1980s and thereafter. Nitrogen fertilizers from nuclear-powered electrical generators, and hydrogen gas for fuel look particularly promising because "off-peak" electrical power can be used to make hydrogen by electrolysis.[32]

Similar nuclear-agro-industrial scenarios have been described elsewhere by the USDA.[33] Figure 1 indicates the general relationship between crop yield and power used in the agriculture of various countries of the world. Remember that this horsepower is accompanied by insecticides, herbicides, and artificial fertilizers, all of which consume large amounts of fossil fuels or other forms of energy. Also, consider that we are encouraging the adoption of this high-energy farm technology all over the world (see Chapter 8).

Agriculture uses more petroleum than any other single industry.[34] In fact, as USDA Secretary Earl Butz has noted, "U.S. agriculture is the number one customer of the petroleum industry." Yet, agriculture is not the only user of energy in our society. In 1970, the United States consumed about 64,000 trillion BTUs of energy, or the equivalent of about 2,000 gallons of gasoline per individual. For instance, a typical American consumes the equivalent of about ten gallons of gasoline annually just to watch a black-and-white television set.[35] By that standard, agriculture's consumption of 116 gallons of gasoline to feed one person does not seem extravagant. Besides, the U.S. uses more than 20 percent of its acreage for exports, which feed citizens of other nations, and some of our crops are used for industrial purposes.

The problem is that agriculture is supposed to be an energy-*producing* sector of the economy. Harvested crops capture solar energy and store it as food or some other useful product. Yet the energy captured is small compared with the energy used in the process. For example, Marvin Harris estimated that traditional Chinese wet-rice agriculture at its best could produce 53.5 BTUs of energy for each BTU of human energy expended in farming it;[36] for each unit of fossil fuel energy we expend (using only energy consumed by tractors, farm chemical production, and electricity) we get about one-quarter in return. Put another way, the energy equivalent eaten by Americans is equal in heat value to

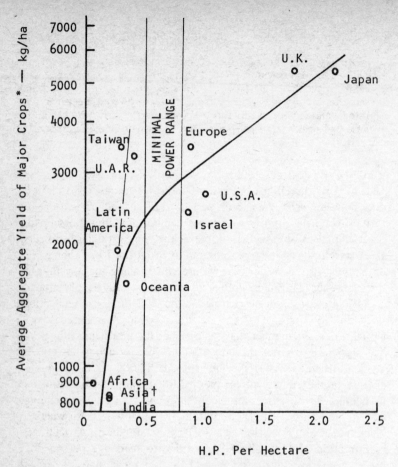

FIGURE 1. Relationship between yields in kilograms per hectare (0.89 kg/hectare = lbs/acre) and power per hectare (1 hectare = 2.47 acres)

SOURCE: President's Senate Advisory Council on the World Food Problem.
 * Crops include cereals, pulses, oilseeds, sugar crops (raw sugar), potatoes, cassava, onions, and tomatoes.
 † Excludes mainland China.

that consumed in the production of fertilizers alone. The question for our time is: How much human energy can we afford to take *out* of agriculture?

With energy as the criterion of efficiency, we can begin to see how efforts to maximize productivity in one food sector can be counterproductive to agriculture as a whole. For example, in the period from 1960 to 1970, tomatoes purchased by processors in California increased on the average of 10 percent by weight, but the increase was all in water content. The effect was to add 450,000 extra tons of water to the 1971 California tomato crop and to the processor's cost of purchase. But 86 percent of tomatoes are processed, the first step of which is water removal. Hence this "productivity improvement" is actually *added* to the cost of drying out the tomatoes for ketchup, tomato sauce, and other processed tomato products. The consumer ultimately pays the water bill.[37]

What I am suggesting here is that in a future with rising energy prices, increasing food needs, and an expanding population, our agricultural priorities will have to change and new technologies will have to reflect these changes. Up to now, mechanization and chemical farming methods have been economically irresistible. But what was irresistible in the past may be unfeasible in the future, for in that future more people will be competing for a fixed or diminishing supply of natural resources. This increased competition will force up the prices of raw materials and lower the value of labor relative to the value of raw materials in spite of the historical tendency for wages to rise.[38] Our technology is based on a historical pattern of falling costs for raw materials. It should be obvious then that the type of technology that is profitable with falling prices of raw material is not the type that will be most profitable when these prices increase. Consider the price of gasoline, for example. As I have noted, thirty gallons of gasoline have a heat value almost sufficient to supply a human with enough calories for a whole year. We pay about sixteen dollars for these thirty gallons of fuel. Even at these low fuel prices, about one-third to one-half the cost of owning and operating a tractor is the cost of fuel. If we paid as much for a calorie of gasoline as we paid for a calorie of corn, the cost of operating many farm machines might cause them to be abandoned as they were in the U.S. during the depression.

To some degree, this may not be so undesirable because, with an increasing population, we need an incentive to discover ecologically sound means of employing people. In fact, our nation already has a labor surplus economy as long as millions of people cannot find work. Yet we call our agriculture efficient while it brags about how few people it employs. We need to think about labor-absorbing agriculture and the resources and research necessary to make it rewarding.

Of course, we could use much more labor to care for our natural resources. But care is unnatural to large-scale farms; it is expensive and uneconomical. It is cheaper to pollute our water with pesticides and nitrates; it is cheaper to destroy our topsoil. As a result, the quality of our food suffers and our natural resources, including people, are wasted.

## EFFICIENCY AND THE SMALL FARMER

In the spirit of Jeffersonian democracy, we have always paid lip service to the family farm. For instance, we can read in the Agriculture Act of 1961 that "It is hereby declared to be the policy of Congress to: . . . recognize the importance of the family farm as an efficient unit of production and as an economic base for towns and cities in rural areas to encourage, promote, and strengthen this form of farm enterprise." Yet in spite of this "official" endorsement, the family farmer continues to disappear. The disappearance of the small farm is a tragedy for several reasons. For one thing, many of the displaced farmers and farmworkers cease to play a productive role in society. Their training is worthless in the city. As a result, they add to the urban welfare rolls and tax burdens.

The small farm also has some very positive human values. A 1947 study by Walter Goldschmidt is illustrative.[39] Goldschmidt studied two farming communities in California's Central Valley. One was dominated by large farms, and the other was a community of small family farms. Where the family farm prevailed, Goldschmidt found a higher standard of living, better community facilities, such as streets and sidewalks, more parks, more stores with more retail trade, and twice the number of organizations for civic improvement and social recreation. Besides, the small-farm community had two newspapers where the other had only one. In short, the small-farm community was a better place to live, perhaps because the small farm offered an opportunity for "attachment" to local culture and care for the surrounding land.

I tried to make a rough estimate of the effect of care on agricultural production in this country, using data from the 1964

Census of Agriculture. I found a pattern that showed that in any state the value of the crops grown on the average acre tends to be larger when the average farm is small. I concluded that, all other things being equal, mechanization tends to decrease yields as well as the farm labor force. Although this drop in yields can be offset by the intensification in the use of other nonfarm inputs such as fertilizer and pesticides, these nonfarm inputs have allowed us to put off the day when we face up to the consequences of diminishing returns and use a more labor-intensive technology.

Will diminishing returns ever catch up with us? If not, then we must make two assumptions: (1) that new technology will be available at a fast enough rate in the future and (2) that the nonfarm resources required by the new technology will be available at a low enough price to make it economically feasible. One piece of fragile evidence relevant to the first assumption is the USDA's index of agricultural productivity, which has fallen from 101 in 1965 to 99 in 1970.[40] Does this statistic represent a straw in the wind or a temporary lull? Answers to questions such as this one can only be matters of faith. However, events of the recent "energy crisis" indicate that the second assumption may be harder to make; we are still in an area dominated by faith.

It seems possible, then, that if, in the future, we have to face either diminishing returns or rapidly rising prices for nonfarm inputs, we will have to develop technologies capable of using more labor, and we will have to adopt policies for making the small and self-sufficient farm a viable reality. Our present technology may not be suitable for these conditions.

Many people grant that the small farm has ecological as well as social advantages over the factory farm, but then they ask if the small farm could feed America. The answer is that it could and would. In fact, small farms have higher yields than larger farms. The yields reflect, in part, the more intensive care which the small farmer gives his land. Moreover, it is logical that we would produce more food by putting more people to work on the land.

But what about our standard of living? The true measure of our standard of living is the happiness of the people, and judged by this standard, America appears to be a poor nation. A reopening of individual opportunity on land might mean a return to sanity.

In fact, we might ask ourselves if the so-called back-to-the-land movement of many people today is simply an incipient stage of a renaissance in land-use patterns in this country.

I am not saying that everybody or everyone must take up a hoe at once. Not at all. But access to the land is imperative. A first step would be to end the government policies that encourage large-scale farming and land monopolies. End the tax and subsidy advantages to large-scale farms. Next we must shift our support and incentives to those farmers whose technology is more ecologically balanced. End the wasteful consumption of our natural wealth, raw materials, and people. And above all, guarantee that people who actually work the land get a decent living. This means land reform, people-oriented rural legislation, and an integration of farm and city. It all depends on which way we want to go.

## Notes

1. Clifford M. Hardin, foreword to *Contours of Change, U.S. Dept. Agr. Yearbook of Agriculture, 1970* (Washington, D.C., 1970), p. xxxiii.

2. Milo Cox, *From Farmer to Consumer, War on Hunger*, U.S. State Dept. Agency for International Development, June 1973.

3. Cited in a speech by Senator Lee Metcalf, 10 March 1969, before the House Ways and Means Committee, printed in that committee's *Tax Reform, 1969* (Washington, D.C., 1969), p. 2070.

4. Charles L. Schultze, *The Distribution of Farm Subsidies: Who Gets the Benefits?* (Washington, D.C.: Brookings Institution, 1971), pp. 15–16.

5. F. E. Faris and D. L. Armstrong, *Economies Associated with Farm Size: Kern County Cash Crop Farms*, Giannini Foundation Research Report No. 26 (1963), p. 9; or Kenneth R. Krause and Leonard R. Kyle, "Economic Factors Underlying the Incidence of Large Farming Units: The Current Situation and Probable Trends," *American Journal of Agricultural Economics* 52, no. 5 (1970): 748–761.

6. Don Paarlberg, "Future of the Family Farm," speech before the 55th Annual Convention of the National Milk Producers, Bal Harbour, Florida, 30 November 1971.

7. Phillip Raup, "Economies and Diseconomies of Large-Scale Agriculture," *American Journal of Agricultural Economics* 51, no. 5 (1970): 1274–1283.

8. "Big Corporations Invest More in Agriculture," *Doane's Agricultural Report*, January 1968. See also Guy Halverson, "U.S. Farmers: What Road Ahead?," part 2, "The Family Farm Faces Corporate Bigness," *Christian Science Monitor*, 29 August 1972.

9. Robert C. Fellmeth, ed., "Power and Land in California," in *Ralph*

*Nader Task Force Report on Land Use in California*, vol. 1. (Washington, D.C.: Center for the Study of Responsive Law, 1971).

10. Cited in "Farm Policy Question," *Agri-Finance*, January–February 1969, p. 42.

11. Said to the Select Committee on Small Business of the United States Senate, Subcommittee on Monopoly, in hearings held in Omaha, Nebraska, 20 May 1968, p. 11.

12. Austin Fox, *Demand for Farm Tractors in the United States*, U.S. Dept. of Agriculture, Economic Research Service, Agricultural Economic Report No. 103 (November 1966): 1.

13. Adam Smith, *An Inquiry into the Nature and Causes of the Wealth of Nations*, ed. Cannan (New York: Random House, 1937).

14. Dowring Folke, *The Productivity of Labor in Agricultural Production*, University of Illinois, College of Agriculture, Agriculture Experiment Station Bulletin 726, September 1967.

15. Ester Boserup, *The Conditions of Agricultural Growth: The Economics of Agrarian Change Under Population Pressure* (Chicago: Aldine-Atherton, 1965).

16. Fred Cottrell, *Energy and Society* (New York: McGraw-Hill, 1955), pp. 138–140.

17. U.S. Dept. of Agriculture, Economic Research Service, *Structure of Six Farm Input Industries* (Washington, D.C., 1968).

18. A. B. Makhijani and A. J. Lichtenberg, "An Assessment of Energy and Materials Utilization in the U.S.A.," Memorandum No. ERL-M310, Electronics Research Laboratory, College of Engineering, University of California, Berkeley, 22 September 1971.

19. C. C. Delwiche, "Nitrogen and Future Food Requirements," in *Research for the World Food Crisis: A Symposium Presented at the Dallas Meeting of the American Association for the Advancement of Science, December 1968*, ed. Daniel G. Aldrich, Jr., Publication 92, (Washington, D.C.: American Association for the Advancement of Science).

20. U.S. Dept. of Agriculture, *Agricultural Statistics, 1971* (Washington, D.C.: Government Printing Office, 1971).

21. Roy M. Kottman, Dean and Director of the College of Agriculture and Home Economics, Ohio State University, points out (personal communication) that the energy costs of fertilizer production might be less than we have estimated. He writes that "the 1967 Census of Manufacturers shows that only about 54 trillion BTUs were used that year to produce and *mix* all of the fertilizers manufactured in the U.S."

22. David Pimentel, "Food Production and the Energy Crisis," *Science*, 182 (1973): 443–449.

23. U.S. Dept. of Agriculture, *Agricultural Statistics, 1972* (Washington, D.C.: Government Printing Office, 1972).

24. Eric Hirst, National Science Foundation, Environmental Program, Oak Ridge National Laboratory, personal correspondence.

25. Barry Commoner and Michael Corr, "Power Consumption and Human Welfare in Industry, Commerce and the Home," to be published

by the American Association for the Advancement of Science, Committee on Environmental Alterations, in *Electric Power Consumption and Human Welfare: The Social Consequences of the Environmental Effects of Electric Power Use*, ed. Howard Boksenbaum et al.

26. Makhijani and Lichtenberg, "An Assessment of Energy."

27. Ibid.

28. USDA, *Agricultural Statistics, 1971*.

29. The average American citizen consumes around 3,000 kilogram-calories daily, which is equivalent to about 12,000 BTUs or 4,380,000 BTUs per year—the heat value of 32.6 gallons of gasoline.

30. U.S. Congress, House, Committee on Agriculture, *Food Costs— Farm Prices: A Compilation of Information Relating to Agriculture*, 92 Cong., 1 sess. (Washington, D.C.: Government Printing Office, 1 July 1971).

31. Kenneth E. F. Watt, *Principles of Environmental Science* (New York: McGraw-Hill, 1973).

32. Perry R. Stout, Agricultural Experiment Station, University of California, Davis, editorial in *California Agriculture* 26, no. 12 (1971).

33. J. Billard, "The Revolution in American Agriculture," *National Geographic* 137, no. 2 (February 1970): 147–185.

34. House, Committee on Agriculture, *Food Costs*.

35. W. D. Brune, Jr., Director of Industrial Services of Pacific Gas and Electric Co., "The Economic Impact of Electrical Power Development: A Talk before the National Engineers Week Symposium," Chico State College, 26 February 1972.

36. See Roy A. Rappaport, *Pigs for the Ancestors* (New Haven: Yale Univ. Press, 1967), p. 262, referring to Marvin Harris, "Cultural Energy," n.d., unpublished.

37. U.S. Dept. of Commerce, National Commission on Productivity, "Productivity in the Food Industry," 1972.

38. Part of the cheapness of some of our raw materials rests on our military and economic influences over the weak nations that export them; that is, some of their cost is hidden in our military and foreign-aid budgets.

39. Walter Goldschmidt, *As You Sow* (Glencoe, Ill.: Free Press, 1947).

40. USDA, *Agricultural Statistics, 1971*.

# 7

# HARD TOMATOES, HARD TIMES: THE FAILURE OF THE LAND GRANT COLLEGE COMPLEX

*Jim Hightower*

## Introduction

Corporate agriculture's preoccupation with scientific and business efficiency has produced a radical restructuring of rural America that has been carried into urban America. There has been more than a "green revolution" out there—in the last thirty years there literally has been a social and economic upheaval in the American countryside. It is a protracted, violent revolution, and it continues today.

The land grant college complex has been the scientific and intellectual father of that revolution. This public complex—composed of colleges of agriculture, agricultural experiment stations, and state extension services—has put its tax dollars, its facilities, its manpower, its energies, and its thoughts almost solely into efforts that have worked to the advantage and profit of large corporations involved in agriculture.

The consumer is hailed as the greatest beneficiary of the land grant college effort, but in fact, consumer interests are considered secondarily, if at all, and in many cases, the complex works directly against the consumer. Rural people, including the vast

Jim Hightower is the director of the Agribusiness Accountability Project.

majority of farmers, farmworkers, small-town businessmen and residents, and the rural poor either are ignored or directly abused by the land grant effort. Each year about a million of these people pour out of rural America into the cities. They are the waste products of an agricultural revolution designed within the land grant complex. Today's urban crisis is a consequence of the failure in rural America. The land grant complex cannot shoulder all the blame for that failure, but no single institution—private or public —has played a more crucial role.

The complex has been eager to work with farm machinery manufacturers and well-capitalized farming operations to mechanize all agricultural labor, but it has accepted no responsibility for the farm laborer who is put out of work by the machine. It has worked hand in hand with seed companies to develop high-yield seed strains, but it has not noticed that rural America is yielding up practically all of its young people. It has been available day and night to help nonfarming corporations develop schemes of vertical integration while offering independent family farmers little more comfort than "adapt or die." It has devoted hours to create adequate water systems for fruit and vegetable processors and canners, but thirty thousand rural communities still have no central water systems. It has tampered with the gene structure of tomatoes, strawberries, asparagus, and other foods to prepare them for the steel grasp of the mechanical harvesters, but it has sat still while the American food supply has been liberally laced with carcinogenic substances.

The land grant complex, as it is known today, has wandered a long way from its origins, abandoning its historic mission to serve rural people and American consumers.

This chapter independently examines America's land grant college-agricultural complex. Its message is that the tax-paid land grant complex has come to serve an elite of private, corporate interests in rural America while ignoring those who have the most urgent needs and the most legitimate claims for assistance.

It is the objective of the Land Grant College Task Force to provoke a public response that will help realign the land grant complex with the public interest. In a recent speech on reordering agricultural research priorities, the director of science and education at the U.S. Department of Agriculture (USDA) said that "the first giant steps are open discussion and full recognition of the need." This chapter is dedicated to that spirit.

## The Land Grant College Complex

As used throughout the report, "land grant college complex" denotes three interrelated units, all attached to the land grant college campus:

1. *Colleges of agriculture*—created in 1862 and 1890 by two separate Morrill acts.
2. *State agricultural experiment stations*—created in 1887 by the Hatch Act for the purpose of conducting agricultural and rural research in cooperation with the colleges of agriculture.
3. *Extension service*—created in 1914 by the Smith-Lever Act for the purpose of disseminating the fruits of teaching and research to the people in the countryside.

Reaching into all fifty states, the complex is huge, intricate, and expensive. It can be estimated that the total complex is approaching an expenditure of three quarters of a billion tax dollars appropriated each year from federal, state, and county governments. The public's total investment in this complex, including assets, comes to several billion dollars in any given year, paying for everything from test tubes to experimental farms, from chalk to carpeting in the dean's office.

## The Research Effort

There is no doubt that American agriculture is enormously productive and that agriculture's surge in productivity is largely the result of mechanical, chemical, genetic, and managerial research conducted through the land grant college complex.

But the question is whether the achievements outweigh the failures, whether benefits are overwhelmed by costs. It is the finding of the task force that land grant college research is not the bargain that has been advertised.

The focus of agricultural research is warped by the land grant community's fascination with technology, integrated food processes, and the like. Strict economic efficiency is the goal, not people. The distorted research priorities are striking:

- 1,129 scientific man-years (smy) on improving the biological efficiency of crops, and only 18 smy on improving rural income.
- 842 smy on control of insects, diseases, and weeds in crops, and 95 smy to ensure food products free from toxic residues from agricultural sources.
- 200 smy on ornamentals, turf, and trees for natural beauty, and a sad 7 smy on rural housing.
- 88 smy on improving management systems for livestock and poultry production, and 45 smy for improving rural institutions.
- 68 smy on marketing firm and system efficiency, and 17 smy on causes and remedies of poverty among rural people.

In fiscal year 1969, a total of nearly 6,000 scientific man-years were devoted to research on all projects at all state agricultural experiment stations. Based on USDA's research classifications, only 289 of those scientific man-years were expended specifically on "people-oriented" research. That is an allocation to rural people of less than 5 percent of the total research effort at the state agricultural experiment station (see Table I).

An analysis of these latter research projects reveals that the commitment to the needs of people in rural America is even less than appears on the surface. In rural housing, the major share of research has been directed not to those who live in them, but to those who profit from the construction and maintenance of houses —architects, builders, lumber companies, and service industries.

Again and again, the point is made that industry needs help because it cannot do its own research and because it is affected by external factors. People, however, are responsible for their own condition. For one, public research assistance is considered an investment; for the other, that assistance is treated as welfare.

### Mechanization Research

The primary beneficiaries of land grant research are agribusiness corporations. These interests envision rural America solely as a factory that will produce food, fiber, and profits on a corporate assembly line extending from the fields through the supermarket

TABLE I

SCIENTIFIC MAN-YEARS OF "PEOPLE-ORIENTED" RESEARCH
CONDUCTED AT STATE AGRICULTURAL EXPERIMENT
STATIONS—1966 AND 1969

| Research Problem Areas | 1966 SMY at SAES | 1969 SMY at SAES |
|---|---|---|
| Food Choices, Habits, & Consumption | 8 | 11.5 |
| Home & Commercial Preparation of Food | 14 | 12.4 |
| Human Nutritional Well-being | 103 | 93.5 |
| Selection & Care of Clothing and Household Textiles | 18 | 15.0 |
| Housing Needs of Rural Families | 11 | 6.5 |
| Family Decision Making & Financial Management | 20 | 16.0 |
| Causes & Remedies of Poverty Among Rural People | 11 | 17.1 |
| Improvement of Economic Opportunities for Rural People | 42 | 27.7 |
| Communication Processes in Rural Life | 17 | 18.3 |
| Individual & Family Adjustment to Change | 28 | 25.6 |
| Improvement of Rural Community Institutions & Services | 29 | 45.3 |
| TOTAL SMY ALLOCATED TO "PEOPLE" RESEARCH | 301 | 288.9* |

SOURCES: USDA science and education staff, *Inventory of Agricultural Research, fy 1969 and 1970*, Washington, D.C., October 1970, pp. 247–278. Also, USDA, USDA-NASULGC, *A National Program of Research for Rural Development and Family Living*, November 1968, pp. 5, 28, and 29.

* This allocation of scientific man-years indicates how meager the commitment to "people-oriented" research really is in comparison with the land grant community's rhetoric of concern. The experiment stations actually were doing *less* people-oriented research in 1969 than they were in 1966. The 289 smy allocated to people in 1969 represents only 4.8 percent of the total of 5,956 smy expended that year at state agricultural experiment stations.

checkout counters. It is through mechanization research that the land grant colleges are coming closest to this agribusiness ideal.

Mechanization means more than machinery for planting, thinning, weeding, and harvesting. It also means improving on nature's design, that is, breeding new food varieties that are better adapted to mechanical harvesting. Having built machines, the land grant research teams found it necessary to develop a tomato that is hard enough to survive the grip of mechanical "fingers," to redesign the grape so that all the fruit has the good sense to ripen at the same time, and to restructure the apple tree so that it grows shorter, leaving the apples less distance to fall to its mechanical catcher. Michigan State University, in a proud report on "tailor-made" vegetables, notes that their scientists are at work on broccoli, tomatoes, cauliflower, cucumbers, snapbeans, lima beans, carrots, and asparagus.

If it cannot be done by manipulating genes, land grant scientists reach into their chemical cabinet. Louisiana State University has experimented with the chemical "Ethrel" to cause hot peppers to ripen at the same time for "once-over" mechanical harvesting; scientists at Michigan State University are using chemicals to reduce the cherry's resistance to the tug of the mechanical picker; and a combination of ferric ammonia citrate and erythorbic acid is being used at Texas A & M to loosen fruit before machine harvesting.

Once harvested, food products must be sorted for size and ripeness. Again, land grant college engineers have produced a mechanical answer. North Carolina State University, for example, has designed and developed an automatic machine that "dynamically examines blueberries according to maturity."

Genetically redesigned, mechanically planted, thinned and weeded, chemically readied, and mechanically harvested and sorted, food products move out of the field and into the processing and marketing stages—untouched by human hands.

Who is helped and who is hurt by this research and development?

*It is agribusiness that is helped.* In particular, the largest-scale growers, the farm machinery and chemical input companies, and the processors are the primary beneficiaries. Big business interests are called upon by land grant staffs to participate directly in the planning, research, and development stages of mechanization projects. The interests of agribusiness literally are designed into the product. No one else is consulted.

Obviously, farm machinery and chemical companies are also direct beneficiaries of this research because they can expect to market products that are developed. Machinery companies such as John Deere, International Harvester, Massey-Ferguson, Allis-Chalmer, and J. I. Case almost continually engage in cooperative research efforts at land grant colleges. These corporations contribute money and some of their own research personnel to help land grant scientists develop machinery; in return, they are able to incorporate technological advances in their own products. In some cases, they actually receive exclusive license to manufacture and sell the product of tax-paid research.

Mechanization of fruits and vegetables has focused first on crops used by the processing industries. Brand name processors (such as Del Monte, Heinz, Hunt, Stokely Van-Camp, Campbell's, and Green Giant) are direct beneficiaries of mechanization research. Many of these corporations have been directly involved in the development of mechanization projects. In addition to the food-breeding aspects of mechanization, processsors and canners also have benefited insofar as mechanization has been able to lower costs of production and insofar as that savings has been passed on to them. Of course, many food processors also are growers—either growing directly on their own land, or growing indirectly, controlling the production of others through contractual arrangements.

Large-scale farming operations, many of them major corporate farms, also are directly in line to receive the rewards of mechanization research. In the first place, it is these farms that hire the overwheiming percentage of farm labor, thus having an economic incentive to mechanize. Second, these are the massive farms, spreading over thousands of acres. This scale of operation warrants an investment in machinery. Third, these are heavily capitalized producers, including processing corporations, vertically integrated input and output industries, and conglomerate enterprises. Such farming ventures are financially able and managerially inclined to mechanize the food system.

*Then there are the victims of mechanization*—those who are directly hurt by research that does not consider their needs. If mechanization research has been a boon to agribusiness interests, it has been a bane to millions of rural Americans. The cost has been staggering.

Farmworkers have been the earliest victims. Again and again

the message is hammered home: Machines are now or are on their way to replacing farm labor. There were 4.3 million hired farmworkers in 1950. Twenty years later, that number had fallen to 3.5 million. As a group, those laborers averaged $1,083 for doing farm work in 1970, making them among the very poorest of America's employed poor. The great majority of these workers were hired by the largest farms, which are the same farms moving as swiftly as possible to mechanize their operation.

Farmworkers have not been compensated for jobs lost to mechanization research. They were not consulted when that research was designed, and their needs were not a part of the research package that resulted. They simply were left to fend for themselves— no retraining, no effort to find new jobs for them, no research to help them adjust to the changes that came out of the land grant colleges. Corporate agribusiness received a machine with the taxpayer's help, but the workers who were replaced were not even entitled to unemployment compensation.

Independent family farmers—at least those who have sales under $20,000 a year (which includes 87 percent of all U.S. farms)— also have been victimized by the pressure of mechanization, and their needs also have been largely ignored by the land grant colleges.

Mechanization has been a key element in the cycle of bigness: Enough capital can buy machinery, which can handle more acreage, which will produce greater volume, which can mean more profits, which will buy more machinery. Mechanization has not been pressed by the land grant complex as an alternative but as an imperative.

Mechanization research by land grant colleges either is irrelevant or only incidentally adaptable to the needs of some 87 to 99 percent of America's farmers. The public subsidy for mechanization actually has weakened the competitive position of the family farmer. Taxpayers, through the land grant college complex, have given corporate producers a technological arsenal specifically suited to their scale of operation and designed to increase their efficiency and profits. The independent family farmer is left to strain his private resources to the breaking point in a desperate effort to clamber aboard the technological treadmill.

Like the farmworker, the average farmer is not invited into the land grant laboratories to design research. If he were, the research "package" would include machines useful on smaller acreages, assistance to develop cooperative ownership systems, efforts to develop low-cost and simpler machinery, a heavy emphasis on new credit schemes, and special extension to spread knowledge about the purchase, operation, and maintenance of machinery. In short, there would be a deliberate and major effort to extend mechanization benefits to all, with an emphasis on at least maintaining the competitive position of the family farm in relation to agribusiness corporations. These efforts do not exist or exist only in a token way. Mechanization research has left the great majority of farmers to "get big" on their own or to get out of farming altogether.

Mechanization also has a serious impact on the consumer, and that impact puts America's "bargain" food prices in serious question. Land grant researchers are not eager to confront the issue of quality impact of mechanization, choosing instead to dwell on the benefits that food engineering offers agribusiness.

The University of Florida, for example, recently has developed a new fresh-market tomato (the MH-1) for machine harvesting. In describing the characteristics that make this tomato so desirable for machine harvest, the university points to "the thick walls, firm flesh, and freedom from cracks." It may be a little tough for the consumer, but agricultural research can't please everyone. The MH-1, which will eliminate the jobs of thousands of Florida farmworkers who now handpick tomatoes for the fresh market, is designed to be harvested green and to be "ripened" in storage by application of ethylene gas.

### Agribusiness Versus Consumers

The colleges also are engaged in "selling" the consumer on products he neither wants nor needs, and they are using tax money for food research and development that should be privately financed. At Virginia Polytechnic Institute, for example, eight separate studies have been conducted to determine if people would like a blend of apple and grapefruit juice.

Another aspect of selling the consumer is "knowing" him. There are many projects that analyze consumer behavior. Typically, these involve consumer surveys to determine what influences the shopper's decision-making. If this research is useful to anyone, it is to food marketers and advertisers, and reports on this research make clear that those firms are the primary recipients of the results. The corporations that benefit from this research should pay for it and conduct it themselves.

The consumer is not just studied and "sold" by land grant research; he is also fooled. These public laboratories have researched and developed food cosmetics in an effort to confirm the consumer's preconceptions about food appearances, thus causing the consumer to think that the food is "good." Chickens have been fed the plant compound xanthophyll to give their skin "a pleasing yellow tinge," and several projects have been undertaken to develop spray-on coatings to enhance the appearance of apples, peaches, citrus, and tomatoes. Following are some other cosmetic research projects that are under way at land grant colleges:

- Iowa State University is conducting packaging studies which indicate that color stays bright longer when bacon is vacuum-packed or sealed in a package containing carbon dioxide in place of air, thus contributing to "more consumer appeal."
- Because of mechanical harvesting, greater numbers of green tomatoes are being picked; scientists at South Carolina's agricultural experiment station have shown that red fluorescent light treatment can increase the red color in the fruit and can cause its texture and taste to be "similar to vine-ripened tomatoes."
- Kansas State University Extension Service, noting that apples sell on the basis of appearance rather than nutrition, urged growers to have a beautiful product. To make the produce more appealing, mirrors and lights in supermarket produce cases were cited as effective selling techniques.

Sold, studied, and fooled by tax-supported researchers, there finally is evidence that the consumer actually is harmed by food engineering at land grant colleges.

Ethylene gas, used to speed up the growth of produce has been shown, when used on tomatoes, to provide lower quality with less vitamin A and C and inferior taste, color, and firmness.

There is strong evidence that DES, a growth hormone fed to cattle, causes cancer in man. Yet DES has added some $2.9 million to the treasury of Iowa State University, where the use of the drug was discovered, developed, patented, and promoted—all with tax dollars. Eli Lilly & Company, which was exclusively licensed by Iowa State to manufacture and sell the drug, has enjoyed profits on some $60 million in DES sales to date.

More and more, chemicals are playing a role in the processing phase. Ohio State University reports that "chemical peeling of tomatoes with wetting agents and caustic soda reduces labor by 75 percent and increases product recovery." One wonders if the consumer will recover. Lovers of catfish might be distressed to learn that this tasty meat now is being skinned chemically for commercial packaging.

Three assumptions are made by the task force. First, if there is to be research for firms that surround the farmer, benefits of that research should flow back to the farmer. Second, no public money should be expended on research that principally serves the financial interests of agricultural input and output corporations—they may be a part of modern agriculture, but they also are very big business and capable of doing their own profit-motivated research. Finally, anything that is good for agribusiness is *not* necessarily good for agriculture, farmers, rural America, or the consumer.

### Failure of Land Grant College Research

Except for agribusiness, land grant college research has been no bargain. Hard tomatoes and hard times is too much to pay. That does not mean a return to the hand plow. Rather, it means that land grant college researchers must get out of the comfortable chairs of corporate board rooms and get back to serving the independent producer and the common man of rural America. It means returning to the historic mission of taking the technological revolution to all who need it, rather than smugly assuming that they will be unable to keep pace. Instead of adopting the morally bankrupt posture that millions of people must "inevitably" be squeezed out of agriculture and out of rural America, land grant colleges must turn their thoughts, energies, and resources to the task of keeping people on the farm, in the small towns, and out of the cities. It means turning from the erroneous assumption that big is good, that what serves Ralston Purina serves rural America. It means research for the consumer rather than for the processor. In short, it means putting the research focus on people first—not as a trickled-down afterthought.

The greatest failing of land grant research is its total abdication of leadership. At a time when rural America desperately needs leadership, the land grant community has ducked behind the corporate skirt, mumbling apologetic words like "progress," "efficiency," and "inevitability." Overall, it is a pedantic and cowardly research system, and America is the less for it.

A change in the focus of land grand research will not happen simply because it should happen. Change will come only if those interests now being abused by the research began to make organized demands on the complex. If independent family farmers, consumers, small-town businessmen, farmworkers, environmentalists, farmer cooperatives, small-town mayors, taxpayers' organizations, labor unions, big-city mayors, rural poverty organizations, and other "outsiders" go to the colleges and to the legislatures, changes can occur. These interests need not go hand in hand, but they all must go if land grant college research ever is to serve anyone other than the corporate elite.

### Making Research Policy

The short-range research policy of the land grant system is the product of the annual budgeting process, and the substance of that research budget is determined by the Agricultural Research Policy Advisory Committee (ARPAC), which reports directly to the secretary of agriculture. Its members are taken from USDA and the land grant community; in fact, they *are* the agricultural research establishment.

The National Association of State Universities and Land Grant Colleges (NASULGC) is the home of the land grant establishment. Their particular corner in the association is under the title of Division of Agriculture, composed of all deans of agriculture, all heads of state experiment stations, and all deans of extension. With eight members on the twenty-four-man ARPAC board, NASULGC's Agricultural Division plays a major role in the determination of research priorities and budgets. The division also represents the land grant college complex before Congress on budget matters.

The top rung on the advisory ladder is USDA's National Agricultural Research Advisory Committee. This eleven-member structure currently includes representatives from the Del Monte Corporation, the Crown Zellerbach Corporation, AGWAY, Peavey Company Flour Mills, the industry-sponsored Nutrition Foundation, and the American Farm Bureau Federation.

Most national advisory structures are dominated by land grant scientists and officials, but whenever an "outsider" is selected, chances are overwhelming that the person will come from industry. A series of national task forces, formed from 1965 to 1969 to prepare a national program of agricultural research, were classic examples of this pattern. Out of thirty-two task forces, seventeen listed advisory committees containing non-USDA, non-land-grant people. All but one of the outside slots on those seventeen committees were filled with representatives of industry, including General Foods on the rice committee, U.S. Sugar on sugar, Quaker Oats on wheat, Pioneer Corn on corn, Liggett & Myers on tobacco, Procter & Gamble on soybeans, and Ralston Purina on dairy. Only on the "soil and land use" task force was there an advisor representing an interest other than industrial, but even there, the National Wildlife Federation was carefully balanced by an advisor from International Minerals and Chemical Corporation.

There are also state and local advisory structures to the land grant complex. Commenting on such groups and their impact on the allocation of research resources, USDA's Roland Robinson wrote: "Many of the advisory groups, similar to those of the Department of Agriculture, are established along commodity and industry lines. Consequently they are oriented toward traditional research needs. The rural nonfarmer, the small farmer, the leaders of rural communities and the consumer are not usually represented on experiment-station advisory committees."

Land grant policy is the product of a closed community. The administrators, academics, and scientists, along with USDA officials and corporate executives, have locked themselves into an inbred and even incestuous complex, and are incapable of thinking beyond their self-interest and traditional concepts of agricultural research.

## The Congressional Failure

Congress holds hearings each year on the appropriations requests for agricultural research. It is here that the public might expect some serious questioning of research focus and some assertion of other than private interests. It does not happen.

Hearings on agricultural research budgets are left pretty much to the land grant community, buttressed by its agribusiness colleagues. The appropriations process falls far short of being a careful, substantive scrutiny—in fact, it is little more than a chance for special interests to press for particular research projects or facilities.

Public witnesses appearing before the agricultural subcommittees overwhelmingly represent agribusiness interests. Technically, anyone can testify, but it is industry that has the resources to maintain Washington representatives and to fly witnesses in and out of the Capital for a day of testimony. There are dozens of agribusiness lobbyists in Washington, ranging from the full-scale operation of the American Farm Bureau Federation to coveys of Washington "lawyers" retained to look out for the special interests of practically every corporate name in agriculture.

The few Washington organizations representing the interests of farmers, sharecroppers, small businesses, the poor, minorities, consumers, or environmentalists either do not have the resources and staff to deal effectively with the agricultural research budget or have failed to perceive their self-interest in that budget. Tax-exempt public interest groups are prohibited by law from lobbying and cannot appear to testify on appropriations unless invited to do so by the committee.

There are hundreds of pages of testimony on the land grant complex each year, but no tough questioning of how those resources are being used. With 2,000 farm families leaving the land each week, with some 800,000 people a year being forced out of rural America, and with all the other stark evidence of rural failure, it seems that some representative of the people would probe a bit into the nature and impact of the land grant complex.

Congress has relinquished its responsibility and authority to narrowly focused officials at USDA and within the land grant community. Like spokesmen of the military-industrial complex, these officials and their allies come to the Capital at appropriations time to assure a docile Congress that its investment in agricultural hardware is buying "progress" and that rural pacification is proceeding nicely.

### Agribusiness Links to Land Grant Campuses

In dozens of ways, agribusiness gets into the land grant college complex. It is welcomed there by administrators, academics, scientists, and researchers who share the agribusinessman's vision of integrated, automated agriculture. Corporate executives sit on boards of trustees, purchase research from experiment stations and colleges, hire land grant academics as private consultants, advise and are advised by land grant officials, go to Washington to help a college or an experiment station get more public money for its work, publish and distribute the writings of academics, provide scholarships and other educational support, invite land grant participation in their industrial conferences, and sponsor foundations that extend both grants and recognition to the land grant community.

Money is the web of the tight relationship between agribusiness interests and their friends at the land grant colleges. It is not that a huge sum of money is given—industry gave only $12 million directly to state agricultural experiment stations for research in 1969. Rather, it is that enough money is given to influence research done with public funds.

But to a larger extent, agribusiness was welcomed into the community because its attitudes and objectives were shared by the land grant communities. Agribusiness corporations wanted help with their new chemical, with their hybrid seed, with their processing facility, or with their scheme for vertical integration. The scientists, engineers, and economists of the land grant community had both the tools and the inclination to deal with those needs.

Industry money goes to meet industry needs and whims, and these needs and whims largely determine the research program of land grant colleges. A small grant for specific research is just good business. In the first place, the grant is tax deductible, either as an education contribution or, if the research is directly related to the work of the corporation, as a necessary business expense. Second, the grant will draw more scientific attention than its value warrants.

One scientist will consult with another, and graduate assistants and other personnel will chip in some time. If the project is at all interesting, it will be picked up and carried on by someone working under a public budget or assigned to someone working on a Ph.D. Finally, not only is the product wrapped and delivered to the corporation, but with it comes the college's stamp of legitimacy and maybe even an endorsement by the scientist who conducted the research. If it is a new product, the corporation can expect to be licensed, perhaps exclusively, as its producer and marketer. Everything considered, it amounts to a hefty return on a meager investment.

There is a long list of satisfied corporate customers. As would be expected, half of industry's research contributions to state agricultural experiment stations in fiscal year 1969 went to just four categories: insect control, weed control, plant and animal biology, and biological efficiency.

Prime contributors are chemical, drug, and oil corporations. Again and again the same names appear—American Cyanamid, Chevron, Dow Chemical, Exxon, Eli Lilly, Geigy, FMC-Niagra, IMC Corporation, Shell, Stauffer, Union Carbide, and the Upjohn Company are just a few of the giants that gave research grants to the University of Florida, North Carolina State University, and Purdue University. Chemical, drug, and oil companies invested $227,158 in research at Florida's Institute of Food and Agricultural Science, for example, accounting for 54 percent of research sponsored there by private industry in 1970.

Where does the corporation end and the land grant college begin? It is difficult to find the public interest in the tangle. These ties to industry raise the most serious questions about the subversion of scientific integrity and the selling of the public trust. If grants buy corporate research, do they also buy research scientists and agricultural experiment stations?

### Land Grant Research Foundations

At least twenty-three land grant colleges have established foundations to handle grants and contracts coming into their institutions for research. These quasi-public foundations are curious mechanisms, handling large sums of money from a wide array of private and public donors, but under practically no burden of public disclosure.

A funding source can give money to a private research foundation, which then funnels the money to a public university to conduct research. By this shell game, industry-financed research can be undertaken without obligation to make public the terms of the agreement. The foundation need not report to anyone the names of corporations that are making research grants, the amounts of those grants, the purpose of those grants, or the terms under which the grants are made.

These foundations also handle patents for the colleges. When a corporation invests in research through a foundation, it is done normally with the understanding that the corporation will have first shot at a license on any patented process or product resulting from the research. On research patents that do not result from corporate grants, the procedure for licensing is just as cozy. At Purdue University, for example, a list is drawn of "responsible" companies that might have an interest in the process or product developed, and the companies are approached one by one until there is a taker.

## Extension Service

The Extension Service (ES) is the outreach arm of the land grant college complex. Its mandate is to go among the people of rural America to help them "identify and solve their farm, home, and community problems through use of research findings of the Department of Agriculture and the State Land Grant Colleges."

Three hundred thirty-one million dollars were available to the Extension Service in 1971. Like the other parts of the land grant complex, extension has been preoccupied with efficiency and production—a focus that has contributed much to the largest producers, but which has slighted the pressing needs of the vast majority of America's farmers, and ignored the great majority of other rural people.

Like their research and teaching colleagues in the land grant complex, extension agents walk hand in hand with agribusiness. To an alarming degree, extension agents are little more than salesmen. A recent article in *Farm Technology*, the magazine for county agents, offers this insight into corporate ties to extension:

> We are impressed with the fact that much time is spent working closely with industry agri-fieldmen and other company representatives. Nearly all states reported that this type of cooperation is increasing.
>
> A good example of this can be found in Arizona, where weed specialists "hit the road" with the chemical company representatives and are involved in cooperative field tests and demonstrations.

Extension Service has not lived up to its mandate for service to rural people. The rural poor, in particular, are badly served by the service, receiving a pitiful percentage of the time of extension "professionals," while drawing temporary assistance from the highly visible nutrition aides program and irrelevant attention from the 4-H program. In 1955, a Special Needs Section was added to extension legislation, setting aside a sum of money to assist disadvantaged areas. Extension has failed to make use of this section.

The civil rights record of ES comes close to being the worst in government. Policy-making within ES fails to involve most rural people, and USDA has failed utterly to exercise its power to redirect ES priorities and programs.

The Extension Service's historical and current affiliation with the American Farm Bureau Federation casts a deep shadow over its claim that it can ever be part of the solution of the problems of rural America.

### Black Land Grant Colleges

In 1862, at the time of the first Morrill Act, 90 percent of America's black population was in slavery. The land grant colleges that developed were white bastions, and even after the Civil War, blacks were barred from admission both by custom and by law. When the second Morrill Act was passed in 1890, primarily to obtain more operating money for the colleges, Congress added a "separate but equal" provision authorizing the establishment of colleges for blacks. Seventeen southern and border states took advantage of the act, creating institutions that still are referred to euphemistically as "colleges of 1890."

The black colleges have been less than full partners in the land grant experience. It is a form of institutional racism that the land grant community has not been anxious to discuss. From USDA, resource allocations to these colleges are absurdly discriminatory. In 1971, of the $76.8 million in U.S. Department of Agriculture funds allocated to those sixteen states with both white and black land grant colleges, 99.5 percent went to the white colleges, leaving only 0.5 percent for the black colleges. Less than one percent of the research money distributed by the Cooperative State Research Service in 1971 went to black land grant colleges. This disparity is not by accident, but by law.

### Public Disclosure

It is difficult to discover what the land grant complex is and what it is doing. For example, most agricultural experiment stations offer an annual report in compliance with the Hatch Act disclosure provisions, but these reports are less than enlightening:

- Some do not list all research projects, but merely highlights.
- Some list research projects, but only by title, without even a brief description.
- Most do not include money figures with the individual projects, and very few reveal the source of the money.
- None contains any element of project continuity to show the total tax investment over the years in a particular investigation.
- Most contain only a very general financial breakdown, listing state, federal, and "other" funds received and expended.
- Few offer any breakdown of industry contributions, naming the industry, the contribution, and the project funded.

These are the basic facts. There is *no* listing of more esoteric items, such as patents developed by the station and held by the college, or advisory structures surrounding the stations.

Data is neither supplied uniformly nor collected in a central location, nor reported in a form that can be easily obtained or understood. Even more significant is the fact that many fundamental questions go unasked and many fundamental facts go unreported.

Millions of tax dollars are being spent annually by an agricultural complex that effectively operates in the dark. It is not that the land grant community deliberately hides from the public. The farmer, the consumer, the rural poor, and others with a direct interest in the work of the land grant complex can get no adequate picture of its work. Congress is no help; it does not take the time to probe the system, to understand it in detail, and to direct its work in the public interest.

The land grant college complex has been able to get by with a minimum of public disclosure, and that has meant that the community has been able to operate with a minimum of public accountability.

### Conclusion

There is nothing inevitable about the growth of agribusiness in rural America. While this country enjoys an abundance of relatively cheap food, it is not more food, not cheaper food, and certainly not better food than that which can be produced by a system of family agriculture. And more than food rolls off the agribusiness assembly line—rural refugees, boarded-up businesses, deserted churches, abandoned towns, broiling urban ghettoes, and dozens of other tragic social and cultural costs also are products of agribusiness.

Had the land grant community chosen to put its time, its money, its expertise, and its technology into the family farm, rather than into corporate pockets, then rural America today would be a place where millions could live and work with dignity.

The colleges have mistaken corporate need for national need. This is proving to be a fatal mistake—not fatal for the corporations or for the colleges, but for the people of America. It is time to correct that mistake, to reorient the colleges so that they will begin to act in the public interest.

### Recommendations

The Task Force on the Land Grant College Complex does not presume to prescribe an agenda for the land grant college complex. That is the proper role of constituencies with a direct interest in the work of the complex.

Rather, the task force seeks, through its recommendations, to open the closed world of the land grant complex to public view and to participation by constituencies that today are locked out.

Generally, these recommendations call for a full-scale public inquiry, both in the Congress and state legislatures, regarding the nature, extent, and national impact of the land grant complex. There should be a General Accounting Office audit of the land grant complex. An immediate reopening of the hearings on the 1972 to 1973 agricultural research budget by the House and Senate is also necessary. Also, the secretary of agriculture should act immediately to restructure the national advisory and policy-making apparatus so that there is a broadened input from "outside" constituencies for research planning.

The task force calls for an immediate end to racial discrimination within the land grant complex, withholding federal money from any state that does not place its black institutions on an equal footing with the white colleges.

Legislation is also needed which would: prohibit land grant officials and other personnel from receiving remuneration in conflict of interests; prevent corporations from earmarking contributions to the land grant complex for specific research that is propitious in nature; and insure that land grant patenting practices do not allow private gain from public expenditures.

Laws requiring full public disclosure from the land grant complex are of crucial importance. Detailed, complete, and uniform reports from each college should be filed annually with the secretary of agriculture, who should compile them and make them easily available to the public.

The land grant colleges must get out of the corporate board rooms, they must get the corporate interests out of their labs, and they must draw back and reassess their preoccupation with mechanical, genetic, and chemical gadgetry. The complex must again become the people's university—it must be redirected to focus the preponderance of its resources on the full development of the rural potential.

*Appendix*

## TERMS, CONCEPTS, AND ACRONYMS

**Agribusiness.** A corporate aggregation that includes: (1) agricultural input firms; (2) agricultural output firms; (3) corporations directly involved in farming; and (4) corporations indirectly involved in farming.

**Agricultural efficiency.** The concept of minimizing agricultural production costs by substituting capital, machinery, chemicals, and other technological and financial inputs for the more traditional farming inputs.

**Agricultural input industry.** An aggregation of firms that supply seed, feed, farm machinery, fertilizer, chemicals, credit, insurance, and other factors of agricultural production.

**Agricultural output industry.** An aggregation of corporate middlemen between the farmer and the consumer, including those firms that pack, process, can, package, distribute, market, advertise, retain, and otherwise handle food and fiber after it leaves the farm.

**Agricultural Research Service (ARS).** USDA's in-house research agency, conducting agricultural research at the federal level, based on USDA's perception of national and regional research needs.

**Cooperative State Research Service (CSRS).** The USDA agency that administers federal research money allocated to state agricultural experiment stations by statutory formula. In addition, CSRS administers a relatively small amount of nonformula funds, expending them through research contracts made with the stations.

**Current Research Information System (CRIS).** A USDA data bank containing computerized information on research projects conducted at state agricultural experiment stations.

**Extension Service (ES).** The national network of extension agents and administrators. The Federal Extension Service (FES) is the USDA agency that administers national funds for extension work. The Cooperative Extension Service (CES) is the usual designation of any state extension service.

**Family farm.** A farm that is controlled and worked by the family that lives on the farm. Financial risk, managerial decisions, and work on the farm are direct responsibilities of the family, which exercises full, entrepreneurial authority.

**Land grant college community.** Includes people directly involved in the land grant college complex at the campus level, in government, and in agribusiness. This is a community of shared interests, involving teachers, researchers, administrators, students, governmental officials relating to the complex, and agribusiness organizations with a proprietary interest in the work of the complex.

**Land grant college complex.** The agricultural component of the land grant university system. The complex includes colleges of agriculture, agricultural experiment stations, and extension services. Engaged in teaching, research, and dissemination of knowledge in all fifty states, the complex accounts for an annual public expenditure approaching $750 million.

**Land grant college system.** The higher educational system created under the Morrill land grant act. It includes sixty-nine land grant universitis, ranging from MIT to the University of California and teaching everything from nuclear physics to Chaucer. Included in this extensive educational system is the land grant college complex, which is focused on agriculture.

**National Association of State Universities and Land Grant Colleges (NASULGC).** A Washington-based organization representing 118 public institutions of higher education, including all sixty-nine land grant colleges. NASULGC's Division of Agriculture represents agricultural college deans, heads of agricultural experiment stations, and deans of extension. The division is operated by and for the land grant complex. The NASULGC division is a powerful spokesman for the complex and is directly involved in the development of agricultural research priorities for the country.

**People-oriented research.** A USDA term referring to research focused directly on people, rather than on production, marketing, efficiency, or some other aspect of agriculture. The term includes twelve research problem areas: food consumption habits, food preparation, human nutrition, clothing and textile care, family financial management, rural poverty, economic potential of rural people, communications among rural people, adjustment to change, rural income improvement, rural institutional improvement, and rural housing.

**Research Problem Areas (RPA).** A series of USDA classifications for agricultural research projects. Allocations of money and scientific man-years are categorized under these RPAs.

**State Agricultural Experiment Station (SAES).** The agricultural research component of each land grant college.

**Scientific man-years (SMY).** A measurement of scientific, technical, and other time expended on research projects. The measurement is based on a standardized formula, and allocations of smy are reported through CRIS.

**United States Department of Agriculture (USDA).** The department with primary federal responsibility for overseeing the land grant college complex.

**Vertical integration.** The movement of agricultural input and output firms into the production stage of food and fiber. The movement can be direct, as when a processing plant buys or leases land to produce commodities for its processing operation. It can be indirect, as when an agribusiness firm contracts with a farmer to produce a certain quantity and quality of a certain commodity at a certain time and for a certain price. In both cases, a degree of control over food and fiber production passes from farmers to agribusiness corporations.

# 8

# THE GREEN REVOLUTION: AMERICAN AGRICULTURE IN THE THIRD WORLD

*Michael Perelman*

The Green Revolution is usually thought of narrowly as the current, accelerated growth in Third World grain production which results from combining the new seeds . . . mostly wheat and rice . . . with heavy applications of fertilizer and carefully controlled irrigation. . . . Yet . . . the Green Revolution is far more than one of plant breeding and genetics. It is woven into the fabric of American foreign policy and is an integral part of the postwar effort to contain social revolution and make the world safe for profits.

—HARRY CLEAVER

The end of the fight is a tombstone white
    With the name of the late deceased
And the epitaph drear!: A fool lies here
    Who tried to hustle the East.

—RUDYARD KIPLING

Michael Perelman, is professor of economics at California State University, Chico.

Norman Borlaug won the 1970 Nobel Peace Prize for his work in developing higher-yielding varieties of wheat and rice. Many observers believe that Third World farmers harvesting these new seeds will soon be producing a surplus of grains and that the resulting agricultural prosperity will set the developing nations on the path to affluence. As the chairman of the prize committee was quoted in the *New York Times* (22 October 1970), "In short we do not any longer have to be pessimistic about the economic future of developing nations." Thus, the Green Revolution has come to symbolize one of the few positive aspects of the world's future.

However, the fact that the United States is exporting its agricultural techniques to the "underdeveloped" countries of the world has serious ecological and social implications for these countries. In fact, it is becoming increasingly clear that the Green Revolution, and the methods by which it is being implemented, could create more problems than it solves. This chapter attempts to describe some of these problems.

### Social Implications

Green Revolutions are nothing new. The Incas selected the best corn seeds from each harvest to plant in the next growing season; as a result, corn evolved from a grass with a tiny kernel to what we have today. We have records of a Sung emperor of eleventh-century China who introduced a rapid-maturing rice from Indochina that could be harvested in 100 days, instead of 180 days. Later, rices were developed which matured within 30 or 40 days.[1]

But not all Green Revolutions have had the desired effect. Sir Walter Raleigh, the Norman Borlaug of his day, introduced the potato to Ireland around 1588. The yield of miracle spuds was far greater than that of cereals, but the benefits went primarily to the landlords, not to the majority of the Irish who were tenants. The standard of living seems to have fallen as yield increased.[2] Irish agriculture, highly concentrated on a single crop, was particularly susceptible to blight, and when the blight finally came in the middle of the last century, it caused the starvation of over a million people and forced another million to emigrate.

Already the Green Revolution is falling in line with the Irish experience; agricultural prosperity is intensifying the poverty of the poor peasants and landless farm laborers who make up the bulk of the population of Asia. One study by the Pakistan Planning Commission shows that mechanization of farms reduces the need for labor by 50 percent. Conservative estimates by the Planning Commission show that by 1985 four million people in Pakistan will be affected by evictions.[3]

AID officials try to argue that mechanization might actually increase the demand for farm labor because machines can plow and harvest a field so fast that two crops can be harvested in a single year. Moreover, the technology of the Green Revolution "can be applied to large and small farms alike and does *not* need mechanized cultivation."[4] But it does require irrigation and the small farmer cannot get the credit necessary for the purchase of wells and the like. Wolf Ladijinsky estimates that it takes 10,000 to 12,000 rupees to equip a 7- to 10-acre holding.[5] Once the farmer owns a large chunk of capital, he can cultivate a few more acres for very little extra cost. So he evicts his tenants and buys out his neighbors. And even in the unlikely event that total employment rises, the land is cultivated by hired landless laborers instead of tenants.

In the Indian province of Punjab, the large farmers are determined to mechanize as quickly as possible in order to be rid of their dependence on agricultural labor.[6] Here the Indians are following a course already taken by American farmers: American farms today employ less than one-quarter as much labor as they did in 1920.[7]

As the demand for farm labor falls, dispossessed farm laborers and peasants go to the city, just as they have in this country. Some economists hope that many of these workers will find urban jobs making tractors and machinery for the farms, but they won't.

Peasants evicted from their land have no place to go, no jobs waiting for them. The International Labor Organization estimates that even with industrial growth rates of 7 to 9 percent a year (a rate which almost no developing nations have achieved), capital requirements in industry are so heavy and the number of new workers per dollar of capital so small, that the industrial sector will have a hard time absorbing the increase in workers created by population growth alone, not to mention displacement from the farms.[8] Robert McNamara told the Columbia University Conference on International Economic Development in February 1970 that "just as the censuses of the fifties helped alert us to the scale of the population explosion, the urban and employment crises of the sixties are alerting us to the scale of social displacement and general uprootedness of population which is exploding not only in numbers but in movement as well."[9]

Arthur MacEwan, a professor of economics at Harvard, who has studied Pakistan, is a little more specific. He writes: "The poor peasants and agricultural workers will see a large increase in output which they are not receiving. In India such a situation has already led to bloody battles between agricultural workers and the landlord's hired 'armies.' Press coverage is so restricted in Pakistan that we cannot be sure what has happened there, but it would seem clear that the same problem exists."[10]

While defenders of the Green Revolution admit the problem of finding jobs for the uprooted peasants and laborers, they counter that the main task of the Green Revolution is to grow more food. But in an economy run by the profit motive, food is grown for markets instead of people; so we should ask if the Green Revolution can eliminate hunger. Perhaps it can in a rational economy where people produce directly to serve the needs of others; but in countries like India and Pakistan larger harvests are little consolation for the poor who cannot afford the price of food.

To begin with, the governments involved in the Green Revolution have had to subsidize it by offering farmers higher prices for their crops. As William Paddock says, "Mexico supports her wheat at $1.99 a bushel or 33 percent above the world's market price for quality grain; Turkey, at 63 percent; India and Pakistan, at 100 percent. But since the quality of the new miracle cereals is low and they are sold at a discount, the subsidies are, in real terms, significantly higher."[11]

Although we can applaud the government's efforts to give farmers incentives to produce enough food to feed the hungry, these incentives cost money. So the government is forced to let the food sell at a very high price or to buy the food from the farmers at the subsidized price and then sell the food at a loss. In countries like India and Pakistan, where agriculture accounts for about half the gross national product, great amounts of money must be used to pay for these subsidies; for instance, in Pakistan, $100 million per year goes to support the price of wheat.[12] Scarce resources must be diverted from social needs like health and education to pay for these subsidies. So we find one of the most influential planners in Pakistan writing that expenditure on the provision of social services such as housing, health, and social welfare is "not economic" and therefore must be resisted. Now remember that many of the uprooted peasants coming to the cities will not be able to find work, thus increasing the need for these "uneconomic" expenditures more than ever.

A symbol of the effects of these subsidies comes from the Philippines. The U.S. House Foreign Affairs Subcommittee held a symposium on the Green Revolution. Many of the participants made a great "to-do" about the Philippines' newly developed self-sufficiency in rice. But much of this self-sufficiency is illusory.

In the Philippines, the support price for corn is less than the support price for rice, so the people have had to replace much of the rice in their diets with corn. As a result the Philippines have become self-sufficient in rice; that is, they do not have to import rice because people can't buy any more rice at the inflated support price. Moreover, the Philippines are importing more and more wheat. So the price supports have helped the Philippines achieve self-sufficiency in rice in two ways: by encouraging the production of more rice and discouraging the consumption of rice. In any case many people are still hungry in the Philippines. The irony is that on the same day as the hearings, the Philippine government asked the United States for an advance of $100 million on its payment due from the United States for our use of military bases in the Philippines. The advance was required to "stave off the emergency" arising from the threatened bankruptcy.[13]

Without these subsidies, the Green Revolution would fall on its face. A USDA report cites two surveys of Philippine farmers who decided to stop growing the new Miracle Rice. Over 50 percent said that their reason was the added expense of growing the new varieties. It also mentions a Burmese village that reduced its planting of the new rice. The villagers were willing to increase their planting of the new rice providing that the government would purchase it at the subsidized price; at free or black market prices, the new rice was not worth growing. The report concludes: "It was widely assumed that the increased returns from growing the new variety would have exceeded the cost. . . . Yet there is little solid evidence on this point."[14]

The subsidies upset government finance in still another way. Because the subsidies make the new varieties so profitable to grow, peasants shift acreage from cash crops or other foods to the new varieties. For instance, in Pakistan many farmers were growing wheat instead of cotton, which used to be grown for export; thus the government lost another of its sources for scarce foreign exchange.

The strategy of the Green Revolution was to launch the new agriculture in those localities where it could be applied most profitably. At present, the Green Revolution is limited to a very small proportion of farm lands in the world. One recent study of a sample area in the Punjab of India found that at a given support price, only 16.9 percent of the area could be profitably cultivated with the new seeds. At a slightly higher support, the percentage was still only 38 percent. *So are we to accept the paradox that the Indian subcontinent can only grow enough food if the prices are so high that the neediest cannot afford the cost? Or does the government have to sacrifice needed projects to subsidize the large farmers to grow low-priced food?*[15]

In spite of these problems, AID officials believe that "pressures are developing for the government to allow prices to decline." But until they do, Third World countries may have to face up to what has been a particularly American problem—bumper crops and hungry people. However, as we shall see later, there are even reasons to doubt that the Green Revolution will continue to produce bumper harvests.

The poor who cannot afford to buy enough food are not the only people who will not be able to bask in the expected prosperity of the Green Revolution: small farmers who rent their land will be singled out for particular hardship. Land values are soaring in Punjab and the Purnea district of Binar, where the new agriculture is in vogue. Rents also are on the rise. In India, rent on the land is rising from the more usual (but illegal) 50 percent of the harvest to as high as 70 percent.[16] Since it is more profitable for the landlord to evict his tenants and hire wage laborers to work in their place, he can consolidate the fragmented holdings of his tenants, allowing him to cut costs by mechanization.

Even those farmers who own their own land cannot always take advantage of the Green Revolution because it costs a lot of money to become a Green Revolutionary: The new seeds increase yields only if they are supported by large doses of fertilizer, pesticides, and water. With interest rates of 20 to 40 percent per year for the small farmers, the new technology lies beyond them. Moreover, as better-off farmers mechanize they will have an incentive to buy up neighboring farms until their holdings reach a size that gives them the full advantage of their machines. Average farm size will increase and machines will replace people on the land; these small farmers will join the landless in the cities to wait for employment, just as they do in America, while a few rich "farmers" accumulate their land monopolies.

### Ecological Implications

Defenders of the Green Revolution admit that it does create social tensions, but they ask critics how the Third World could feed its growing population without this modern technology. Ironically, this same technology might worsen the nutritional patterns of the Third World while stimulating its population problems.

It is true that harvest of wheat and rice have increased. But higher yields are bought at the cost of a lower protein content. This leads two officials of the USDA to report: "The new cereals they are now eating may have a lower protein content than the varieties from which they were derived. This fact is inadequately documented, but whenever we have examined the relationship between yield and protein content, the suspicion about lower protein content was confirmed. . . . And less legumes are being grown; their acreage is being displaced by higher yielding grains."[17]

This last point is crucial because legumes are an important source of protein in Asia. In India, for example, they account for 25 percent of all protein consumed. They also contain necessary amino acids not found in grains. Furthermore, the legumes build up the fertility of the soil by fixing nitrogen from the air. And in other parts of Asia much of the protein comes from the fish that are raised in the rice paddies. With the intensive use of pesticides these fish may die or be unfit for human consumption. Thus the Green Revolution might actually lower the protein production of Asia.

Less protein in the diet caused by the technology of the Green Revolution would tend to produce an increase in child mortality; and over and above its human cost, this higher child mortality would mean a higher rate of population increase. This apparent contradiction is explained in the following way:

> A study last year [1969] in a village in the western state of Gujarar concluded that "families continued to have children until they were reasonably certain that at least one boy would survive. Once they had this number, they attempted to stop having more.
>
> A recent computer simulation reveals that with current estimated infant and adult death rates in India, a couple must bear 6.3 children to be 95% sure that one son will be surviving the father's 65th birthday. The average number of births in India per couple is 6.5, which tends to support the thesis that parents continue to bear children until reasonably sure of the survival of at least one son.[18]

The effect of high infant mortality can be devastating to an economy. One study estimated that 22.5 percent of the national income of preindependence India was required to rear children who would not live long enough to make a productive contribution to the society.[19]

There are other ecological problems created by the Green Revolution. Already the new rice technology of the Philippines has produced a disease which was never before a serious problem in the Philippines.[20] Ecologists and agronomists know that as the local varieties of disease-resistant crops are replaced by a few higher-yielding but more susceptible strains, the possibilities for epidemic plant diseases (such as the potato or corn blights, or the wheat stem rust) become much more likely.

The Green Revolution with its singular focus on a few crop strains also has serious environmental consequences. The major famines of the late nineteenth century, led to research into crop varieties that were disease-, insect-, and drought-resistant. The difference between contemporary research involving genetic engineering and the selection experiments of the past is more than just the pursuit of higher yields. As noted by W. David Hopper: "The new varieties that are beginning to emerge are a significant break with traditional agriculture for they are being selected for response to inputs that are not part of conventional [Third World] farming. Most notable among these, of course, is fertilizer. . . . As plant stands thicken with high fertility conditions, insect population and the opportunity for the spread of disease will increase."[21]

The same concern has been put another way: "[Traditional] Japanese rice has a resistant quality against rice stem-borers. Both the number of egg-masses and larvae that actually bore into the plant is small and the damage is little. However, with the [new varieties], the number that actually bore into the plant is larger and the damage greater. . . . Immunity against this insect decreases with the more abundant use of nitrogenous fertilizer."[22]

The irony of this situation is that the only way to correct the genetic weaknesses of high-yield crop strains is by future incorporation of genes from the very varieties that the new hybrids were designed to replace. Up to the present time, developed countries have been able to return to areas of genetic diversity, usually located in Third World nations, to collect genetic wealth for future breeding programs. Clearly one of the top agricultural priorities of Third World countries in the future should be the establishment of national "gene banks" for preserving endemic strains of cultivated crops.

In addition to these problems is the fact that the new varieties also demand amounts of water that go far beyond those available with irrigation systems designed for drought protection.[23] *Thus, in a variety of ways, the Green Revolution may endanger the environment in those countries where sound natural resources are most needed.*

### Political Implications

The Green Revolution does look promising to some classes. A new set of farmers is emerging in the developing nations to reap the harvests of the Green Revolution: retired military officers and civil servants, doctors, lawyers, and businessmen, for many of whom farming is a tax dodge (another sign of the Americanization of farming). Landlords too have been reaping the benefits of the Green Revolution as noted before. But the real beneficiaries of the Green Revolution will be the corporations that supply the pesticides, fertilizers, and farm machinery. In the early part of the 1960s oil companies saw fertilizer manufacturing as a perfect outlet for their petroleum. Around the same time, the M. W. Kellog Division of Pullman, Inc., developed a new process that cut the costs of fertilizer production in half. As W. L. Smith, president of Kellog has said, "When companies saw the production costs they knew they could *not* afford to buy the new plants."[24] Kellog alone sold more than $400 million worth of fertilizer plants; between 1963 and 1968 United States fertilizer companies invested more than $4 billion to build fertilizer mines, plants, and distribution networks. By the time the scramble for new investment was over, these firms had severely overinvested. In the view of one oil company executive, "A lot of blood will be shed before we have a stable [fertilizer] industry."[25]

There is little doubt that the world chemical fertilizer market is quite vast. For example: "Using the rule of thumb of one pound of plant nutrients for each 10 pounds of grain, the current yearly fertilizer consumption of 7 million tons in the less developed countries must climb to 47 million tons in 1980. At $150 per ton of fertilizer this prospective market could well expand from the present one of $1 billion a year to at least $7 billion 15 years hence. This volume of fertilizer, averaging about one-fourth the Japanese rate of usage, would still be far from optimal."[26]

Consider also that if the Indians applied fertilizer at the same per capita rate as Holland, then Indian agriculture alone would require one half the world's existing capacity.[27]

Of course, the Third World fertilizer imports began before the Green Revolution. For instance, from 1960 to 1967, the percentage of India's total export earnings required to finance fertilizer imports alone rose from 2.5 to 20 percent.[28] However, Indian fertilizer consumption was encouraged by our State Department by lending the Indians money to import fertilizer; then came the droughts of 1965 and 1966 during which the Indian harvest suffered terribly. At that time, the United States government ceased signing annual food sales with India and began to dole out food a few months at a time. Meanwhile, the United States and the World Bank put a great deal of pressure on both India and Pakistan to "encourage" multilateral corporations like the combine of Standard Oil of California and International Minerals and Chemicals to invest in fertilizer plants. This "encouragement" took the form of giving these corporations advantages not usually available to domestic firms. At the same time the Indian government also agreed to maintain the high agricultural prices mentioned earlier.

By this time the stage was set for the launching of the Green Revolution. Dr. Borlaug and his associates from the Rockefeller and Ford Foundations agricultural research centers had perfected the packaging of the Green Revolution. The new seeds could only out-perform traditional varieties when combined with pesticides and fertilizer. So the men who designed the Green Revolution created a package with the optimal mix of seeds and other inputs; then the package was sold at a very low cost and with easy credit terms to the most progressive farmers in the best of areas to create an agricultural industry dependent on these manufactured inputs.

The Green Revolution means more good news for the petrochemical industries because the new Miracle Seeds also require much heavier use of pesticides. The president's Science Advisory Committee of 1967 estimated that if the Third World is to double its food production by 1985 using modern agricultural techniques, then it will require a 600 percent increase in the rate of pesticide application.[29] In response to this potential market, Standard Oil of New Jersey (Exxon) has some 400 "agro-service centers" where the farmer can buy his seed, pesticides, and chemical fertilizers from his local Exxon dealer.[30]

In effect the Green Revolution makes the Third World dependent on the United States. Once the farmers come to rely upon the foreign-built machinery or petrochemicals, then an embargo represents a powerful threat. Of course, India and Pakistan were dependent on the United States for massive food shipments even before the Green Revolution. In return, the United States has received Indian and Pakistani money to the extent that the United States now owns about 15 percent of all Pakistani currency.[31] The United States has not been above using these funds to apply political pressure when it feels so inclined.

At the same time, the United States is dependent upon the Third World for food, but our dependence is of a different kind. In the first place, the United States, like most of the other countries of the developed world, is a net protein importer; the developed world ships about 2.5 million tons of gross protein to the Third World and receives about 3.5 million tons of higher-quality protein in the form of fish meal, press-cakes or oil-beans, and soybeans.[32] Second, we are dependent on the Third World for our coffee, tea, chocolate, and other nonessential foods and nonfoods. We spend more than $3 billion per year on food imports, which is more than any of the Third World countries spend, in spite of the widespread belief that the United States is feeding the world.[33]

This pattern of trade did not always exist. Thirty-five years ago India was a major exporter of grain, and as recently as 1955 exports of basmati rice totaled more than 100,000 tons per year.[34] Now the developed nations have little need for Third World grain imports; we have more than we need. So the Third World is left to earn its scarce supply of foreign exchange by selling nonessential foods or nonfoods like rubber and tobacco. But it is getting harder and harder to earn foreign exchange by selling such export crops. George Borgstrom noted in 1968 that the Third World nations were delivering a tonnage of cash crops 33 percent greater than the tonnage of 1952: in return for this increased volume of exports, the Third World earned only 4 percent more income.[35] Certainly they need all the foreign exchange they can get because in the next few years countries like India and Pakistan will need about 50 percent of their export earnings just to pay off the interest on their outstanding loans.[36] For instance, India had an exchange gap of $1.5 billion in 1970 and a $0.5 billion charge on her loans outstanding.[37]

What, then, is the best course for India and the other Third World nations? How are they to feed their growing population? First, a better distribution of income could go a long way toward limiting the problem of infant mortality. Second, small farms in India as well as in other nations can produce higher yields than larger farms. So yields can be increased through land reform. In India for example, 33 percent of the land could be allocated in units of three acres to 45 million households including all of those currently landless, which would still leave two-thirds of the land to be divided into farms of an average size of twelve acres.[38] Finally, there needs to be a reevaluation by Third World countries of the concepts and desirability of development itself. In the past, world leaders have debated over the best strategy for bringing the "have-not" nations up to the level of the "have" nations. Often the problem is treated as if the attainment of technological affluence were a goal in itself, urbanization and industrialization being its own reward. However, it is becoming increasingly obvious that this may not be the case. In China, for example, agricultural production has more than doubled since 1949 without any appreciable change in the level of technology. Such "agricultural output . . . has been achieved largely by the intensification and improvement of traditional techniques within the new institutional framework."[39] This point has been made in more general terms by Kasum Nair, an agricultural expert with the Center for Asian Studies at Michigan State University. In a 1971 interview at the University of Michigan's Conference on Asian Environments, Mrs. Nair noted that

> Asian countries must improve their agricultural output by refining existing methods of organic farming rather than depending on Western fertilizers and insecticides. . . . The methods of modern technology must be adapted to the realities of Asia's environment, both economic and ecological. . . . Asian attitudes failed to protect the environment from pre-industrial technologies and cannot be expected to deal effectively with the far more powerful pollutants of modern technology. . . . Mainland China is trying to [avoid] that pattern now by employing green and organic manure and compost, allowing the farmer to improve his techniques by better use of his traditional resources rather than investing in new products.[40]

Clearly, land reform and social reorganization have produced positive results in China. India and the rest of the Third World could follow a similar course by harnessing their used and underused labor to farm and improve the soil. By making it profitable and culturally desirable to do so on a broad basis, the proper relationship between people and land can be maintained so that people can be fed and pursue an acceptable standard of living.

### Notes

1. Keith Buchanan, *The Transformation of the Chinese Earth* (New York: Praeger, 1956); also, Ping-ti Ho, "Early Ripening Rice in Chinese History," *The Economic History Review*, December 1956.

2. Redcliffe Salaman, "The Influence of the Potato on the Course of Irish History," The Tenth Finlay Memorial Lecture, delivered at the University College, Dublin, 27 October 1943 (Dublin: Brown & Nolan).

3. "More Landless," *Pakistan Forum* 1, no. 2 (December 1970-January 1971).

4. Robert J. Muscat, chief of the planning division of the Bureau for Near East and South Asia of the Agency for International Development, in a letter to the *New Republic*, 21 and 28 August 1971, pp. 33–35.

5. Wolf Ladijinsky, "Ironies of India's Green Revolution," *Foreign Affairs*, July 1970.

6. Francine Frankel, *India's Green Revolution: Economic Gains and Political Costs* (Princeton, N.J.: Princeton Univ. Press, 1971).

7. U.S. Dept. of Agriculture, "Changes in Farm Production and Efficiency: A Summary Report," Statistical Bulletin No. 233, June 1970.

8. Max Millikan, "Population, Food Supply and Economic Development," *Technology Review*, pp. 43–58.

9. Cited in William C. Thiesenhausen, "Latin America's Employment Problem," *Science* 171, no. 3974 (5 March 1971): 868–874.

10. Arthur MacEwan, "Contradictions of Capitalist Development in Pakistan," *Pakistan Forum*, October-November 1970.

11. William Paddock, "How Green Is the Green Revolution?," *BioScience*, 20, no. 16 (15 August 1970): 897–902.

12. Walter Falcon, "The Green Revolution: Second Generation Problems," *American Journal of Agricultural Economics*, December 1970.

13. Paddock, "How Green Is the Green Revolution?"

14. Dana G. Dalyrmple, "High Yielding Varieties of Grain," from *Technological Change in Agriculture: Effects and Implications for Developing Countries*, USDA, April 1969, in *The Symposium on Science and Foreign Policy: The Green Revolution*, Proceedings, Subcommittee on National Security Policy and Scientific Development, House Committee on Foreign Affairs, 5 December 1969.

15. S. M. Hussain, "Price Incentives for Producing Mexican Wheat," *Pakistan Development Review* 10, no. 4 (Winter 1970): 448–468.

16. Ladijinsky, "Ironies of India's Green Revolution,"

17. Aaron M. Altschul and Daniel Rosenfield, "Protein Supplementation: Satisfying Man's Food Needs," USDA, Foreign Economic Development Service No. 3, reprinted from *Progress, the Unilever Quarterly* 54, no. 305 (March 1970).

18. Alan Berg, "The Role of Nutrition in National Development," *Technology Review*, February 1970, pp. 45–51.

19. Ibid.

20. E. Wayne Denney, "Typhoons and Tungro Cause Philippines to Import More Rice," *Foreign Agriculture*, 31 January 1972, pp. 8, 12.

21. W. David Hopper, "The Mainsprings of Agricultural Growth," Rajendra Prasad Memorial Lecture, 18th Annual Conference of the Indian Society of Agricultural Statistics, January 1965.

22. Takekazu Ogura, ed., "Agricultural Development in Modern Japan," Tokyo, FAO Association, 1963, Appendix to Part III, pp. 611–612.

23. Berg, "Role of Nutrition."

24. Thomas O'Hanlon, "All That Fertilizer and No Place to Grow," *Fortune,* 1 June 1968, pp. 90–95.

25. Ibid.

26. Lester Brown, "The Stork Outruns the Plow," *Columbia Journal of World Business,* January-February 1967, pp. 15–21.

27. Paul R. Ehrlich and John P. Holdren, "Population and Panaceas: A Technological Perspective," *BioScience* 19, no. 12 (1969): 1065–1071.

28. Statement of Arthur Moser before the House Subcommittee on National Security Policy and Scientific Development, printed in *The Symposium on Science and Foreign Policy: The Green Revolution.*

29. H. F. Robinson, "Dimensions of the World Food Crisis," *BioScience* 19, no. 1 (1969): 24–29.

30. Lester Brown, *Seeds of Change* (New York: Praeger, 1970).

31. "O.L. 480," *Pakistan Forum,* February-March 1971, p. 11.

32. Paul R. Ehrlich and Anne H. Ehrlich, *Population, Resources and Environment* (San Francisco: W. H. Freeman & Co., 1970).

33. Frederick D. Gray, "That Coffee from Brazil, and Other Food Imports," *Food for Us All, USDA Yearbook of Agriculture, 1969* (Washington, D.C., 1970), pp. 18–19.

34. John J. Parker, Jr., "India Struggles to Increase Exports of Agricultural Items," *Foreign Agriculture,* 14 September 1970, p. 3.

35. George Borgstrom, "The Dual Challenge of Hunger and Health," in *Man and The Environment,* ed. Wes Jackson (Dubuque, Iowa: William C. Brown, 1971).

36. Paddock, "How Green Is the Green Revolution?"

37. *New York Times,* 19 January 1970, p. 68.

38. Wyn F. Owen, "Implications of the Green Revolution for Economic Growth," *American Journal of Agricultural Economics,* December 1970.

39. Michael Allaby, "Green Revolution: Social Boomerang," *Ecologist* 1, no. 3 (1970): 19–21.

40. Anon., "India Expert Says Asia Farms Must Refine Existing Methods, *Allentown Morning Call,* 17 June 1971, p. 66.

## References

Allen, Robert. "A New Strategy for the Green Revolution." *New Scientist* 63(909): 8 August 1974, pp. 320–321.

Anderson, Alan. "Farming The Amazon: The Devastation Technique." *Science,* 30 September 1972, pp. 61–64.

Berg, Lasse and Lisa. *Face to Face: Fascism and Revolution in India.* Berkeley, Calif.: Ramparts Press, 1971.

Boerma, Addeke. "A World Agricultural Plan." *Scientific American,* August 1970, pp. 54–69.

Borlaug, Norman. "The Green Revolution: For Bread and Peace." *Bulletin of the Atomic Scientists*, June 1971, pp. 6–9; 42–48.

Cleaver, Harry. "The Contradictions of the Green Revolution." *Monthly Review* 24 (June 1972): 80–111.

Falcon, Walter. "The Green Revolution: Generations of Problems." *AAEA Annual Meeting*, University of Missouri, Columbia, Mo., August 1970.

Frankel, O. H. "Genetic Dangers in the Green Revolution." *World Agriculture*, 1971, pp. 9–13.

Gomez-Pompa, A., Vázquez-Yanes, C., and Guevera, S. "The Tropical Rain Forest: A Nonrenewable Resource." *Science* 177: 762–765.

Harris, Marvin. "How Green the Revolution." *Natural History* 81 (1972): 28–30.

Janzen, Daniel. "The Unexploited Tropics." Bulletin, *Ecological Society of America* 51 (1970): 4–7.

White, Gilbert. "The Mekong River Plan." *Scientific American*, 1966, pp. 49–59.

# PART III

# Rural Struggles and Alternatives

# 9

# THE NATIONAL SHARECROPPERS FUND AND THE FARM CO-OP MOVEMENT IN THE SOUTH

*Robin Myers*

In 1963, as the civil rights revolution was building dramatically toward its historic march on Washington, and as the pressures that germinated the war on poverty were pushing to the surface, the National Sharecroppers Fund (NSF) was sponsoring a series of conferences across the South. Not very exciting, and they didn't attract attention outside of very concerned circles; but this was, in fact, the heart and essence of the Southern rural cooperative movement.

The initial conference, in Bricks, North Carolina, November 1962, was NSF direct response to poverty and the civil rights struggle. In 1960 when some seven hundred sharecroppers and tenant farmers in Fayette County, Tennessee, received eviction notices because some had been active in a voter-registration drive, NSF moved in to help with more than charity. NSF insisted that government farm programs were designed by Congress to benefit all farmers, not just the commercially successful and politically powerful, and that the Fayette farmers were entitled to assistance. The year 1961 saw the first NSF fieldwork, to bring technical assistance to poor farmers who did not know how to secure the

Robin Myers is a former research consultant for the National Sharecroppers Fund.

help to which they were entitled. The same program was needed across the entire South, and within a year plans were made for an initiating conference at the Franklinton Center in Bricks.

What made this conference so different from most was that sharecroppers, tenants, and small farmers themselves came from all parts of the South; it was not just professional people talking about rural problems; the desperately poor were also there, knowing only the land, and seeking the resources and tools to stay on it. They were encouraged enough by what they learned, and by the technical assistance available, to plan two more conferences, for Frogmore, South Carolina, and Mt. Beulah Conference Center, Edwards, Mississippi. Other state conferences followed. The Frogmore and Mt. Beulah conferences discussed many things, since rural needs were so deep and so varied. But the emphasis had shifted from Bricks, where it had been brought out that federal programs should serve the poor instead of discriminating against them. The moving idea now was that of people themselves in action; the reports of both conferences centered on rural cooperatives.

Many reasons lay behind this. The time was ripe; people trusted themselves rather than outsiders. The technique of cooperatives had been tested and proven in rural areas, and the idea had federal support. NSF's long background in southern work bridged many gaps to bring people, techniques, and organizations together.

## The Roots of the NSF

The Sharecroppers Fund had been working in the South since the thirties, although not organized in its present form until 1943. In the depths of the depression a Southern Tenant Farmers Union had been organized in Arkansas by black and white sharecroppers who were being forced off the land while landowners accepted government subsidies that should have helped the poor. By 1943 it was clear that theirs was not a depression phenomenon but permanent misery embodied in a feudal system of land control. As a result the National Sharecroppers Fund took on a full-time job.

Agricultural needs and manpower scarcity brought on by World War II focused attention on the intolerable conditions of farmworkers and many farmers who had lost their land. Some attempts at union organizing were made; the workers might have been able to improve wages and working conditions except for the increasing use of foreign contract workers. As NSF and, later, the National Advisory Committee on Farm Labor (NACFL) moved to help migrant farmworkers, it was understood that *the problems of all who worked the land were interrelated.* The decline and failure of small farmers in the South, especially black farmers, was due more to lack of political power than to economic inefficiency. They couldn't compete and weren't good loan risks. Discrimination against them was indirect, the result of compounded policies favoring ever-larger commercial farmers. The pressure against the small black farmers was traditional and deliberate. The local-control structure of agricultural programs placed committees such as the Agricultural Stabilization and Conservation Service (ASCS) in the hands of the plantation owners; they used allotments and other programs for their own benefit. The local system endured because the seniority system of Congress placed power of appropriations and agricultural committees in the hands of the Southerners, who were always reelected by their limited (white) electorate.

Part of the New Deal attempt to alleviate rural poverty, was to support the formation of cooperatives through the Farm Security Administration (FSA) established in 1937. More than 25,000 were organized in the South before FSA was limited (1943) and ended (1946) by the continuous conservative opposition in the Department of Agriculture and Congress. Funds were lent for securing land and housing as well as for buying and marketing operations, and the purchase of machinery and stock. Two FSA communities, Gees Bend, Alabama, and Holmes County, Mississippi, survived as core centers when the movement of the sixties came along.

Thousands shared in at least temporary FSA benefits, though far fewer blacks than whites. But the program was under constant attack and it never had money or scope enough to meet the problem: there was too much ignorance, too little capital, and, above all, no local political power. Some co-ops survived, though FSA didn't. Most successful were the credit unions, which were less subject to the hazards of agriculture. *Not until the local power of the landowners was challenged by the civil rights revolution could the popular basis for economic change be found.*

Through all of this NSF had been there, pressing for legislative change in Washington, contributing money to help keep alive the union and other people's organizations in the South, educating relentlessly against the national apathy, and joining with the farm organizations in warning that one of the largest forced migrations in history was taking place without planning, recourse, or any assessment of human cost. *The nation boasted of its agricultural revolution, heedless of its victims.*

## NSF and the Civil Rights Movement

The mass rural emigrations of the fifties brought a million people a year off the land into the cities. Because of the high birthrate, rural population remained generally stationary despite the migration. Large numbers of those who left the farms simply became part of the nonfarm rural poor. The decade of the sixties saw a 26 percent decline in the number of farms but a 27 percent increase in the average size of farms, although the amount of land remained about the same. The number of farms continues to decline at a fairly steady rate of about 100,000 a year; today there are around 2.8 million farms left in rural America.

Black-operated farms, almost entirely in the South, are disappearing nearly twice as fast as white. Between 1950 and 1970 their number dropped from 560,000 to 98,000. Farms with at least $10,000 in yearly sales grew in number by 33 percent from 1960 to 1970, but those that grossed less than $10,000 declined by 42 percent; 1.8 million of the remaining farms are in this category, and they will go, too, unless they rescue themselves.

When the civil rights movement (the Student Nonviolent Coordinating Committee, CORE, and the Southern Christian Leadership Conference) entered the South in the sixties, it came up against the grinding poverty of the people. Obviously, the struggle for life was basic to the struggle for liberty; they took it on and looked for tools useful in economic action. They found the cooperative idea and technique; and they found that the National Sharecroppers Fund (with its new tax-exempt Rural Advancement Fund) had a knowledge of the South and the other organizations that would help. The conferences, starting with Bricks, brought together the idea, the people, and the technique.

Credit, which had been difficult to obtain under the old Department of Agriculture outlook, became available through Office of Economic Opportunity (OEO) loans. Even this was a struggle. Although most of the poverty was in the rural areas, most of the programs were in the urban areas, thanks largely to the presence of experienced institutions and personnel. Although some credit unions had been established among the poor, and among blacks, most successfully in northern Alabama, they had accumulated far too little capital to help finance the agricultural co-ops.

Marketing was always a problem for poor farmers who were usually tied to a single white broker for sale of their cash crops. To break away from these brokers was to risk all hope of income and debt payment.

Farm advisers were more apt to think first of improving farming methods and crop diversifications, so that the poor farmers would have something to offer in a highly competitive market. Economic freedom, however, was usually the farmers' first thought. Since it was easier to secure government loans for specific things like machinery, the first Southern co-ops tended to be machinery co-ops that shared tractors and other equipment among poor farmers. But there was no "instant income" in a co-op and it was hard for people so often deceived and exploited to work together.

From the Mt. Beulah conference it was clear that education would be essential if the poor were to develop and control their co-ops themselves. Simple literacy came first, then bookkeeping. One of the earliest of the co-op developments, in Louisiana, was due to the enthusiasm of the Rev. A. J. McKnight, a participant in the Bricks conference who had also studied the cooperative movement in Nova Scotia. With members of poor black parishes, he formed the Southern Consumers Cooperative and the Southern Consumers' Education Foundation early in the sixties. The agricultural part of the Louisiana co-op, the Grand Marie Vegetable Producers Cooperative, Inc., broke a sweet potato marketing monopoly with an organization that was interracial until local pressures forced white farmers out. The co-op brought processing machinery and a storage shed with an FHA loan, and received OEO funds to pay staff to work on recruitment and education.

The history of the various contemporary Southern co-ops was quite similar, and seemed to repeat the experiences of the New

Deal FSA co-ops. After an initial spurt, lack of management and marketing skills plus a membership that was unstable because it needed immediate gains threatened the very existence of the co-op.

The largest of all the new low-income co-ops was SWAFCA, the South West Alabama Farmers Cooperative Association. SWAFCA was an outgrowth of the voter registration drives of the mid-sixties. Six of the ten counties included were among the hundred lowest income counties in the United States. Forty percent of the 1,800 black family members farmed with mules and 38 percent with only hand tools.

To help the members raise their incomes, SWAFCA not only had to introduce group buying and marketing, but it had to break dependence on cotton allotments and cotton brokers, and to substitute raising peas, cucumbers, okra, and other vegetables. Along with NSF, interested organizations included the Southern Christian Leadership Conference, the Southern Regional Council, and the Cooperative League of the United States.

Small but concerned organizations provided seed money. The initial loan of $25,000, from the International Foundation for Independence, went to buy seed, fertilizer, and similar supplies for the co-op to sell to members on credit; $5,000 from Operation Freedom in Philadelphia became operating capital. OEO grants were used to pay administrative and technical staff, to establish a loan guarantee fund to secure additional credit, and for general operating expenses. The OEO loan was protested in Washington by local officials and politicians. Similar pressures gave the co-op a hard time with the Farmers Home Administration before funds were made available for receiving and distributing sheds in the ten counties.

SWAFCA was beset with delays, deliberate obstacles, and attacks—conditions especially frustrating to people living marginally. When outside troubles were compounded by poor management, members lost hope. It was next to impossible to train members in the use of new equipment and the management of new crops and at the same time ensure that their incomes were rising enough to strengthen their loyalty to the co-op and provide confidence in its future. Both crop volume and membership dropped and the co-op was rift by factionalism as the sixties ended.

Fortunately, by this time help was available from the new

Southern cooperative movement itself. Both a Federation of Southern Cooperatives (FSC) and a Southern Cooperative Development Program (SCDP) could offer short-term loans and field representatives to aid recruitment and membership education. The very survival of SWAFCA was a victory that changed attitudes and procedures. Greater economic independence meant that farmers received respect and better treatment locally. Agencies of the Department of Agriculture had to get used to working with black farmers, and they did.

Both FSC and SCDP owed much to Rev. McKnight's Southern Consumers' Education Foundation, an idea that preceded the necessary resources. The Mt. Beulah conference decided on a permanent link among the co-ops for mutual strength and education, and the federation developed out of this base. The Ford Foundation agreed to fund the SCDP, to help the low-income co-ops secure credit and technical assistance. The program soon decided to concentrate its efforts in Louisiana, Alabama, Mississippi, and Tennessee, and on about two dozen existing co-ops. Most of these same co-ops had met in Atlanta in 1967 to form the (FSC) but the federation was open to co-ops in seventeen states and has had as many as several hundred members. In 1970 SCDP and FSC were merged. Today, the field staff is assigned to various co-ops but is also available to concentrate in areas of trouble.[1]

The West Batesville Farmers Cooperative had been organized in 1965 by SNCC fieldworkers and an NSF staff member. West Batesville used an FHA loan (secured with NSF help) to buy three automatic cotton pickers, three combines, four trucks, and other equipment; today, these are used by 250 to 300 members. The co-op was able to encourage members to diversify from cotton to vegetables. With management and marketing help from the interested organizations and SCDP and the federation, it overcame early difficulties and started meeting its machinery payments.

Most of the co-ops succeeded in raising the incomes of members a few hundred dollars a year—a great deal when usual annual earnings were around $1,000 to $2,000. But co-ops continue to depend heavily on outside financial support and technical assistance[2] and to have insecure marketing arrangements. To take the further step into food processing would have been hazardous for organizations not yet fully stable as producers; the history of such experiments is not encouraging. Yet what security could be found often resulted in a new chattel relationship: contracting with a large cannery or freezer operation that controlled everything, which, in effect, made the independent farmer a contract worker again.

### NSF and the Organic Food Market

It was NSF that pioneered a new direction. One offshoot of the movements of the sixties was a concern for the whole quality of life. This included the environment, and it also included food. Some consumers reacted against the poor quality of mass-produced food available in the food chain monopolies; others went further and tried to grow their own. The idea of natural, organic farming became the basis for a consumer movement; health food stores spread, and their products even entered the chains as consumer knowledge and independent buying habits grew.

NSF's historic middleman role thus found new groups to connect: consumers in search of organically grown food; farmers in search of a market that would not try to exploit or control them.

In 1972 it was estimated that there were 3,477 organic food stores in the United States, which sold products valued at more than $100 million. About a thousand farms were registered as being "organic" by the Organic Gardening and Farming Association. The organic movement had reversed a basic farm problem; here was a growing market in search of crops that small farmers might be especially qualified to produce.

The first NSF-assisted county came to the organic experiment through difficulties that were great even for the rural South. Three-quarters of the county was farmland with a scattering of nonunion, low-wage industries. And even as the sixties ended, Halifax County, Virginia, was a microcosm of the Old South—still basically a row-crop monoculture dependent on Department of Agriculture allotments for tobacco. Most farmers were small and half the farms were tenant operated; sharecropping was still common with a $500 annual income in a good year. The threat of dispossession was immediate, and the need for capital for chemicals, mechanization, land expansion, which had driven people from the cotton fields earlier, now extended itself to tobacco farming. Isolation deepened the poverty, and even the great years of the civil rights movement passed by Halifax. Only in 1967 did the local anti-poverty agency suggest the possibility of farm cooperatives when NSF agreed to provide technical assistance through its field staff.

Learning from the difficulties of other co-ops, Halifax started with a two-day workshop in February 1968, from which temporary officers and committees emerged. The marginal farmers themselves formed a survey committee, including two NSF-paid aides, to contact some 300 farm families. They found interest in growing potatoes, tomatoes, and cucumbers.

A training grant from OEO was used for a feasibility study of the area and technical training for members. Outside consultants studied land, resources, soil, climate, and markets. Ten farmers were trained for "outreach" work to carry the message to people who had never received government or other benefits.

### Halifax and Burke: Models for Cooperation and Organic Farming

In the early years, the co-op (the Southern Agricultural Association of Virginia) proved again that the poor could work together, benefit by training, and grow crops other than tobacco. But there was also recurrent drought, jobbers' trickery on prices, and the complexities of marketing. By the end of 1969 the co-op faced failure.

That was when NSF (now with the tax-exempt Rural Advancement Fund) was able to help with two new ideas. The first was concentration of effort. While maintaining its general program and political pressure, NSF decided to concentrate most of its funds available for grants and loans in just two geographical areas instead of spreading thinly over many; this was partially because these were areas where struggling people had organized co-ops and had faith in themselves and their own efforts. There was action, not just to save a few farms but, starting with the co-op, to rebuild a whole area with the kind of health, housing, and community services that make life worthwhile. Halifax was one such county, and Burke, in Georgia, was the other.

The second idea was to link the co-op with the growing organic market in a nonexploitive relationship that could benefit both farmers and consumers.

Already, despite difficulties, the Halifax co-op had a combination office, grading shed, and loading platform. Co-op families had organized a $20,000 credit union, so at least small farmers could get loans that banks had refused to give. And some land had been purchased for future developments. Under the new plan, farmers were given alternatives. They could continue to farm independently, buying seed and fertilizer at a discount through the co-op; they could also let the co-op use the land and split the profits with them fifty-fifty; finally, the land could be leased to the co-op for rent, with the co-op hiring the labor and retaining the profits. This would increase the amount of land under cultivation so that a larger crop would give the co-op a better break in the market. It would also provide a demonstration farm where the organic techniques of growing specialty crops like strawberries and tomatoes could be tested.

NSF/RAF put in $100,000, which went to buy both land and tools. It also paid for experts in organic farming and the fertilizer they recommended: feather meal, bone tankage, seaweed, leached langbenite (a double sulfate of potassium and magnesium), and dolomite (a double carbonate of calcium and magnesium). Prior to this, crops had been limited to what the poor soil and even poorer farmers could handle. Cucumbers, squash, and peppers were planted first, with turnips for a winter crop. With that success behind them, farmers tried even more diversified crops like field tomatoes, carrots, beets, early peas, string beans, sweet corn, and white and sweet potatoes. Money was distributed around the community since there was work for others in picking and grading the crops.

Despite a drought, the forty acres cultivated organically outproduced the others. Moreover, the vegetables were superior. So the next year a hundred acres were planted organically, and farmers were willing to put still more of their acreage into vegetables.

At the present, farmers growing "organic" vegetables in the co-op say that the organic crops pay double what tobacco would bring per acre in a good year. Sales are being made locally, in Washington, D.C., and in Chicago. The three hundred co-op members are beginning to see small income gains, but what they feel most is the relief from perpetual debt. Tackling the dilapidated housing is high on the agenda now, and a community family health council has been set up. A survey of medical and sanitary needs has led to plans for a better water supply and more adequate plumbing.

The story in Burke County, the second area of concentration, was much the same. Burke County was even poorer than Halifax, and it was cotton, rather than tobacco, that determined the debt cycle. The 170 farmers who joined the co-op farmer between one-half and eighty acres. They were still struggling for existence when NSF/RAF was able to pump in $100,000, as it had in Halifax. To start, $35,000 was used for an urgently needed irrigation system. Farmers learned to stagger crops so that all could use the pea picker and tote-lift box cooperatively. The funds also provided a warehouse, a home-crafted cucumber picker, a well, a two-and-one-half-ton truck, a pickup truck, a large tractor, and twenty-five acres of land.

As in Virginia, the farmer-members learned to diversify by growing vegetables, to plant more acreage, and to work together. At the same time, the co-op itself started organic farming experimentally.

Burke is now busy planning a whole new community. NSF lent $4,000 toward the initial purchase of fifteen acres and helped co-op members to file FHA loan applications for eight homes as a start. Through NSF staff efforts, engineers from Georgia Tech built a mock-up of what the community might become: 2,500 families on 10,000 acres of land. There is a creek that could become a lake for irrigation and recreation activities. Burke County has visions of a great future, as co-op members work their way out of debt toward small increases in income.

In so concentrating, NSF has not forgotten the rest of the South and the rest of the rural poor. In South Carolina[3] funds were provided for a farmers' co-op, an oyster co-op, a cucumber co-op, a crabmeat co-op, and, in Florida, a farmworkers' co-op through which migrants are developing self-help programs. It also gave financial support to Education and Training for Cooperatives, a grass-roots group of people developing community-controlled industries.

The major new direction of NSF/RAF, however, was made possible by a special gift for the purchase of five hundred acres of land in Anson County, North Carolina. This is equidistant from

Halifax and Burke and near more than thirty other low-income co-ops. It will become the training-in-action center that has been sorely needed.

There will be two kinds of schools. In one, co-op managers and members will learn management techniques, financing and accounting, crop selection, fertilizers, processing, marketing. In the other, the land itself will be a school for farmers who come for a day or a semester. They will learn how to bring depleted fields back to life; how to produce specialty vegetables and fruits; how to use new mechanical equipment; and how organic production is linking farmers and consumers in new markets that serve both.

These all are co-ops that have developed, in one way or another, to supplement and undergird the great civil rights and anti-poverty movements of the sixties. It is thus natural that out of similar circumstances in other parts of the country, similar developments are arising. La Raza, an organization of Mexican-Americans, has reported a hog-raising cooperative in Uvalde, Texas, and undoubtedly it is one of many. There are now more than thirty Mexican-American farmers in Cooperativa Compesina near Watsonville, California. The four who started it were migrant workers who had learned, through OEO courses, how money could be made in specialty crops. At present, they are growing strawberries and expecting to net $4,500 an acre. Because it is a labor-intensive crop, the co-op has a good chance to succeed despite the fact that it must compete with some of the largest agricultural corporations.

The outlook for co-ops was promising when the exploitive policies of the huge landholders (of land, labor, and consumers) were exposed and the public was sympathetic. However, it is clear that in order to continue, the co-ops need help. Title VII of the OEO bill, which would have authorized broad financial and technical assistance to the low-income co-ops, was vetoed by President Nixon in 1971. His subsequent dismantling of OEO on the one hand and of farm programs on the other means that the rural poor must look elsewhere, especially to their own strength.

The loss of 100,000 farmers a year has meant, also, a loss of rural political influence. *Part of that can be made up by the growing consumer and organic movements, but only if they understand how fundamental to their own goals is control and use of the land.*

Alliance with the cities who have had to take in the refugees from the land, and whose people are also turning in some measure toward cooperatives can help too. Co-ops can't do it all. But "save the land, save the people" is a message that ought to get response if it is put to the nation loudly and clearly. There isn't much time left.

## Notes

1. About a dozen handicraft or producers' co-ops are affiliated with the Federation of Southern Cooperatives. Though not directly part of the agricultural story, they are relevant as part of the developing sense of cooperative community which is as vital as direct economic gain. Often these are formed by wives of farmers active in the agricultural co-ops, and take the form of sewing or quilting co-ops. The most successful of the quilting co-ops have been in Gees Bend, Alabama, the old FSA community, and several in Appalachia where there is a broader crafts tradition. Most interesting as a new departure is the Poor People's Corporation (PPC) of Jackson, Mississippi, with thirteen affiliated producers' co-ops, and Liberty House, their marketing outlet. PPC is wholly independent of government funds although it has received some help from the federation. All the producers' co-ops are marginal; their workers often put in long hours and net less than the minimum wage. But they are self-directed, and they do help to alleviate the crushing poverty of the rural areas.

2. Lest it be said that the low-income farms and their co-ops are not viable because of this, remember the continued indebtedness of the largest commercial farms to the government for many subsidies.

3. In South Carolina the Coastal Plains Human Development Coordinating Council is turning the tide of the out-migration by aiding self-help and cooperative efforts. The council was organized in 1971 to operate in twenty-six South Carolina counties. It was funded by small grants from the United Methodist Church and NSF. Late in 1972 it finally received $30,000 from OEO which benefited, among other things, three local co-ops in which NSF had been interested: the Bluffton Oyster Co-op for transportation; the St. Helena Tomato Marketing Co-op (Frogmore) for equipment purchases; and the United Communities for Child Development, for training day-care-center personnel.

## References

Keyserling, Leon H. *Agriculture and the Public Interest: Toward a New Farm Program.* Washington, D.C.: Conference on Economic Progress, 1965.

London, Joan, and Anderson, Henry. *So Shall Ye Reap.* New York: Thomas Y. Crowell, 1970.

Marshall, Ray, and Godwin, Lamond. *Cooperatives and Rural Poverty in the South*. Baltimore: Johns Hopkins University Press, 1971.

National Sharecroppers Fund. *A Better Life for Farm Families*. New York: National Sharecroppers Fund, 112 E. 19th St., N.Y. 10003, 1968.

National Sharecroppers Fund, Board. *Annual Report: 1970–1972*. New York: National Sharecroppers Fund. Reports of other years, however, summarize the story as well.

"Only Six Million Acres: The Decline of Black-Owned Land in the Rural South." Report sponsored by Clark College, prepared under the direction of Robert S. Browne. New York: The Black Economic Research Center, 112 W. 120th St., N.Y. 10027, 1973.

Shannon, Fred. *The Farmer's Last Frontier: Agriculture 1860–1897*. New York: Harper & Row, 1968.

U.S. Commission on Civil Rights. *Equal Opportunity in Farm Programs*. Washington, D.C.: Government Printing Office, 1965. An appraisal of services rendered by agencies of the Department of Agriculture.

U.S. Dept. of Agriculture. *Fact Book of U.S. Agriculture*. Washington, D.C.: USDA, 1972.

U.S. Dept. of Agriculture. *Farm Policy in the Years Ahead*. Report of the National Agricultural Advisory Commission. Washington, D.C., USDA, 1964.

*Win*. Special issue. "The Death of the American Farmer." July 1972.

# 10

## THE INVISIBLE WORKERS: LABOR ORGANIZATION ON AMERICAN FARMS

### George L. Baker

The story of farm labor organizing in America is not the kind of tale we will retell to our grandchildren. Consigned to the social scrap heap, the nation's 2.8 million farmworkers have been denied the basic rights accorded workers in every other industry. This dark chapter of social history encompasses as much human neglect and exploitation, greed, legislative complicity, and outright violence as any to be found in our history books.

For it is a history that reaches far beyond the confines of a traditional struggle between labor and management; indeed, it goes to the heart of such varied issues as racial discrimination, minority oppression, political overkill, and social equality. In the background always has been America's food industry, a $130-billion-a-year chain that provides one out of every seven jobs in our country.

One who drives leisurely through the agricultural fields and vineyards of, say, California, hardly senses the proportion of the farm labor question. The fields are deceptively still, partly because the problems of working conditions, housing, and wages are set back away from America's main byways, and away from the

---

George L. Baker is a correspondent for a California newspaper.

bustling cities. As they have been hidden from history, they have been hidden from public view.

For years no one could see how a veritable flood of foreign labor was ruthlessly used to harvest crops, how each succeeding minority group was set against another to discourage unionization and to drive down wages, how farmers, loyal politicians, and local governments—including the courts and law enforcement agencies —banded together whenever workers threatened to organize themselves.

Periodically, stories illuminating farmworker problems would flash across newspaper front pages, stir the public conscience briefly, and then recede once again into the anonymity of the green fields. Only with the emergence of César Chávez in Delano, California, in the early 1960s, did the American people begin fully to comprehend the plight of the farmworker. For all his accomplishments in building the first viable farmworkers union in America, Chávez's longer-lasting achievement may be that he focused the nation's attention not on a labor movement but on the despairing lot of the vast majority of the nation's agricultural workers.

At this writing there is grave doubt whether the United Farm Workers (UFW) can withstand the dual onslaught of the growers and International Brotherhood of Teamsters who have combined in a naked attempt to put Chávez out of business. By signing up nearly half of California's grape growers—formerly under UFW contracts—the Teamsters have delivered a devastating blow to the cause of the farmworkers union. The AFL-CIO and its president George Meany have vowed to give Chávez whatever money and manpower is needed to turn back the Teamsters.

But that issue aside, there is no question that California agriculture and eventually the rest of the nation will someday be unionized. The tidal wave of discontent unleashed by Chávez is too strong to be turned back. But what shape will that unionization take? Agriculture, whose reactionary philosophy is rivaled only by that of the South's textile industry, needs a thorough housecleaning from top to bottom. That certainly would not be accomplished by the Teamsters, who prefer cozy arrangements with growers to militant unionism. America's labor movement has grown fat and complacent, satisfied and content with the status quo. The Teamsters, who have added a dash of corruption to their movement, symbolize all that is failing within the stagnant labor movement.

On the other hand, Chávez is starting from a position that many labor leaders occupied thirty-five years ago. Consequently, he is not content to talk about only "bread-and-butter issues" of wages, fringe benefits, and vacations. His concern broadens to include some of the more fundamental problems of farmworkers—simple dignity, equality, and opportunity. He would create a force that could compete on even terms with the accumulated power of the agribusiness industry. In a word, Chávez would force the growers to confront their past injustices. That is something the Teamsters surely wouldn't do.

It is too early to predict the outcome. But it's possible to raise a rhetorical question: Would farmworkers be better off without any union instead of being forced into the Teamsters? Or put another way, would a program of land reform—returning to the workers the land they have cultivated for so many years without reward—be a sensible alternative to unionism? It is a question that perhaps has no answer, but in light of the current struggle and the following discussion it is something to ponder.

### Corporate Invasions and Rural Workers

That it has taken so long for unionism to take hold is not surprising, given the array of forces farmworkers have had to face. It has been more than the simple stubbornness of farmers, anyway now becoming a diminishing political force in America. The entire agribusiness chain—from the grower to the banks, to agricultural chemical companies to the chain stores—have bridled at the thought of dealing with unionized farmworkers.

Always the most vocal opponent was the corporate farmer, whose posture was that of a family farmer but whose operations were like an open-air assembly line.

Clearly, corporations are turning more and more to farming as an easy way to gain profit and income tax write-offs. As the shadow of this corporate invasion spreads ominously across the Great Plains, the Corn Belt, and the South, the notion that farming is a "unique and special industry," a phrase the American Farm Bureau Federation lavishes about, is false and misleading.

These corporate farmers—Tenneco, Boeing, Dow Chemical,

Pacific Lighting Corp., the Getty Oil Co., and scores of others—
are the same companies that treat land as if it were an inexhaustible
resource. They pile it high with chemical fertilizer, lace it with
pesticides, and spray it again for good measure. The same approach
is taken with labor. Like land, labor is seen as simply another cost
factor added to the ultimate cost of goods. The social ills of farmworkers are left for someone else to remedy.

As the trend of rural emigration continues, control of land
naturally falls into the hands of fewer and fewer people. In California, for example, 5 percent of the corporate farmers control
60 percent of the state's prime farmland. Land monopoly is a
particularly vicious monopoly for it means virtual absentee control
of people, social institutions, and the economy. This kind of
power—concentrated among a few corporate executives in New
York, Houston, or Los Angeles—places both farmworkers and
small farmers at a disadvantage. Both are at the mercy of cost-
conscious officials and neither has political or economic leverage
with which to fight back. Thus, family farmers and farmworkers
have a common enemy and it would seem reasonable that they
could join forces to forge an alliance for mutual self-preservation.
For reasons that will be outlined later, this alliance has not yet
materialized, but as trends develop it could.

Despite the posturing of the agribusiness complex, every farmer
does not have an equal stake in unionization. In 1964, for example,
88 percent of the nation's 3.1 million farms used little or no farm
labor, according to U.S. Department of Agriculture statistics. The
remaining 12 percent accounted for 77 percent of the wages paid
nationwide. A miniscule 919 farms, each grossing more than $1
million annually, averaged 130 man-years per farm, a figure that
means far more than 130 people working on the farms since most
workers are employed only a few months out of the year.

Such employment figures hardly paint the picture of a pastoral,
genteel country of family farmers, imbuing the nation with the
virtues of the soil and Jeffersonian democracy. In truth, agriculture
is an industry dominated by the few. The Farm Bureau, Council
of California Growers, and various chambers of commerce have
perpetuated a fraud on the American public.

Only when it suits their purposes are they happy to acknowledge
the "importance" of the small farmers. One instance of this

occurred in late 1972 when the Farm Bureau sponsored a ballot initiative in California, Proposition 22, that would have made it impossible to organize the state's farmworkers. (The proposal was similar to Farm Bureau-sponsored legislation that had passed in Kansas, Idaho, and Arizona.) To give the proposition the appearance of widespread support, small farmers were trotted out across the state to announce their support of the measure while the Bank of America, corporate farmers, wine companies, and other anti-labor organizations channeled money to the campaign. Nevertheless, Proposition 22 was soundly defeated by the voters who saw through it as a sham.

### National Labor Relations Act and Farmworkers' Power Base

The 1930s—a time of widespread depression and great social ferment—was also the time when labor unions finally received the legislative assistance they needed to organize many of the nation's biggest industries. The vehicle was the 1935 National Labor Relations Act, which established ground rules for representational elections among workers, collective-bargaining procedures, and the National Labor Relations Board to oversee the law. It was dubbed the "Magna Carta" of labor.

Yet agriculture was specifically excluded from the law because farm groups succeeded in passing off farming as a way of life, not as the industry it truly was. Besides, they said, the law would not suit farming, an industry too vulnerable to strikes and the other economic measures used to unionize other industries.

The sponsor of the bill agonized only a little, when he said that "The committee discussed this matter carefully in executive session and decided not to include agricultural workers. If we get this bill through and get it working properly, there will be opportunity later, and I hope soon, to take care of the agricultural workers."

They are still waiting to be taken care of, because four decades later there is no such legislation. While this discrimination has resulted in an enormous economic benefit for farmers, it has ensured that the power to control farmworkers' lives will continue to rest with the farmers. Social, political, and economic power are the natural outgrowths of unionization, at least in its infant stages. The power of numbers, coupled with strong organization that can speak with one voice at election time, is a very real threat to the agribusinessmen who have come to look upon rural America as their own little fiefdom.

If there is one thing more disturbing to farmers than having to pay increased wages, it is having to confront a power structure of the poor that can compete on an equal footing at school boards and city councils. One of the Delano table-grape growers most troubled by the Chávez movement, Jack Pandol, was fond of telling newsmen, "He's trying to replace my power structure with his." Chávez is hardly trying to take over agriculture and rural America, as many farmers complain. Rather, his intent is only to achieve a measure of equality between the rich and poor, the powerful and the powerless. It is this gap that gave birth to his movement and those before him and it is this gap that will provide the impetus for continued agitation for change.

In their history of the farm labor movement, Joan London and Harry Anderson provided some insight into this aspect of the farmworkers' struggle (see References).

A successful farm labor union will by its very nature "subvert" the present rural California social order under which elected representatives and administrators have served the minority who own the land rather than the majority who work on it; under which justice is not blind, but peeks out from under her hankerchief to see if a plaintiff or defendant has brown skin and calluses on his hands; under which schools, newspapers, police departments, churches and every other social institution covertly or overtly perpetuates the premise that landowners are a class above; land-workers a class below.

Power is what has allowed farmers to dominate the farmworkers. Chávez made that point when he told Mark Day, in the book *Forty Acres*: "It is necessary to build a power base. To change conditions without power is like trying to move a car without gasoline. If workers are going to do anything, they need their own power. He [the grower] has more power because he has economic power. If we had economic power, our thousand votes would count a thousand times more than any individual's votes."

## History of Farmworker Exploitation

To write about the history of farm labor is to write about discrimination on a scale nearly unmatched in any other American industry. It is a history of one foreign minority after another brought to the orchards and vineyards of California, exploited thoroughly, and then cast loose when the growers' purposes were fulfilled. Always there was one firm rule: Find the most tractable, compliant group that will work for the lowest possible wages.

In California it all began with the Indians, first brought under domination by the Spaniards and missionaries who pressed them into service to construct missions and forts, and to cultivate the few agricultural crops then being grown. Disease and war wiped out much of the Indian population, and after California became a state in 1850 agriculture changed from a subsistence existence to a profit-making venture.

This time, the farmers turned to the Chinese who were just then "liberated" from working on the transcontinental railroad. They were good, industrious workers who complained little, the type growers looked upon most favorably. But their frugality and oriental nature, combined with their willingness to work for below-standard wages led to a vast outpouring of prejudice, particularly from labor union elements in San Francisco.

As the Chinese were excluded from California, growers turned to another minority—the Japanese. Their importation began in the 1890s and by 1909 there were an estimated 30,000 Japanese working on California farms. Once again they were an ideal labor force—they underbid other workers, depressed farm wages, and labored without rancor or agitation. It was an ideal system for the growers until a strange thing began happening. Japanese—frugal, ambitious, and talented farmers themselves—began leasing and buying their own land to compete with the Anglo growers.

That was enough for the growers. Anti-Japanese agitation soon began, culminating in the passage of the Alien Land Act of 1913 forbidding Japanese to own land. By 1924 the agitation had reached such a peak that the Japanese presence in California agriculture had been ended almost entirely.

Perhaps not coincidentally, farmers by this time had begun looking south to Mexico where political upheaval had unleashed a huge labor force that was desperate for work and could easily be imported into the United States. And so began the use of Mexican labor to supplant the Japanese. The transition was so swift that by 1920 there were an estimated 100,000 Mexicans working on California farms.

Like the Japanese and Chinese before them, the Mexicans were uniquely suited to California agriculture. They were willing to work for what amounted to slave wages, they didn't remain in the state when the picking season was over, thereby easing the social and political burden on rural communities, and they labored silently.

Except for a period during the 1930s when "Dust Bowl" migrants from Oklahoma, Texas, and Arkansas descended on California by the hundreds of thousands, Mexican-Americans have remained the backbone of California's farm labor force. World War II with its shipbuilding, munitions, and aircraft work in the cities, drew off many of the "Okies" and left farmwork once again to the Mexicans.

"The Mexican has no political ambitions; he does not aspire to dominate the political affairs of the community in which he lives," one grower remarked. Early on the growers became aware that their positions of preeminence could be threatened by a cohesive organization that would vote itself into power. The object, then, was to ensure that the Mexicans would leave the country when their job was done. Consequently, they were denied any social benefits from county governments, nearly disenfranchised at every election, and encouraged to get back to their native land.

It was left to the federal government to institutionalize the use of cheap foreign labor. In 1951, under pressure of Korean War labor shortages, Congress enacted Public Law 78 which governed the "temporary" importation of Mexican workers. From 1944 until that time they had been imported from year to year. But P.L. 78 transformed it into a formal practice.

P.L. 78 was so temporary that it was extended year after year until 1964 when urban and labor pressures led to its abandonment. This, the bracero program, was conceived purely as a labor subsidy to growers, already growing fat from substandard wages. Ernesto Galaraza in his book *The Merchants of Labor* estimated that from 1950 to 1960 more than 3.1 million Mexican nationals were imported to work on more than 270 crops. At its high point, more than 500,000 were brought in during one year at a cost to the taxpayers in 1945 of $21.6 million.

Aside from the wages, the other effect of the bracero program was to further discourage unionization. Mexican nationals who talked of higher wages or better working conditions could be summarily sent back to Mexico. And conversely, if domestic workers did the same thing and walked out on strike, they could easily be replaced by braceroes, eager for the work and money.

With the bracero program, farm labor had come nearly full-circle, from the use of contract Japanese and Chinese to the use of contract Mexicans. When the bracero program ended, growers began to rely on illegal Mexican aliens and greencarders who crossed the border to find work in the fields. The number of illegal aliens in California alone has been estimated at 500,000, about 70,000 of whom are working on farms. Farmworkers themselves say that 10–15 percent of the labor force in many agricultural counties is composed of illegals.

This nearly invisible influx of workers has had much the same effect as the use of previous minorities, from the Chinese to the occasional use of Filipinos, Hindus, and Armenians. Wages were depressed, worker rights were shackled, and nationality was set against nationality to insure obedience.

An oversupply of labor meant that no strike could be effective, for there would always be someone ready to step in when workers walked out of the fields. Keep enough people hungry and someone will always be ready to work, the theory went. Equally important, the use of foreign labor made the task of unionization doubly difficult. The labor force was highly mobile, spoke little English, was unaware of the nuances of organized labor movements, and was politically impotent.

The legacy of this bitter harvest of exploitation is still being reaped today. For the most part, farmworkers remain a silent, poverty-stricken group, floating from one crop to another, making enough dollars to sustain themselves during the "cropping season" and then returning to the barrios or ghettoes of southern Florida, the lower Rio Grande Valley in Texas, or California's San Joaquin Valley, to collect welfare and food stamps until the fruit ripens again.

Consider that in 1972, according to the USDA, there were 2.8 million farmworkers who worked an average of eighty-eight days and earned $1,160. Two years earlier, for an average of eighty days farmwork, hired farmworkers earned $887 in cash wages. The hourly composite farmworker's wage rose from $1.33 in 1969 to $1.42 in 1970, yet that was still only 42 percent of the average factory worker's wage.

There is a minimum wage for farmworkers of $1.60 an hour— but it is twenty-five cents less than the minimum wage for comparable work in factories. An exemption from federal-state unemployment insurance saves California growers over $20 million a year in payroll taxes. (In 1971 and 1972, the California legislature passed compulsory unemployment insurance for farmworkers, only to have it vetoed by Governor Ronald Reagan, who said it would cost the growers too much money.)

Wages are only one aspect of the farmworker's dilemma. Farming is the third most dangerous occupation behind mining and construction, according to the Department of Labor. Yet, agricultural pressure groups continue to press for exemptions from the poorly enforced Occupational Safety and Health Act of 1970. More than 2,400 accidental deaths and 200,000 disabling injuries occurred on American farms in 1970, the National Safety Council estimated, and a Senate subcommittee was told there are 800 pesticide poisonings a year, most of which go undetected and unreported. While Congress appropriates $20 million a year for pesticide research, it spends only a tiny fraction, $120,000, on pesticide safety.

Then there is the matter of the health of migrants. The migrant's life expectancy of forty-nine years is twenty years less than the national average. The average person pays $300 a year for health care, while the federal government expends only $15 a year for migrant health care. Infant mortality for migrants is 25 percent higher than the national average.

An even worse legacy is child labor, a form of involuntary servitude thrust upon children because of the migrant's nomadic search for work and the need to supplement family income. It is largely a matter of guesswork as to how many children work on American farms. The American Friends Service Committee places the number at 800,000, a figure disputed by most farmers. What they can't dispute is the terrible toll it takes on the child's health, schooling, and self-image.

One of the most intriguing aspects of the farm labor force is that no one knows how many migrants there are. A migrant is defined as a worker who moves from one county to another within the year to work in agriculture. The USDA estimates the total at 196,000, or 7 percent of the total farm labor force. It is an interesting and probably inaccurate figure for there is no official census of migrants. While USDA calculated that farm labor increased from 2.6 million in 1970 to 2.8 million in 1971, the number of migrants remained the same, an implausible occurrence, given the depressed state of the economy during that period. UFW places the figure at 500,000.

Farm leaders are fond of citing USDA migrant statistics to show that only a fraction of workers are forced to move from state to state to earn a living. The below-average wages are justified on the basis that farm work is an unskilled labor that any child tall enough to grab an orange can learn in a matter of hours. In truth, farm labor is a back-breaking, strenuous skill that in most instances requires months and sometimes years to acquire, particularly in fruits and vegetables. These crops are heavily dependent on hand labor and without it the farmer would be unable to stay in business.

"The wages paid migratory farmworkers are relatively low because the value of their services as measured by the volume of their output and farm prices is relatively low to their employers," William J. Kuhfuss, president of the American Farm Bureau, explained rather lamely to a Senate subcommittee. "Neither a farmer nor any other employer could afford to pay as much per hour for relatively unskilled labor," he said, neglecting to mention the great disparity between farm and nonfarm wages. Farmers never have had to pay competitive wages because history was kind enough to provide them with an abundant pool of cheap labor.

## History of Farm Labor Organization

If farmworkers were so maltreated and underpaid for so long, the logical question arises: Why didn't they rise up long ago to form their own unions? In part it was because the growers' consistent use of imported workers made that task impossible. Also, there are the sheer physical dimensions of organizing and striking on the farm. Unlike an auto plant or similar factory there are no gates or parking lots where workers can congregate. The factories, in this case, are wide fields that spill across the horizon, out of sight and sound. Many employees live in grower-owned labor camps that are off-limits to union organizers. It is like trying to organize a factory whose location changes every day and whose workers change nearly as often.

Additionally, there was never any indigenous leader to emerge from the ranks of the workers. Always, the organizing was done from the top down, with predictable results. Organized labor, for all its breast-beating, exhibited little imagination or interest in tackling agriculture. Periodically, organizing efforts were made but they generally were half-hearted attempts.

Though the AFL discussed organizing workers as early as 1889, it did not make a concerted effort until after World War II, a push that was largely unsuccessful until it affiliated with the Chávez union. Instead, one of the earliest efforts was carried out by the International Workers of the World, whose concept of unionization was not so much to secure workers better wages, but to get them a piece of the farm. Formed by the merger of the Western Federation of Miners and the American Labor Union in 1905, the IWW concentrated on the migrants and "fruit tramps," who moved from town to town following the crops.

At every turn, the IWW was confronted with the kind of establishment "law and order" that so often has been called into play whenever unions have threatened the delicate rural social balance. Often this involved the formation of vigilante groups, described by Carey McWilliams in *Factories in the Fields*: "During the period when vigilantes were in action, they completely usurped the functions of governmental officials, defied the governor of the state, conducted their own trials, equipped and drilled an armed force, and operated in effect as an insurrectionary junta . . ."

The IWW's loose, decentralized structure was not the kind of organization able to bring unity to the huge migratory labor force. But through a bizarre incident called the Wheatland Riot in August 1913, it was able to galvanize national attention for a time on the farm labor question. The incident took place in Wheatland, California, where more than 2,800 workers—men, women, and children—were crowded into a farm labor camp that needed only 1,500 workers. At a mass meeting, a riot was touched off in which the district attorney, a deputy sheriff, and two workers were killed.

The IWW leaders were convicted of murder and sentenced to life imprisonment, though they later were released. The ranch owner, D. B. Durst, escaped implication in any of the violence. But magazines and newspapers throughout the U.S. and Europe were intrigued by the incident and began probing the question of migrant labor.

The IWW efforts in the West, however, were all but buried in that volley of gunfire at Wheatland, so it soon moved into the Midwest and enrolled more than 20,000 members. By 1917, there were an estimated 70,000 IWW farmworker members. But as happened all too often, the government intervened on the side of the growers during the Red Scare following World War I. IWW leaders were arrested and convicted of violating California's criminal syndicalism law which made it a felony to "teach, advocate, aid or abet acts of violence to effect political change."

The IWW's back was broken and it was not for another seventeen years that a serious effort was mounted to organize workers. This time it was the Cannery and Agricultural Workers Industrial Union (CAWIU), backed by the Communist Party, that began staging sporadic strikes in California's agricultural valleys to protest the depression wages then prevailing. By one estimate it carried out twenty-five strikes alone in 1935, involving 37,550 workers. And it was at center stage in the most successful strike of 15,000 cotton pickers in Corcoran, California.

Tainted by charges of communist influence—a verifiable charge —the CAWIU, like the IWW, was disbanded after many of its leaders were jailed on similar criminal syndicalism charges. For a so-called radical organization its demands were moderate enough. They included a minimum wage of seventy-five cents an hour, an eight-hour day, time-and-a-half for overtime, decent homes and sanitary conditions, elimination of contract labor, equal pay for women, and no child labor. Modest though the demands were, only a few have been met to this day, further evidence of the shabby treatment of agricultural workers, especially when contrasted with industrial workers, who have all of the above benefits and much more.

While all this was going on out West, Southern sharecroppers who lived in a latter-day form of human bondage were rebelling against their landlords who were reaping money from the 1933 Agricultural Adjustment Act. The act was designed to decrease cotton production to stabilize prices. This was done by paying farmers to not grow cotton on all their productive land. As the land was idle, so were the tenants. Angered by the growers' almost total domination of their lives, the tenants—white and black— formed the Southern Tenants Farm Union in 1934.

It had only limited success in organizing strikes and eventually was merged into the AFL with the new name of the National Farm Labor Union (NFLU), but only after first leaving the Congress of Industrial Unions. On 30 September 1947, NFLU Local No. 212 launched one of its most famous and publicized strikes against the DiGiorgio Corporation of Kern County, California, a prototype of the vertically integrated corporations which had interests in growing, packing, shipping, and selling fruits and grapes. Again a labor strike was met by violence, punitive court orders, cries of "communism," and foreign-imported strikebreakers. Thirty-two months after the strike began—May 1950—the strike was called off.

From that time until the early 1960s there were scattered efforts to organize workers in California, backed primarily by the AFL. But the drives were underfinanced, poorly directed, and generally ineffective. The central problem—as with most previous attempts —was the absence of a leader who could both organize and command the respect of workers. There was, in short, no leader from among the people who understood what was at stake. All this began to change, imperceptibly at first, and later with the force of an erupting volcano.

## Chávez and the United Farm Workers

The catalyst was Chávez, a man admirably suited for his long-held ambition to organize farmworkers. So much has been written about Chávez that he has taken on larger-than-life qualities. It is well known that this son of a Mexican national tramped the orchards and vineyards of California and Arizona, met the very prevalent radical discrimination in such towns as Delano, engaged in religious fasts, bespoke a code of nonviolence in an industry noted for its violence, and finally prevailed against odds so long that no one cared to compute them.

Above all, Chávez is an organizer—not of just farmworkers but of politically and economically oppressed people throughout the United States. He is the symbol of hope for millions who have not been made a part of the American dream.

Chávez is uniquely suited for his role. He spent nearly a decade with the Community Service Organization, an urban-oriented Mexican-American organization that headed up voter registration drives and economic boycotts in California's cities. Chávez headed the organization in the late 1950s but quit in 1962 when CSO leaders did not match his enthusiasm for leaving the cities and moving into rural areas among workers.

With only a dream and $800 in savings, Chávez formed the National Farm Workers Association (NFWA) in Delano in 1963. From the outset, Chávez made it clear that it was not a union in the traditional trade-craft sense. The NFWA was an organization of mutual self-help, where roles were blurred and farmworkers could join a credit union, buy gasoline for their autos at a cooperative, and generally assist one another.

But above all, Chávez used the tactic of nonviolence, of massive civil disobedience to laws he felt were either unconstitutional or immoral. Mindful of the violence that had attended every strike before him, Chávez knew that repetition of it would surely distend his organization.

He spent hundreds of hours with farmworkers, exploring their strengths and weaknesses, desires and resentments about the growers, and in the process, finding leaders for his movement. He crisscrossed California farmlands fully aware that it was necessary to have a strong organization before staging a strike, a premise lost on most of the earlier strike leaders. His patience was rewarded when the first Delano strike began in September 1965 over the issue of wages, then $1.20 an hour. It was a strike that came three years premature for the NFWA, Chávez has recounted, but he had little choice when the Agricultural Workers Organizing

Committee—a Filipino organization headed by Larry Itliong and backed by the AFL-CIO—walked out of the Delano vineyards first. The two later merged to form the United Farm Workers Organizing Committee (UFWOC).

It was no mere accident that Chávez chose to settle in Delano. Delano is the center of California's $40 million table-grape industry, run by second- and third-generation farmers whose political and economic philosophies are not unlike the rest of California agriculture. But more important, table grapes are a labor-intensive crop that must be handled carefully by skilled workers and consequently is immune for some time to mechanization.

The cry of *huelga* (Spanish for "strike") was heard across the fields. As strikes go, it was about as successful as those that had preceded it; in other words, the growers imported strikebreakers, mostly illegal aliens, to pick the crop, sometimes at great expense. But the crops were harvested.

After all, California history was littered with the remnants of farm-union leaders who had called strikes, pulled a few workers out of the fields, and then vanished into thin air. But there was something different about this one. In retrospect, it is easy to pinpoint some of the things that were changing.

The 1960s was a time when America's social conscience was stirred from a generation of lethargy. Causes became America's life-style and here was a cause both meaningful and tangible— the deprivation of minorities oppressed by one of the most powerful and conservative industries. Chávez, in his quiet, understated but powerful manner, did much to focus the nation's attention on a problem too long neglected. He was the master of the media, a classic labor leader who talked not simply about paid vacations and health plans, but about workers' rights as humans. That was one part of the equation.

Additionally, there was the reaction of the growers who played right into Chávez's hands. They dismissed him as "just another dumb Mex," and a spokesman described the NFWA as "another fly-by-night operation, something that will be over as soon as the outside agitators that started it get tired of the game they're playing and go home."

A California Assembly committee was rushed in to hang the tag of "communism" on the movement, just as that tag had been used to demolish the IWW, CAWIU, and others. But this time it didn't stick. Clergymen of all faiths from across the nation rallied to the cause. The National Council of Churches dispatched its National Migrant Ministry to help Chávez, Catholic bishops in the San Joaquin Valley spoke out in favor of *"la causa,"* and several Catholic priests traded in their cloth to join the movement.

The labor movement, which had shown only casual interest in the problem for decades, pitched in with unprecedented fervor. The national AFL-CIO began giving the union $10,000 a month to support the organizing effort. And Walter Reuther, the late president of the United Auto Workers, spoke out often and articulately in favor of the farmworkers and gave the union $5,000 a month. (The AFL-CIO has estimated it spent more than $2.75 million in direct cash grants from 1959 to 1972 to organize farmworkers. This does not count manpower contributed or other services.)

Senator Robert F. Kennedy and others were attracted, partly out of the evident political possibilities and partly out of conscience, lending their names and considerable political muscle. Committee hearings were staged in Delano, focusing the issues of worker deprivation even more. Surely, such a vast array of support was convincing evidence that social commitment, not communism, permeated the movement.

The first break for the union came in 1966, at the end of a 244-mile march from Delano to the state capital in Sacramento, when the Schenley Corporation agreed to a contract, under the threat of a nation-wide boycott of its other products. The intractable DiGiorgio followed, but only after first allowing entry to the Teamsters and then agreeing to a representation election, in which the Teamsters were soundly beaten by the farmworkers union.

But most of the other Delano growers remained unmoved. Then Chávez launched what proved to be the ultimate weapon, a consumer boycott of all California table grapes. The growers called it "illegal and immoral," an interesting bit of rhetoric since farmworkers were not covered by any federal labor laws. It was a piece of irony the growers found most unpalatable, especially when the boycott began to have a devastating effect. Grapes rotted in cold-storage houses, prices plummeted as demand dropped, and things began to look quite gloomy.

The growers filed multi-million-dollar lawsuits that gathered dust in courthouses, they enlisted the help of politically powerful advertising firms, used Senator George Murphy as a congressional shill, and even formed company unions—all to no avail. Only when a group of ten growers, clearly sensing that to continue their intransigence meant bankruptcy, did the Delano growers sign with the UFW.

In an emotionally charged ceremony in Delano on July 29, 1970, the twenty-six growers who had reviled Chávez and his movement signed the contracts in the presence of Chávez, 300 farmworkers, and the Catholic Bishops Farm Labor Committee that had been instrumental in bringing the two warring sides together. The terms of the contracts—something no farm labor movement had ever obtained—tell how far the movement had advanced. The standard wage for field work was $1.85 an hour, up 65 cents an hour from when the strike started. Employees could no longer be fired at the whim of a supervisor or grower. Growers had to contribute to a medical plan and union building fund. There was a strict prohibition against the use of some pesticides and regulations covering the application of others. Farm labor contractors, despised by workers for years, were displaced by a union hiring hall. But most important, the workers had wrested control of their lives from the growers and attained a measure of self-determination.

The boycott marked another turning point for the farm labor movement. Previous organizing efforts had operated in a political and social vacuum, devoid of any outside assistance. This made the job of crushing dissent in rural areas extremely easy. But the boycott linked the rural Mexican-American community with the urban Chicano community along with liberals and others who sensed the plight of the farmworker. This coalition, though fragile at times, remains an important element in the strength of the farmworkers movement.

The ink was hardly dry on the grape contracts, however, when the Teamsters walked in through the back door to sign up 30,000 lettuce workers in Salinas, who were the next objective of the UFW. It was clear that the growers were taking the easy way out, a fact confirmed by the California Supreme Court in December 1972 when it found that the contracts were tantamount to formation of a company union.

Chávez called a strike in August, pulling five-thousand workers out of the fields and paralyzing the Salinas Valley at peak harvest. It was the most successful farm strike in history, a clear indication that the workers favored the UFW over the Teamsters. To avert any further bloodletting, the two unions late in the year reached a jurisdictional agreement that left fieldworkers to the farmworkers union and processing workers to the Teamsters. Additionally, growers were "invited" to give up their Teamster pacts and sign with the UFW. The accord didn't work. Only five growers moved over to the UFW while most others spent nearly a year going through the motions of negotiating with Chávez.

That the jurisdictional agreement had no meaning became evident in late 1972 when the Teamsters announced it would renegotiate the lettuce grower contracts, two years before they expired. As soon as that was completed, they moved into the same vineyards that Chávez had opened up to unionization and began signing up grape growers in another attempt to dismantle the only legitimate farm union in history.

### The Future for Farmworkers

Confronted with insurmountable odds before, Chávez as of this writing faces perhaps his most critical test. In a way, Chávez may be wishing that workers were covered under the NLRA because it would force growers to hold elections to determine which union is favored by the workers. Chávez would most probably win the elections. But since 1969 he has opposed coverage under the act because of the Taft-Hartley and Landrum-Griffin amendments that severely circumscribe union activities. In particular, his most effective weapon, the secondary boycott, would be outlawed.

"Under the complex and time-consuming procedures of the National Labor Relations Board, growers can litigate us to death," Chávez has explained. "Forced at last by court order to bargain with us in good faith, they can bargain in good faith—around the calendar if need be—unless we are allowed to apply sufficient economic power to make it worth their while to sign."

The current Teamster move into the fields, which this time appears to be permanent, may force Chávez to alter his position. It would be difficult to give up the boycott as a club to be used against the growers, but in the end, Chávez may have no other choice. It is ironic that the growers, who for years contended their workers were happy and satisfied and consequently opposed inclusion under the NLRA, have recently been clamoring for it. They know the boycott would be eliminated as any threat to their pocketbook.

Though the Teamsters profess great concern for the farmworkers, their true position was outlined by the president of Western Conference, Einar Mohn, in April 1973, when he said it would take at least two years for the farmworkers to have a part in union activities. That hardly fits the picture of a democratic union hueing to the cry of the rank and file.

As of May 1973, the Teamsters had signed up more than thirty-five grape growers with close to 18,000 workers. The wages were good—$2.30 an hour as well as the fringe benefits: a pension plan, unemployment insurance, medical benefits, paid holidays, and paid vacations. But there was no protection for workers against pesticide misuse or contract language that ensured the worker protection against grower abuses. If one contract provision sums up the growers' reprehensible stance it is the clause that permits growers to use farm labor contractors. The labor contractor is a body merchant who supplies farmers with workers for a price.

To eliminate the contractors' exploitive practices, the UFW uses a hiring hall where workers are dispatched to fields on the basis of seniority. That approach has had its problems but it is far superior to allowing the contractor to deal with workers as ciphers rather than as human beings.

The Teamsters have admitted that they are organizing the fields not because of their compassion for the worker, but to protect their packing house, warehouse, and truckdriver members. They reason that a farm strike or boycott by the rival UFW would cut the flow of agricultural products and perhaps put some of their members out of work. It is logical, but it doesn't say much about their ideological commitment.

Their compassion for and understanding of the farmworkers is on a par with the growers. As Chávez says, the farmworker would be lost in the two-million-member Teamsters because "they would be too busy worrying about their truckers."

Chávez remains confident that his union will prevail, just as it has in previous battles that seemed nearly as hopeless. In this regard he was buoyed by the announcement in mid-May by AFL-CIO president George Meany that the labor federation would be giving the UFW $1.6 million over a three-month period to carry out strike activities. It is an open-ended commitment.

If it wins, what will the UFW have won? Agricultural mechanization continues apace, aided by fat subsidies to private corporations provided by the USDA and land grant colleges. Mechanical devices to harvest wine grapes, lettuce, broccoli, asparagus, olives, and scores of other crops are being developed. If and when they are perfected, the bitter harvest will be the elimination of thousands of jobs for farmworkers. In two years (in the mid-1960s), a mechanical harvester for processing tomatoes was in full use, displacing over 25,000 workers in California alone.

Chávez and his associates are fully aware of the problem, having written antimechanization clauses into some contracts, but they are so wrapped up in the day-to-day problems of survival that down-the-road problems are given only a quick glance.

### Alternatives to Unions

Given all this, is there a possibility that farmworkers can develop economic and political power, independent of a union?

One alliance that has great potential is one between farmworkers and family farmers. After all, both groups are used by corporate farmers, exploited mercilessly, and stripped of any power that could be used to change things. If the Democratic Farm Labor Party could be so successful in Minnesota, why can't that alliance be transferred to California and other industrialized agricultural states?

One of the reasons is the absolute fear that small farmers have of the farmworkers. They have read so much Farm Bureau propaganda that they are convinced the end result of unionization would be the expropriation of their land. Additionally, the growers continue to look at their workers with equal feelings of contempt and paternalism.

On the other hand, as was told to a Senate subcommittee by UFW vice-president Dolores Huerta: "It is unfortunate that small growers who are not unionized are so blinded by bigotry against unionization because we do have many, many problems in common—the lack of bargaining power and the lack of political power. But their attitude prohibits our working with them. . . . So we will have to wait until we find ways to unionize them, then we can start talking to each other."

As Ms. Huerta points out, it matters little if the worker "is being underpaid and has terrible working conditions from a grower that has 40 acres or one that has 40,000." Corporate farmers, in fact, are easier to organize than family farmers because of their fear of a boycott and a more sophisticated understanding of labor relations.

But the fact remains that despite some recognition on the part of the National Farmers Organization and others of mutual problems, a joint family-farmer/farmworker effort is a long way off.

There remains one other alternative to unionism, one that on the surface offers great opportunities, but upon closer examination has some serious flaws. It is a farmworker cooperative like the Cooperative Compesina organized in Watsonville, California. Some twenty-five farmworker families, with no previous farm managing experience, organized together to grow 140 acres, primarily strawberries.

The idea is to allow the families to use their own skills of tending the land and to develop new managerial skills, thereby offering them an avenue of escape from the dead-end job of picking crops. With each acre producing 3,000 boxes of strawberries selling for $3, the families can expect to make some $9,000 a year. Even with expenses taken out, the income is two to three times what the farmworkers were making before.

Rather than trying to retrain workers with industrial skills for the time when they will be uprooted by mechanization and dropped in urban ghettoes, the cooperative has the aim of stabilizing a work force willing to remain in rural America.

Moreover, such projects offer the chance of achieving real land reform, which would have the dual effect of at least holding back corporate land monopoly and equalizing the distribution of wealth in rural areas. Yet one of the biggest drawbacks of these cooperatives was what held back its formation for so long—the lack of money. A grant of $100,000 was given by the Office of Economic Opportunity (OEO) to the Central Coast Development Company, and a $150,000 government-guaranteed loan was obtained from the Wells Fargo Bank after others balked.

As Chávez has learned, anyone who gives away money can just as easily take it back (e.g., the dismantling of the OEO in early 1973). Without money for the initial investment, meeting the high initial costs of cooperative farming is extremely difficult. Even more important in the long run is the economic base for farming. Cooperative Compesina can succeed because strawberries are a labor-intensive crop; that is, labor is the most important economic input for cultivation and harvesting. But such a program would be difficult to apply to such crops as cotton, wine grapes, peaches, and scores of others because of the imminence of mechanization.

Given such pessimism for these two alternative forms of changes, what then of the future of farmworkers? It is bleak but not hopeless. The Teamster raids on the UFW have imperiled the movement for radical change. But even if the growers and Teamsters succeed in putting the union out of business, they will not bankrupt Chávez. "Organizing farmworkers is our life," says Chávez, and thus only the weight of public opinion can improve the lot of these impoverished workers. It has taken nearly 100 years to move agriculture to the point where unionization is even given serious consideration and it may take another 100 years to move realization to reality. But in the end it will be done and we will all be better for it.

## References

There are a great many books on the farmworkers movement in the 1960s, but sadly the history prior to that is found wanting.

### California Agriculture and Social Problems

Carey McWilliams, *Factories In the Fields* (1939; New York: Peregrine Press, 1971). The best book on the subject, it is a telling and heartbreaking account of migratory labor force in California.

Henry Anderson and Joan London, *So Shall Ye Reap* (New York: Thomas Y. Crowell Co., 1971). Another fine history of farm labor, this book also deals with Chávez and other men who preceded him in attempting to organize workers.

*Chávez: The following books are "pro-Chávez":*

Peter Matthiessen, *Sal Si Puedes—Escape If You Can* (New York: Random House, 1970; paperback edition: New York: Dell, 1973), the most insightful book on the subject.

Mark Day, *Forty Acres: César Chávez and the Farm Workers* (New York: Praeger, 1971).

John Gregory Dunne, *The Story of the California Grape Strike* (New York: Farrar, Straus and Giroux, 1967).

Paul Fusco and George D. Horwitz: *The California Grape Strike* (New York: Macmillan, 1971).

Jean M. Pitrone, *Chávez: Man of the Migrants* (New York: Alba House, 1971).

*Chávez: The following two books depict Chávez in a less favorable light and are less accurate than the above:*

John Steinbacher, *Bitter Harvest* (Orange Tree Press, 1971), a right-wing tract against unions in general.

Ralph de Toledano, *Little César* (Anthem Books, 1971), a more readable book but no less derogatory.

*National Sharecroppers Fund/Rural Advancement Fund*

Several valuable pamphlets have been published by this fund:

*Report on Farm Labor, 1959; Agribusiness and Its Workers, 1963; Farm Labor Organizing, 1905–1967; The Condition of Farm Workers and Small Farmers*, annual publication (New York: The National Advisory Committee on Farm Labor, 112 East 19th Street).

*Congressional Hearings*

United States Senate Committee on Education and Labor, *Hearings on Violations of Free Speech and Rights of Labor, Pursuant to S. Res. 266, 74th Congress.* These are out of print but may be available in university libraries.

Migratory Labor Subcommittee of U. S. Senate Committee on Labor and Public Welfare, 1969–1970, 91st Congress, 11 vols. *Hearings on Migrant and Seasonal Farmworker Powerlessness.*

Migratory Labor Subcommittee of U. S. Senate Committee on Labor and Public Welfare, 1969–1970, 92nd Congress, *Farmworkers in Rural America*, 7 vols.

*Farmwork Problems*

Ernesto Galaraza, *Merchants of Labor* (1964), a valuable account of the bracero program; *Spiders In the House and Workers In the Field* (1970), a history of the 1949 DiGiorgio strike. Both can be obtained by writing to Dr. Galaraza, 1031 Franquette Avenue, San Jose, California.

C. E. Bishop, ed., *Farm Labor In the United States* (New York: Columbia University Press, 1967).

Robert Coles, *Children of Crisis: Migrants, Mountaineers, and Sharecroppers*, vol. 2 (Boston: Litte Brown and Company, 1972).

Carey McWilliams, *Ill Fares the Land and North From Mexico*, two books that bear on the question of migratory labor.

Julian Samora, *Los Mojados: The Wetback Story* (Indiana: University of Notre Dame Press, 1971).

Ronald B. Taylor, *Sweatshops In the Sun: Child Labor On the Farm*, a book that destroys the myth that work is good for children.

Dale Wright, *They Harvest Despair: The Migrant Farm Worker* (Boston: Beacon Press, 1965).

*Other Writings*

*Reader's Guide to Periodical Literature.* Scores of magazines have highlighted the story of the farmworker. Among the best coverage for the past forty years are *The Nation* and *New Republic*.

# PART IV

# Cities and Farms

# 11

## LAND DISPUTES AT THE
## URBAN-RURAL BORDER

*Paul Relis*

> Physically incoherent, socially disparate, [these new metropolitan districts] are at best statistical collections. Here and there in the mass one may partly trace the outline of a city: but the mass itself is not a city, in a functional sense, any more than the immediate countryside that surrounds it is a rural area.
>
> —LEWIS MUMFORD
> *The Culture of Cities*

Over the past five years a land-use struggle has emerged in America that promises to be profoundly important to the future viability of our cities and our agricultural communities. This struggle has pitted our country's land speculation and development interests, as well as its banking institutions against a powerful grass-roots political movement whose aim is to contain urban sprawl at the existing urban-rural border.[1]

Paul Relis is director of the Community Environmental Council in Santa Barbara, California.

While the struggle is still being fiercely contested there are now many hopeful signs that suggest that containment of such sprawl may be achieved within the decade. Certain local and state governments have already implemented policies that have greatly curtailed the uncontrolled spread of urban development, and many more will no doubt follow their example in the very near future.

Five years ago no one would have predicted when or how urban sprawl could be controlled. It had such tremendous economic and political momentum that its spread seemed inexorable. Yet the past three years have seen the birth of an environmental revolution in our country. Americans have awakened to the fact that their resources are finite and that personal action is necessary to effect policies that will carefully conserve and distribute those resources that rcmain. In the process of their awakening, large segments of the public have come to identify sprawl as a major culprit in the deterioration of their environment and the decline of the quality of their lives.

This chapter discusses briefly some of the major problems created by urban sprawl into rural areas and some of the difficulties and successes citizen groups and governments have had in trying to contain it. I will be drawing heavily on examples from California—since it is an area of our nation where the problem has been very acute and the public response most dramatic.

### California: A Golden State Turning Gray

Nowhere in history have a people consumed natural resources as voraciously and rapidly as they have in the state of California. In barely one hundred years much of this fabled "golden state," noted for its rich and broad valleys, its vast timber resources, and seemingly endless coastline has been transformed into a depleted gray wasteland of spent opportunities.[2]

Already the landscape between the Mexican border and northern Los Angeles County (an area of about 700 square miles)— once an abundant stretch of agricultural lands, small towns, and metropolises—has disappeared under a wave of suburbs and shopping centers.

This tendency to expand the metropolis into the country has brought about a general surrender of human values and an undermining of the genuine values of the metropolis itself. The loss of any inner cohesion has left California cities, and indeed most other American cities, without the means to draw to themselves the rich and varied resources of their surrounding regions.

Consider the fate of the Los Angeles–Orange County area of southern California. As late as the postwar 1940s the city of Los Angeles was surrounded by prime agricultural lands. Within a radius of fifty miles from the civic center was enough agricultural productivity to satisfy most of the city's produce, dairy, and poultry needs and enough citrus to supply most of the nation. This fertile agricultural environment was the site of many small towns (generally with populations of less than 10,000) that provided goods and services for the farms and orchards of the region and acted as the supply, packaging, and shipping centers for the foodstuffs destined for Los Angeles and elsewhere in the nation.

Yet in the span of less than fifteen years this reciprocity between urban and rural functions was lost under a flood tide of urban expansion. Only a few scattered fields of croplands and orchards remained—more museum pieces than functional components of a once-healthy urban-rural economy.

This change in the landscape of southern California was so alarming to many northern California residents that they began to suggest that California be divided into two separate states. But it wasn't long before the same forces that had virtually destroyed southern California were at work on the rich lands surrounding San Francisco and other communities on the coast and interior.

To the south of San Francisco lay the Santa Clara Valley, otherwise known as the "Valley of the Heart's Delight." Back in 1940 this valley was a fine example of a fully integrated agricultural community. The cities were functionally related to the entire agricultural matrix. San Jose, with a population of about 50,000, was the county seat and the center of a food processing industry. Six other towns with about 5,000 persons each were distributed around the valley, each performing specialized but interrelated agricultural functions.

These towns were the service centers for roughly 100,000 acres of orchards and 8,000 acres of vegetable crops, situated on two alluvial fans. The topsoil was a fine loam, thirty to forty feet deep in places, and underlaid by a tremendous underground water storage basin with a capacity of roughly one million acre-feet.[3] But within twenty years this California treasure, like its counterpart in the southern half of the state, became a victim of uncontrolled and unplanned urban expansion.

The stories of Santa Clara Valley and Los Angeles are only part of the continuing tragedy of California's (and the nation's) settlement pattern. For example, current estimates by the U.S. Soil Conservation Service, with respect to California, indicate that:

> A total of 1,173,656 acres are expected to be urbanized and a total of 808,871 acres are expected to be devoted to recreation subdivision between 1967 and 1980 for a gross of 1,982,527 acres. This calculates out to be an average of 388 acres per day that will be converted during this period (1967–1980).
>
> Approximately 25 percent of the total acreage that is expected to be converted is prime agricultural land (land capability classes I and II). This amounts to 528,600 acres which is 7.5 percent of the total 7.2 million acres of prime land available for agriculture in California.
>
> It is largely on these prime lands that the specialty crops are produced that make a substantial contribution to the total agricultural economy of California and the nation.[4]

If this rate is allowed to continue it has been projected that within thirty years the remaining three-fourths of California's existing farmland will be gone.[5]

Most of these changes are destined to take place around California's major metropolitan areas. Already the amenities that originally brought people out of the cities have vanished. Even within the suburbs there is precious little open space or wild area for recreation; there are no effective public transit systems to link suburbanites to the cultural benefits of the cities (the car is rapidly becoming too expensive to operate on a day-to-day basis), and the congestion and polluted air make a second migration from the metropolitan areas almost a certainty.

Even in areas that are currently committed to California's multibillion-dollar agricultural industry, the same suburbanization is becoming a threat. Agriculturally oriented cities like Fresno, Modesto, and Sacramento are now large metropolises, swelling and spilling over into the heart of the prime farmlands that surround them (see Figure 1).

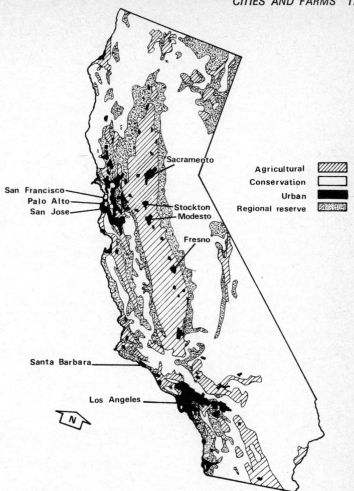

FIGURE 1. The four state zones of California showing the major urban areas described in the text.

## The Tax Relief Strategy for Farmlands

Conservationists battling sprawl have historically embraced the arguments put forth by farmers that new housing developments drive the farmland taxes so high that the farmers are forced to sell out. Farmers have pointed out that since developers prefer the flat lands of prime agricultural areas, there is a real danger of much of the country's food supply being paved over. In unison farmers and conservationists have pleaded that the only way of preserving agricultural land and open space at the edge of the cities is to lower the taxes on farmlands.

Partially in response to this combined plea for agricultural land preservation by farmers and conservationists, the state of Maryland enacted a tax relief law to farmers in 1956. Since then some twenty-seven states, including California, have adopted similar measures of their own.

Dr. Thomas Hady of the USDA's Economic Research Service has divided these tax relief laws into three categories: "preferential assessment" for farmland, which is pure tax relief with no strings; "deferred taxation," under which varying percentages of back taxes come due when the farmland is developed; and "restrictive agreements," under which the landowner agrees to certain restrictions in return for tax relief.

But none of these programs which have aimed at preserving farm lands and open space near the city has really been effective. Maryland's "preferential assessment" legislation has had no apparent effect on slowing down the exodus of farmers from the land or encouraging the "development" of open space. A 1962 study of six Maryland counties showed that 71 percent of rural land buyers were not farming, and 79 percent had not previously farmed. More important, the study indicated that these buyers paid much higher prices for the land than their "farmable" value would justify. All of which suggests that Maryland's "preferential assessment" legislation has become more of a tool for land speculation than land preservation.

California's "restrictive agreement" has fared no better. Known both as the "Land Conservation" or "Williamson" Act, it requires farmers, in return for tax relief, to sign contracts to keep their land undeveloped for at least ten years, though a landowner desirous of developing his property earlier can usually abdicate the agreement by paying a substantial penalty. Land under contract is taxed in proportion to its agricultural productivity rather than its potentially higher real estate value.

By 1972, some 9,562,706 acres of California lands had come under Williamson Act contracts: almost one-tenth of California's privately owned land, 25 percent of private farmland, and 50 to 75 percent of the land in some counties. But a 1970 survey revealed that less than 20 percent of the land then under this act was "prime agricultural land"; that only 1.33 percent of it was within a mile of any city, and less than 5 percent was within three miles of any city.[6]

The fact is that the Williamson Act has failed to preserve agricultural lands adjacent to the cities. In addition, it has cost California's local governments a staggering $40 million a year because of the tax breaks it provides.[7] Among the largest benefactors of these reduced taxes are gigantic landholders whose properties are under little threat of urban encroachment (see Chapter 3).

Preservation of farmland and open space against urban sprawl has not been accomplished by the Williamson Act or any other state farm tax relief program because the legislations misinterpret the fundamental reasons for urban sprawl and because they are so full of tax loopholes that developers are able to manipulate them to their own advantages.

Consider the Williamson Act in California. The act was conceived on the premise that prime agricultural lands around cities should be preserved because they are vital to the maintenance of our nation's food supply. Yet the fact is that neither California nor the United States is currently short of agricultural land, nor do they expect to be in the future. For, despite the U.S. Soil Conservation District's alarming figures, which indicate rapid depletion of California's existing farmlands,[8] the state continues to increase its total farm acreage. With the aid of enormously costly water aqueducts like the California Water Project and the massive energy support systems consisting of farm machinery, chemical fertilizers, and pesticides, thousands of new acres—mostly in marginal areas—have been added to California's farmlands. For example, when metropolitan development in Los Angeles moved into orchard lands, the oranges and lemons that succumbed to the bulldozers moved up to the southeast San Joaquin Valley and over to Arizona. And when the fruit orchards of the Santa Clara Valley likewise fell before the bulldozers, the orchards were similarly relocated.

Thus the argument for urban containment in prime agricultural areas is not justifiable in terms of simply preserving our food production potential. A stronger argument would be made if the act had stressed the importance of the location value of prime soils near the cities. Maintenance of farm production adjacent to the cities ensures a food supply that is not dependent on the high-energy support system required to make marginal lands fertile. The precariousness of this dependence becomes apparent once the energy system underlying it begins to falter—as our petroleum economy is now doing. When that happens, the real costs of removing agriculture from the urban periphery to marginal areas begin to surface, in the form of greater production and distribution changes and increased ecological damage.

Paul Shepard, Dartmouth professor of environmental perception, has powerfully described these costs.

> Unless our course is altered, the years from 1970 until the end of the century will see the construction of huge pipelines running thousands of miles into the interior of all the continents. Besides being irrigated by reactor-operated desalination processes, soils will be fumigated. The night sky will be illuminated; gigantic electric-generating plants will empower fertilizer factories so that several multi-crops will be grown per year. On the surface this seems to be the epitome of progress and human betterment. However factory mono-culture—the dream of engineers and politicians—is a dependent enterprise, requiring heavy industrial support for sifting the planet for materials and energy. . . . The economic and social superiority of modern farming as a way of life is a fiction; as technology it is destructive; as a Green Revolution, hopeless.[9]

The Williamson Act and the other state farm tax relief programs have only remotely touched on this scenario raised by Professor Shepard, which is shared by others in this book, yet it may be the strongest argument conservationists and planners have in favor of preserving agriculture at the urban-rural border.

Another strong argument in support of saving prime agricultural lands near the city is that although 25 percent of the nation's land is capable of sustaining agriculture, only 0.5 percent is capable of producing specialty crops. The problems of agricultural over-production (which has already been cited as an argument used against prime land preservation near cities) is irrelevant to the question of prime land, since such overproduction is limited primarily to grain crops, which do not require prime land. If future generations are to have available a wide variety of produce at reasonable prices, then prime lands adjacent to cities must be preserved. According to Robert Goodier, chief of the California Division of Soil Conservation, destruction of the prime agricultural land adjacent to one southern California city will force transfer of production to poorer lands that require an additional $1,500 per acre of capital investment to bring under production. Finally, it is worth mentioning that the mounting problems of organic waste disposal in the cities will sooner or later have to be alleviated through a massive program of recycling into rural areas. The advantages of having farmlands near the cities under these circumstances is obvious.

Other examples of misinterpretation in most state farm tax relief programs include: (1) the erroneous assumption that open space can best be achieved through the preservation of prime agricultural lands and (2) the belief that farmers want tax relief because they want to remain farmers.

One of the major motivations behind measures such as the Williamson Act is that preservation of agricultural land around the city will serve as badly needed open space. This is the planners' old green-belt idea of ensuring at least a ring of open land between urban settlements.[10] The concept, of course, is a good one provided that one doesn't confuse the purposes of agriculture and that of open space, which is apparently what most farm tax relief programs have done. While it is true that the farmer is our greatest land resource custodian it does not follow that farmland is, in the public sense, "open space." Unfortunately, prime agricultural land does not resemble the urban bucolic idea of a recreational retreat. The working farm of today means toxic fertilizers and pesticides, noisy farm machinery, foul smells, and contaminated runoff water from fields and feed lots (which incidentally suggests that non-chemical or "organically" oriented farming is a more desirable form of agriculture around urban areas).

And as for the Williamson Act's implied assumption that farmers want to remain farmers if given a tax break, a few brief remarks need only be made to dispel this illusion. Conservationists and legislators have overly sentimentalized the farmer as the committed steward of civilization's soil. Given a tax break it is assumed that he will stand fast in pursuit of his ancient art. But this is the exception rather than the rule. Today's farmer is basically a businessman and, therefore, is often satisfied to abandon his fields for a good price. And farm tax relief measures like the Williamson Act help him to get that good price. They provide him the tax relief that enables him to keep farming until urban pressures force the land values to a lucrative point. This real estate option, inherent in all farmland tax relief programs, helps to explain why so few acres of agricultural land have been preserved near our cities.

The only effective way to ease the urban growth pressures on farm and open-space lands near the cities is to attack the economic incentives that make farmers want to sell and speculators want to buy. These incentives, which include tax and service subsidies, are the real loopholes that cause sprawl to continue in spite of land conservation measures.

The fact is that federal, state, and local taxpayers have massively subsidized urban sprawl. A number of tax policies have made it profitable for speculators to hold land that should be developed, forcing other developers to leap-frog over these holdings to cheaper land farther from town. This has resulted in the familiar pattern of larger and looser urban masses thinly spread across vast expanses of otherwise scenic and ecologically important land.

While the list of land income shelters that promote sprawl run into the dozens, there are a few that are particularly abused. Probably the most familiar is the federal capital gains tax that allows a "person" to make an investment in land and be taxed at the capital gains rate of only 25 percent when the land is sold. An individual or a corporation in a high income tax bracket finds the capital gains tax on land investments particularly lucrative. Instead of having his investment in a savings account taxed at a rate of 50 percent or more annually, he puts his investment in land and has the investment taxed at 35 percent or less. Moreover, until he actually sells the land he pays no income tax at all, although he can borrow against the land's appreciated value and deduct the interest payments to boot. Farmers are also permitted to deduct against their income improvements to land, such as planting an orchard or building up a herd of cattle. Deduction of this sort enables persons and corporations in high income tax brackets to operate orchards or breed herds at an economic loss, while at the same time making a handsome profit on saved taxes.[11]

On top of income tax loopholes are local property tax assessors who tend to assess raw land at a smaller fraction of market value than developed land.

> Farmland tax break laws legitimize and greatly extend this practice. The resulting undertaxation makes it even cheaper to hold land out of development, increasing still further the value of land to the speculator.
>
> While vacant land is undertaxed, buildings are often overtaxed, particularly in slum areas. Sources of slum overtaxation include: lagging reassessment as the buildings decline in value, numerous tax breaks for homeowners which do not apply to the rental units typical of slums, and political gerrymandering that leaves the areas most in need of services with the poorest property tax base. Any tax on buildings, as opposed to the land under them, penalizes the owner who develops and maintains his property. But the slum landlord faces a disproportionately whopping tax increase should he try to renew his buildings or construct new housing. . . . As a result, the development that might have occurred in the rundown areas instead sprawls onto new land.[12]

Compounding the problems caused by tax loopholes are federal, state, city, and county boons to sprawl that come in the form of subsidies for services such as streets, sewers, and utilities. Roads, for example, are frequently built in undeveloped subdivisions at public cost. The highway lobby made this possible through highway trust funds, which earmark gas tax money for new highway construction. Because of this policy, developers have been able to rely on state and local governments to run roads to their subdivisions.

Complementing this form of subsidy has been the prevailing policy among growing communities that they should provide services such as sewers and water to peripheral or leap-frogged subdivisions at low prices to encourage further development and thus expand the "heralded community tax base." Proponents of this form of subsidy have been the community chambers of commerce, realtors' associations, banks, and lending institutions. They have been campaigning for decades on the platform that new housing and commercial development would expand the tax base and thus freeze or drive down individual property taxes. But recent studies by several planning consulting firms and city and county planning agencies have proved these beliefs to be misleading. Among the most widely proclaimed of these studies was the Foothills Environmental Design Study for the city of Palo Alto[13]—a suburban community on the San Francisco Bay Peninsula. The report's economic analysis revealed that it would actually be cheaper for the city to buy the foothills and conserve them for open space than it would be to permit them to be developed.

Studies such as Palo Alto's, together with a growing concern over the *total* costs of suburban life have given opponents of sprawl strong political ammunition. Conservationists in particular have seized upon the information provided by these studies and used it as the basis for successful land reform campaigns across the country. Open space or antidevelopment measures spearheaded by conservationists won voter support in the fall of 1972 in New York, Florida, Washington, and California and brought into office a wave of new environmentally oriented politicians.

In some cases the political fledglings of land reform have embarked on bold efforts to create land use policies and plans that would help to reestablish the role of agriculture at the city's edge. Employing such tactics as down-zoning, land acquisition bonds, and land trusts, these policy makers have begun to contain the ominous urban flood.

But the strength of this embryonic land reform movement remains to be seen. Containment and slow-down measures may prove short-lived unless their advocates address the root causes of sprawl. People who can afford it will continue to buy land and thus inflate rural space in order to pursue life-styles no longer possible in urban settings as they now exist. And why not? For the prosperous American who feels crowded or rootless, tradition suggests that what he needs is land of his own. Thus he buys ten acres of farmland in Maine, or ranchland in California, retreats to it for three weeks a year, and thereby perpetuates the sprawl syndrome.

### Santa Barbara: A Community Struggles for Land Reform

Because of an absence of effective federal and state land settlement policies, cities and counties seeking to contain sprawl have had to develop their own regulatory measures. Borrowing from the containment experiments of one another, a handful of communities have already managed to successfully implement no-sprawl policies.

One such community is Santa Barbara, California. Better known for its infamous oil disaster of January 1969, Santa Barbara has been engaged in a fierce political struggle to contain sprawl for the past five years.

In the mid 1950s, this sedate coastal community began to experience intense growth. While farmworkers quietly went about tending to the truck crops and the bountiful lemon, avocado, and walnut orchards, land speculators visited the homes of their employers offering them tempting prices for their land.

These speculators had advance knowledge that the quiet little agricultural community of Goleta, just north of Santa Barbara, was on the verge of a development boom. They had learned that the University of California was about to expand greatly its campus just a few miles from the town of Goleta, and that a number of large research and development industries were interested in locating branches near a growing campus. Correctly viewing the university as the magnet that would attract development and create an accompanying need for housing and services, speculators from all over the country began buying up Goleta Valley lands as fast as they could. At the same time they also began demanding protective zoning ordinances to secure or increase the value of the lands they had purchased.[14]

With the Chamber of Commerce fanning the flames of anticipation and excitement, land values soared, and finally, when supplemental water was brought to the valley, a building frenzy began.

Confronted with unprecedented pressure for development, the local government faced the unenviable task of determining the future shape of the Santa Barbara area landscape. How was growth to be accommodated? What was to become of agriculture, which was then the area's primary economic base? These were hard questions demanding immediate answers. Two opposing viewpoints soon emerged.

On one side people argued that new development should be guided into the foothills behind the valley, thereby sparing the prized agricultural lands. Opposed to this view was a large and powerful force of speculators and builders who argued that the valley floor was the place for development and that agriculture should make way for housing and commercial sites. The county's planning director, siding with speculator and developer interests, suggested that the entire valley be zoned for urban purposes and that agriculture should retreat into the foothill canyons.

But by early 1970 there were some indications of opposition to this position. Santa Barbara's oil blowout in January 1969, which had precipitated a feeling of fear and outrage through the community, had the secondary effect of awakening otherwise complacent people to the plight that was affecting their local environment, land as well as sea.

Subdivisions were going up at a phenomenal rate (Santa Barbara County was then the second fastest developing county in the nation); the newspapers were filled with announcements of subdivision openings; and the agenda of local agencies were glutted with requests for more building permits.

Alarmed by the speed of the transformation of the valley floor, citizens began to organize and seek out courses of action to curb the urban onslaught. Their only apparent handle on the situation appeared to exist in the Santa Barbara County General Plan. It prescribed a boundary beyond which urban development was not to spread and established planning goals such as "protecting the fertile lands of the Valley for the growing of crops" and "providing for the sound growth of cities through careful balancing of and proper relationship between residential, commercial, industrial, and open uses." Though the plan itself was not legally binding in its prescriptions for land use, and did in fact recommend the urbanization of the entire valley floor, it did give citizens a reference point from which to base their goals and actions.

The importance of having such a document surfaced when a major Los Angeles developer announced plans to develop some 3,638 acres of virgin Santa Barbara coastland. The proposed development would have extended upper-class suburbia into the middle of agricultural and grazing lands about eight miles beyond the county's General Plan boundary, and opened the intervening area to rapid development. Citizen opposition to the proposal was unprecedented, but the combined campaigning by the Santa Barbara Building and Trades Council, the Chamber of Commerce, the Board of Realtors, and the local television station made it relatively easy for the county administrators to nod their approval of the scheme.

Infuriated by what seemed to be a flagrant act of ignoring the popular sentiments of the community, the citizen opposition united under the auspices of "Citizens for the General Plan" and initiated referendum action. The struggle that followed was one of California's first and certainly most bitter efforts to subdue urban sprawl. A classic land use battle—"El Capitan," as it was known— pitted citizen suburbanites against the unified forces of the banking and lending institutions, the Building and Trades Council, a highly successful developer, the Chamber of Commerce, and the county government.

But all the publicity tactics they could muster failed to stimulate voter support for the project. When the votes finally were counted, "El Capitan" went down to a smashing defeat—marking one of California's first successful attacks against urban sprawl.

Inspired by the decisiveness of the vote and the political awakening which "El Capitan" had stimulated, participants of Citizens for the General Plan began to direct their actions toward a more extensive land control campaign. This program took the form of educational forums in land-use planning plus political campaigns for local government offices.

Through a series of planning forums sponsored by Santa Barbara's continuing education division of the city college, entitled "Last Call for Santa Barbara," citizens learned of the fallacies inherent in the argument that urban sprawl is good for the tax base and for taxpayers. A panel of lay planners exposed their fellow citizens to the wide array of subsidies that were available to the speculator and showed them a blueprint of what the future of the valley would be if these subsidies continued to be provided by their county government.

The citizen panel suggested that a new plan for the valley be developed that would preserve open space and agricultural lands at the peripheries of suburbia and would direct all future development to select areas that had been skipped over in the sprawl process.

The panel suggested that the county employ such controversial land use measures as down-zoning and open-space bonds and that it develop a land bank as a way of realizing the objectives of the new plan.

Concurrent with this educational program were a series of "minor" political campaigns over control of the valley's water and sanitation districts. Armed with the information developed in the "lay planning forum," a coalition of land reformers challenged and defeated a series of pro-sprawl incumbents who had reigned over the water district for more than a decade. The new water board immediately initiated policies prohibiting the extension of water services beyond the General Plan's prescribed urban boundary, and also banned future water service to areas not already supplied with a water meter. In addition, it authorized a research program designed to explore alternatives to importation of supplemental water from the California Water Project.

Following the election of the water board coalition a political campaign was mounted to elect land reform candidates to the board of supervisors, the most powerful political body of county government. Once again citizens faced the combined forces of the Chamber of Commerce, the banking and lending institutions, and the Building and Trades Council. After another bitter campaign the land reformists scored another decisive victory.

At the time of this writing it appears that Santa Barbara is well on its way to ending further urban encroachments into the countryside. It has initiated severe water conservation policies that all but preclude further urban development. It has also initiated downzoning policies, and has started a comprehensive land use review for the entire county.

Hopefully, Santa Barbara's present policies will mature and bear fruit in the form of a community that blends the values of agriculture with urbanity. Perhaps its experiences will serve as a small example for other cities and communities of the country who seek to integrate in a more positive way the values and economies of a rural society with those of the city.

Certainly the climate in America is ripe for embarking on land settlement programs that redirect our energies away from the social, economic, and environmental ravages of sprawl. The urban crises of our time compel us to seek and realize a more symbiotic relationship between the city and the rural lands that surround it.

## Notes

1. Mason Gaffney, "Containment Policies for Urban Sprawl," Univ. Kansas Publ., Gov. Research Series #27, Lawrence, Kansas.

2. Raymond Dasmann, *Destruction of California* (New York: Macmillan Co., 1965); Robert C. Fellmeth, ed., "Power and Land in California," in *Ralph Nader Task Force on Land Use in the State of California*, vol. 1 (Washington, D.C.: Center for the Study of Responsive Law, 1971).

3. Jay Thorwaldson, "The Palo Alto Experience," *Cry California* (681 Market St., San Francisco), Spring 1973, pp. 4–17.

4. U.S. Dept. of Agriculture, U.S. Soil Conservation Service, California Soil and Water Conservation Need Inventory, California Conservation Needs Committee, 1970.

5. Dean Eckbo, Austin & Williams, "State Open Space and Resource Conservation Program for California," prepared for California Legislature, Joint Committee on Open Space Lands, April 1972 (from Library and Courts Building, Room 509, Sacramento, California).

6. Polly Roberts, "Farmland Tax Breaks: How Not to Stop Sprawl" (unpublished manuscript), Berkeley, California, 1972.

7. Don Demain, "Tax Shelter Costing Millions," *Oakland Tribune*, 2 January 1973.

8. Eckbo et al., "State Open Space Program."

9. Paul Shepard, *The Tender Carnivore and the Sacred Game* (New York: Charles Scribner's Sons, 1973).

10. F. J. Osborn, *Green Belt Cities*, (New York: Schocken Books, 1969); Ebenezer Howard, *Garden Cities of Tomorrow*, ed. F. J. Osborn (Cambridge, Mass.: M.I.T. Press, 1965).

11. Fellmeth, "Power and Land."

12. Roberts, "Farmland Tax Breaks,"

13. Livingston and Blaney, Foothills Environmental Design Study for the City of Palo Alto, Land Use Study, Department of Planning, City of Palo Alto, California, 1971.

14. Paul Relis, "Contemporary Planning: Evolution or Repression?," Part 2, "The Search for an Alternative," *Survival Times* 2, no. 6 (1972): 19–23, Santa Barbara Community Council, Santa Barbara, California.

# 12

# URBAN-RURAL FOOD ALLIANCES: A PERSPECTIVE ON RECENT COMMUNITY FOOD ORGANIZING

*Darryl McLeod*

The late sixties saw small groups of people in every major North American city begin organizing food cooperatives. Their motives and methods were diverse: Some opened small stores ("storefront co-ops") while others organized federations of neighborhood buying groups ("food conspiracies"). Riding on a subsiding wave of political and cultural activism, the co-ops formed quickly and initially depended completely on volunteer labor. Not a few organizing efforts were ill-defined and short-lived. In many cities, however, the new groups flourished and common patterns of organization began to emerge. Co-ops in Minneapolis, Madison, and Seattle, for example, rapidly developed networks of self-managed restaurants, bakeries, warehouses, grain mills, numerous storefronts or neighborhoods, and even nearby farms.

This chapter attempts to document the growth and to understand the direction of these new food distribution systems. How do these organizations compare with traditional American cooperatives and how do they differ from the conventional corporate food markets?

Darryl McLeod, a graduate student in agricultural economics at the University of California, Berkeley, was affiliated with the Berkeley Food Project, 1971 to 1973.

The answer to these questions involves more than a discussion of a few stores and neighborhoods. An urbanized society depends upon a complex food and agriculture system. The organization and regulation of food marketing determines the fate of the farm economy as well as the quality, cost, and type of food distributed. In order to deal with the problems of this complex marketing system many urban groups have found it necessary not only to alter handling and processing techniques within the cities but to encourage diversified ecological farms and compatible rural communities outside the cities. In many regions farmer-consumer associations have formed to deal with marketing, food quality, and food production.[1] These "urban-rural alliances" are necessary if an urban group hopes to provide any sort of viable "alternative" to the American food industry.

The development of this economic link between city neighborhoods and rural settlements is of particular interest to us. The federation of small farms and urban co-ops qualitatively changes the potential and dimension of both groups' efforts. Aside from the obvious advantages of farmer consumer cooperation, people are able to get beyond the narrow self-interest of individual farmers and shopkeepers, which handicapped earlier cooperative movements. When workers begin to view the food economy in its totality rather than in terms of their own niche within it, seemingly contradictory problems (i.e., high food prices, shortages of surpluses, rural poverty) find their common root.

### Farms and Markets

The old idea of trying to solve the farm problem on the farm is outmoded . . . modern agriculture is inseparable from the business firms which manufacture production supplies and which market farm products.—John Davis, 1956.

This wholistic approach acknowledges the high degree of interdependence and specialization that characterizes the modern industrialized food complex. Today's food economy depends on more than a few farms, stores, and middlemen; it includes food processors, farm machinery companies, petro-chemical and transportation complexes, packaging and advertising firms, government agencies, university laboratories, banks and many other institutions that surround the farmer. Attempts to reform (or substitute for) one of these components without altering the others are doomed to failure. One example is the long battle by reformers to preserve the dispersed family farmer in the face of growing centralization and corporate domination of the institutions surrounding him.

One of the first to comprehend the reality of a broad unified corporate food complex and to predict its effects on the farm economy was John Davis, assistant secretary of agriculture under Eisenhower. Back in the fifties, when farm leadership was still painting a picture of independent family farms, farmer-controlled cooperative marketing, and consumer sovereignty in the cities, Davis was talking about contract farming and "vertical integration." He and his colleague Earl Butz went against secretary Benson and all of the farm organizations when they advocated "corporate control to 'rationalize' agriculture production."[2] The interdependence and the power of this emerging government and corporate agriculture bloc so impressed Davis that he devised one word to describe its totality—*agribusiness*.

Davis's analysis also points to the crucial and dynamic role the food marketing structure plays in determining the way food is grown, the size and location of farms, and, in short, the type of rural economy that will emerge. If we explore the historical interaction of farms and markets in the United States, the importance of this relationship becomes even more striking. Today, one could almost say that the type of agriculture we have is determined not in the countryside, but in the offices and factories of the city. It is not surprising that the changes in conventional marketing structures made by the new cooperative networks have begun to foster new experiments in small-scale, ecological farming.

### An American Cooperative Heritage?

Some observers have loosely labeled the appearance of the new food groups a "cooperative revival" in the American tradition.[3] However a quick comparison of the form and functions of the newer cooperative networks with those of established consumer and farm marketing cooperatives reveals little common ground. Aside from a few experiments with joint farmer-consumer buying organizations during the thirties and perhaps the Farmer Consumer Associations formed recently in many states to certify organic food, real rural-urban cooperation is unprecedented.[4]

The thought of today's huge farm marketing "cooperatives" (e.g., Sunkist, Land o' Lakes, Asociated Milk producers) working with any urban consumer group stretches the imagination.[5]

Even the more progressive "Twin-pines" consumer cooperatives have often viewed the unorthodox urban co-ops with hostility. Art Danforth, leading idealogue of the national "Cooperative League," once likened food conspiracy organizers to "agents of Safeway," working to subvert successful cooperative supermarkets.[6]

It is not quite fair, though, to damn past cooperative movements solely on the practices and attitudes of the surviving institutions that call themselves co-ops. A brief look at the development of nineteenth-century cooperatives reveals more subtle distinctions and similarities.

### Farm Cooperatives in Perspective

The American farmer has never glorified the ideal or accepted the condition of self-sufficiency. He has always seen himself as a creature of the marketplace. . . .—W. A. WILLIAMS, 1969

Both American consumer and farm cooperatives trace their origins to the small English industrial town of Rochdale. By 1844, this small community of flannel weavers had long been battered by fluctuating world markets and economic crises. Facing a broken strike and a severe depression, these twenty-eight producers felt compelled to abandon the illusory comforts of the "invisible hand." They set down a successful three-stage plan for the development of a "Cooperative Society."[7] A consumer purchasing co-op was to be set up immediately and run on certain principles (modified to become the famed Rochdale principles). Soon thereafter "production cooperatives" would be organized (the weavers reopened their mill and ran it cooperatively). Finally, their town would become one of many "cooperative settlements": ". . . as soon as practicable this society shall proceed to arrange the powers of production, distribution, education and government; or in other words establish a self supporting colony of united interests or assist other Societies in establishing such Colonies."[8]

American farmers began organizing a few years later in response to similar economic pressures, but their co-ops would play a somewhat different role in the rural economy. To a certain extent, the Rochdale weavers wanted out of the international marketplace, while in the United States cooperatives worked to ease the farmer into the larger economy. This inversion of the Rochdale model was possible because the emerging farm businessman rarely went beyond Rochdale's first stage of group purchasing and selling. It is true that the Grange and other farm associations often tried to own or control related industries and suppliers.[9] But there is little evidence that the farmers ever organized real cooperative production units or had a "self-supporting colony" vision in their cooperative efforts.

American farmers became unhappy with their marketing situation as early as the 1850s. An 1858 issue of *Homestead* complained: "There is a growing dissatisfaction with our present way of making sales of all kinds. Under our present system of city marketing our producers lose on the average at least a quarter by selling to speculators (traveling town buyers)." *Homestead* suggested "using town fairs to bring farmers, butchers and retailers together for mutual trade, thereby avoiding the middleman."[10] Several of these "farmer's markets" were established. Earlier, in 1851, in what may have been the first agricultural co-op, a small group of New York farmers got together to market cheese.

The real wave of farm cooperative organizing began in the 1870s, after new transportation and communication systems had brought international markets to the farmers' doorstep. At first, it seemed that marketing had ceased to be a problem. A family could grow whatever it wanted, put it on a train, and let the market system do the rest. But the promise of new markets soon proved illusory. The new commercial farmers were taken on a roller-coaster ride of fluctuating commodity prices and farm exports. Rural depressions were frequent and severe. ". . . at the precise moment when farming became dependent on the market, the agricultural market of the Union was suddenly turned from a local one into a world market and became prey to the wild speculations of a few capitalist mammoth concerns."[11]

The Agrarians responded in two ways. First, as the rural majority, they put pressure on the central government to improve and regulate transportation, adjust tariffs and money markets, and, most importantly, to expand the farmers' and the country's marketplace through exports or conquest. The implications of this "agrarian discontent" for U.S. foreign policy and economic development are well documented elsewhere.[12]

The farmers' second response to the economic instability was to accelerate group efforts to increase the producers' share of the market dollar. Another development was reinforcing this trend toward group economic action. This was the appearance of the "bonanza farms" and the introduction of their exploitative land and labor techniques.[13] These early precursors of industrialized agriculture would seed thousands of acres of virgin land, harvest huge wheat crops with the newest of machines, and be gone a few years later. In order to compete with these new city farmers and pay postwar taxes, whole rural communities turned to a few cash crops. The farmers were divided in their response to the new methods. The Populists fought the domination of the city financiers and industrial monopolies, while other farmers retained their faith in the market system and technical progress.[14] Few could foresee the vast changes the new methods and machines would bring to agriculture—and to society—save one 1867 critic:

> In the sphere of agriculture, modern industry has a more revolutionary effect than elsewhere. . . . Capitalist production, by collecting the population in great centres, and causing an ever increasing preponderance of town population, on one hand concentrates the historical motive power of society; on the other hand it disturbs the circulation of matter between man and the soil, i.e., prevents the return to the soil of its elements consumed by man in the form of food and clothing; it therefore violates the conditions necessary for the lasting fertility of the soil. By this action it destroys at the same time the health of the town labourer and the intellectual life of the rural labourer.[15]

The economic strategy of the farm co-ops was centered on two closely related objectives. The first was simply to give members immediate savings on production costs, to provide services otherwise unavailable to rural households, and in general to improve the material, "bread and butter" status of the average farmer. Cooperatives, like the Grange, were an important part of the local social life. It is often implied that farmers were unable to cooperate or that cooperatives were bad businesses. A closer look at these assumptions by USDA researcher L. S. Tenny revealed that cooperative business failures had not been "unduly great." In fact: "The same factors working in a privately owned organization would have brought about the same disastrous results."[16]

This was an era of rapid centralization of corporate power and regular, economic depression; few small businesses found survival easy. The farmers tried to keep their co-op enterprises small, controllable, and within the local community. In a world of big and expanding urban markets, this tactic meant slow death. It also contradicted the second broad objective of the farm cooperative movement—to stabilize the food and agriculture system in such a way that the economic status of the independent family farmer was preserved. The local associations were faced with a dilemma. In order to improve their position in the marketplace, they would have to become large enough to bargain with and supply the new factory processors and shipping trusts. The farmers would have to become a very small part of a large, centralized organization which extended far beyond their local community and the limits of real cooperative control.

An attempt to reconcile this contradiction between size and local membership were the cooperative federations. Starting in 1893 with the California Fruit Exchange, many locals organized together to use larger processing plants or to make long distance shipments.[17] But organizing federations was slow going, and they were limited in their urban activities and effective size.

It is conceivable that the local associations could have been more effective and still have remained within the community had they expanded in a different way. The cooperatives could have assimilated a variety of related functions and services within the rural towns. Experience in other countries has demonstrated that "multipurpose cooperatives," organizations that provide credit, supplies, machinery, research, and other related services, are most effective in building a strong economic community.[18] Partial self-sufficiency at the local level was a form of bargaining power never widely used. As it was, most nineteenth-century co-ops were formed around narrow functions and had few ties with other local institutions (today cooperative functions are made narrow by charter laws). One notable exception was the SVEA community in Minnesota.[19] Starting in 1896 with a Rochdale-plan creamery, this Swedish farm community took over the phone system, the local bank, an insurance company, the village store and, by 1910, had their own grain elevator. The members of this cooperative, most of them old Farmer's Alliance members, soon ran the county government. This type of rural cooperation may have been carried to its logical conclusion in the "agrarian socialist" government of Saskatchewan, Canada, and for a time in North Dakota.[20]

Beginning in 1912 with the Sun Maid Raisin Growers, the cooperative federation movement was eclipsed by another form of organization—the Centralized Cooperative.[21] These associations were much larger, with tens of thousands rather than thousands of members. By 1925, sixty-one of these centrals had a larger total membership than the five thousand remaining locals and federations combined.[22]

If the old associations were formed to serve expanding new markets, these new "co-ops" promised to create and control them. Often organized by outside businessmen, farmers would join or subscribe to the marketing co-ops' services as individuals rather than through the old locals. There was debate in cooperative circles as to whether these organizations could really get farmer support or whether they were even cooperatives.[23] The management of these associations aimed at becoming the bargaining and promotional agency for the crop of an entire region.

In 1922 the Capper-Volstead Act gave federal endorsement to these questionable "cooperative" tactics of the centralized co-ops. The law gave the cooperative managers extraordinary powers to enforce future production contracts on farmers and to regulate and standardize farm produce. These central bureaucracies were even exempted from various antitrust statutes.

Business and government leaders hoped the centralized co-ops would act as "bargaining agents" for the thousands of farm operators, thereby bringing the troublesome farmers in line, much as the unions did for labor. Special powers and government endorsement were not enough however, to enable the central association either to gain the loyalty of the average farmer, or to really distinguish co-op marketing tactics from those of their agribusiness counterparts. Just two years later (1924) farmers essentially gave up on any sort of "cooperative" farm program and returned to their familiar cry for expanded export markets and tariff equality under the new McNaury-Haugen Bill.[24]

Today the Agricultural Cooperative story hardly needs telling. Out of the seven thousand surviving cooperatives, the one hundred largest account for half the total sales and assets. Several "cooperatives" are among the nation's five hundred largest corporations.[25] [Today, talk of a "return to farmer control" is good for a few cynical chuckles from old-time farmers.]

### Consumer Cooperatives: From Twin Pines to Cooperative Supermarkets

The development of *urban* consumer cooperatives followed a pattern similar to that of the agricultural associations. The first real city co-ops got their start in the tight ethnic communities of European immigrants who brought the cooperative heritage from the old country. As the surrounding minority communities melted into the great American majority, the small isolated stores were left to fend for themselves in the world of business. These stores might have faded away with their original communities had this not been the "Age of Reform." Early in the twentieth century, a few reformers saw consumer cooperation as a "scheme for social reorganization," a way to check growing corporate producers' domination and replace the "profit motive with the service motive."[26]

One of the most influential of these progressives was Dr. James Peter Warbasse, founder of the Cooperative League. The twin pines symbol Warbasse designed and the league he financed have become synonomous with the idea of consumer cooperation in the U.S. Unfortunately, Warbasse epitomized some of the worst traditions in American Cooperative organizing. He began a one-man, Owen-like crusade to organize cooperatives for "the common American." Like the many cooperative idealists who followed him, his singular means precluded genuine collective ends. Saddest of all, Warbasse and his league began a tradition of separation of producer and consumer cooperatives.[27] He had no faith in the ability of farmers or workers to organize their own enterprises. Warbasse taught that without expert management or something called "consumer control" the production cooperative would always become either a business failure or a profit-seeking private business. As a result the league became a barrier to the development of a "full cooperative,"[28] or any workable alternate food or agriculture system. Today, while they still promote the idea of "a new cooperative owned and controlled by farm and urban cooperatives,"[29] the league has never developed a complete critique of the food marketing system to which they claimed to offer an alternative.

Left with a few stores that had lost their communities, and a program with no broad appeal, the league began to advise cooperatives on business practices and to lobby the federal government for official sanction. The prescription for the ailing co-ops included expert management, centralized administration for efficiency, and competitive expansion. The Rochdale principles were diluted even further. Membership control became a meaningless voting ritual; patronage dividends became a token sales gimmick and the Rochdale home settlement expansion was replaced by an ethic of corporate growth and merger.[30] The twin pines co-ops were swept up by the same chain-store revolution and supermarket concentration that changed private food retailing and the eating habits of America.

Though twin pines leaders advocated and endorsed prevailing business methods of food marketing, they could not have comprehended the importance of the changes that mass-merchandising and corporate concentration were to bring to the food and agricultural system of America. The industrialization of farming and food processing that followed the chain-store revolution came at the expense of the very farmers and consumers the cooperatives represented. Chain stores began in the late twenties as a very profitable method of integrating back to suppliers and circumventing the traditional "terminal markets" and the thousands of small middlemen.[31] Unfortunately, this antiquated system of small distributors and "ma and pa" stores was the only access to urban markets the small farmer had. It was "a marketing system shaped to fit this [small-farm] agriculture," and its obsolescence signaled the decline of a dispersed agriculture system in the U.S.[32] When the real concentration in retailing began in the late forties and fifties, most small stores and farm co-ops were left in the dust. Today three-quarters of all the food sold in this country is sold in "super-markets" (chains with sales of over $500,000) with the four largest accounting for nearly one-third that total.[33]

Today, twin pines literature discusses the tough competitive food retailing business, the narrow profit margins, and modern, efficient supermarkets. Cooperative spokesmen offer pathetic plots "to convert a national chain to a cooperative."[34] Supermarkets are efficient if one accepts the "silent violence" of chemical food technology[35] and the ecological nightmare of frivolous food packaging.

The modern supermarket depends upon the ability of science and factory to alter some very fundamental natural qualities of food. Freshly harvested food has certain well-known tendencies. Chemical enzyme activity continues within the harvested fruit or vegetable and its taste, appearance, and nutritional value change rapidly with handling and the passage of time. As with all living things, harvested food is susceptible to bacteria, fungus, and the attacks of other living organisms. Using their arsenal of sprays, fumigants, additives, and processing tricks, food technologists have been able to duplicate or preserve the appearance and even the taste, but rarely the nutritional value of food products.[36]

Food processing technology has only given food the appearance of durability and freshness, while sacrificing essential nutritional and living qualities. It has attempted, at a very high price, to tailor food to its own marketing schemes.

As a recent Agriculture Extension Service publication declares: "The old simple concept of food as a staple, basic commodity and of a sovereign consumer with a clearly defined demand for food is being cast aside. Industry seeks to create consumer demand for differentiated products and then to tailor the marketing and production process accordingly. This is a process that has been effectively used in the production and marketing of automobiles, television sets and numerous other consumer items. . . . The demands of the supermarket for specific quantities and qualities are being transmitted to the processors and producers and are being acted upon."[37]

There is little doubt as to what kind of food the supermarket demands; between 1929 and 1958, per capita consumption of fresh fruits and vegetables declined 30 percent while processed food consumption increased 152 percent. The value added to raw produce by food processing increased 325 percent between 1939 and 1957.[38] The same modern industry that strives to make potentially durable appliances and automobiles quickly obsolete has tried to transform perishable, healthy food into a durable, lifeless commodity.

Interestingly enough, supermarket technicians make no pretense of efficiency. A 1966 government technical study, "Food from Farmer to Consumer," concludes that when it comes to food retailing: ". . . efficiency is secondary to merchandizing. The paramount aspect of retailing is affecting the consumer's decisions—the decision to enter the store and the decision to select from the items displayed."[39] In 1970 the food industry spent well over $3 billion on food advertising and another $9 billion on food packaging[40] (compared to $300 million on nutritional research in 1967). As for profits, that small 1 to 2 percent sales margin always translates into a healthy 15 to 20 percent real profit (return on investment).[41]

In retrospect, twin pines cooperatives were a classic case of confusing community service with business survival. Had they not chosen to compete and expand by the supermarkets' rules, cooperative managers reply, there might be no consumer cooperatives left at all. But meager "extra" services that twin pines co-ops offer are not enough to compensate for the damage their presence does to any remnants of a popular "alternative" consumer cooperative. Not since the early fifties have twin pines supermarkets let social concern or political strife interfere with business profits. As the manager of Berkeley's "cooperative" natural foods supermarket declared, while ignoring the decisions of the store's membership council, "My first duty is to see that the store survives."[42] Words guaranteed to discourage any beleaguered believer in cooperative consumer control or economic democracy.

### The New Community Food Cooperatives: The Urban-Rural Link

American cooperatives began as something a lot of farm and city people needed, and became something a few organizers wanted. If, in fact, we are experiencing a "cooperative revival" in the American tradition, we have little to look forward to. After the initial wave subsides, the most we can expect is a few more cooperative businesses and perhaps a couple of zealous cooperative crusaders. Cooperatives have done little or nothing to reverse trends toward larger farms, ubiquitous supermarkets, deteriorating food and environmental quality, and the general loss of initiative and self-determination on the part of the common American.

Fortunately, there are numerous characteristics that tend to distinguish today's new community co-ops from the old cooperative movements in particular, and from conventional food marketing in general. In order to understand the most important of these distinctions, it may be useful to think in terms of three roles a cooperative or any economic "alternative" institution might play in a given situation.

I

Cooperatives get their start because they are sensitive to a need in a certain area. . . . They survive if they stress survival rather than ideals of individual members' participation as realistic goals.
—Howard Adelman, 1969

Initially, the purpose of the cooperative simply may be to provide immediate economic benefits for its members. In this first role, group economic action can increase member savings, provide new services or create additional social motivation or bargaining power. These initial benefits have been the primary focus of nearly all American food and farm cooperatives. Conventional cooperatives often have been transitional, partly because these limited needs are soon filled and partly because they could be replaced by private firms.

Most of the new food associations differ from the old even in the immediate needs they were created to meet. Conventional cooperatives were usually started in times of material hardship and economic uncertainty. The sixties, on the other hand, were times of material prosperity, but cultural upheaval and political unrest. The neighborhood and storefront co-ops are still very much concerned with food prices and nutritional quality (some more than others), but community economic sovereignty and changes in interpersonal relationships (e.g., man to woman, worker to manager) merit equal if not greater emphasis. Some may view the new cooperatives as a base from which to continue their attacks on capitalist institutions and others as an opportunity to implement more satisfactory interpersonal relationships, but all seek more than just cheap food.[43]

The new groups are not concerned with business survival so much as with continuing to meet the basic needs of their own communities. If one of the new storefronts or neighborhoods loses local support or internal consensus, it is not "saved" by business reorganization or consolidation as were the twin pines stores; it simply closes down for a while. The working groups stress the goods and services they provide each other rather than those a private wage could buy. Paying themselves uniformly low "subsistence wages," the workers have expanded into many food-related processes and produce a wide variety of goods and services consistent with their low-consumption, ecological life-styles. Storefronts in Minneapolis operate on a gross margin of 10 percent across the board, less than half that of the largest chain stores. Yet the co-ops have no exclusive membership or any other method of discriminating benefits and obligations other than the neighborhood one lives in. Their efforts have begun to produce unique social institutions. In Minneapolis, there is a restaurant with no prices—for two years, people have paid what they could for good, well-prepared meals.

This is definitely not the model for cooperative organizing recommended by the Cooperative League or by government community development projects. A closer parallel might be the original vision of the Rochdale Weavers. Or, perhaps, the municipal economic sovereignty documented by Peter Kropotkin in his idealized, *Mutual Aid* description of the medieval "free cities."[44] Of course, one limitation is that, despite the best efforts of food cooperatives, free clinics, and tenants unions, living in America's cities is not "free" in any sense of the word. Unless the cooperatives are able to grow rapidly and meet a wider range of human needs, the self-sacrifice that working in a co-op requires might become only a phase in a person's development—sadly analogous to the Peace Corp, community lawyers, Vista, etc. Such a transient workforce would undermine the continuity the groups and community need in order to grow and develop.

II

> . . . consumer cooperatives alone have measurably lightened the
> burden of existence for a great many people. Their weakness con-
> sists in the fact that they have not altered the contents of the modern
> social order even when they have altered the method of distribu-
> tion.
>
> —Lewis Mumford

A second role of an "alternative" co-op may be to develop and
implement new forms of organizations and alternate techniques of
production or distribution. Traditional cooperatives promised to
cure the ills of the industrial food system, yet they failed to alter
its basic organization of production or its technical methods. Their
leaders were enthralled with miracles of the new technology, were
blinded by specialization, and were infatuated with the large centralized firm. The new cooperatives, on the other hand, have
experimented with neighborhood councils, planning and accounting methods, work-teams, preorder food systems, recycling, bulk-
packaging, and ecological farming. Today people seem to be less
intimidated by the mythology of markets and capitalist efficiency
and entertain fewer utopian fantasies of quick improvements without fundamental restructuring.

The new community food cooperatives differ in another way
from traditional cooperative businesses in that the latter have
usually failed to "decentralize" individual responsibility for community economic growth and for day-to-day management decisions.
A key trait of traditional cooperatives (as well as capitalist and
state capitalist institutions) is that they make the investment and
management decisions of a large group the responsibility of a few
individuals.[45] The ability of people to be leaders in some situations
and followers in others, to be both teachers and learners, or to
contribute their varied talents in the work place is stifled by these
business bureaucracies. Broader-based management techniques
would not only be more consistent with cooperative ideals, but
would foster greater individual freedom and creativity while making
small, powerful management groups superfluous.

In the food cooperative communities of Seattle, Minneapolis,
and Berkeley a conscious effort has been made to keep all decision-
making in the hands of the working groups and neighborhoods.[46]
With the aid of newsletters, informal planning, and regular meet-

ings, a workable consensus management process has evolved in Seattle and Minneapolis. Successful restaurants, bakeries, warehouses, and stores are run on this basis day to day, while broader issues are discussed at all co-op meetings. The people who attend these central meetings are not elected representatives; rather, their "liaison" role is similar to that of European workers councils' "revocable delegate."[47] This process contrasts with the "one man, one vote" system and the "patronage dividends" investment technique of earlier cooperatives. This organizational form places limitations on the size of the economic or neighborhood unit (about twenty people is the most that can be expected to agree), but it usually means that when an important crisis arises it will be resolved, not voted away. The consensus process puts the responsibility for every economic decision upon each individual, for one objection can stop the works.

This increased individual responsibility and activity works in part because such techniques as visible markups, preordering, rotated jobs, open books, and source identification make the food distribution process almost transparent. A relatively simple idea like preordering has a wide variety of implications and has greatly improved certain aspects of alternative food distribution.[48] Preordering almost naturally establishes a rhythm more suited to the various forms of perishable food. Preordering also discourages compulsive patterns of buying so typical in traditional supermarkets where the buyer is faced with a wide variety of choices. Through some relatively simple planning at the group level, "impulse-buying" and distribution costs are greatly reduced. Purchasing decisions are based on household planning rather than on the confused quagmire of supermarket packaging and advertising.

An important by-product of the preorder system is a stock of information that can be used to form a more direct link between farmers and urban consumers. Ever since the Homestead settlers stopped producing for nearby tables and became raw material sources for food processors and shippers, one "farm problem" has been to reconcile supply with the fluctuating demands for foodstuffs. When combined with good communication links, a few joint handling facilities, and diversified urban-fringe farms, last year's ordering information can be used to coordinate production with consumption and encourage a steady flow of fresh food to urban communities. A good example is the Common Market in Madison, which serves several hundred neighborhood groups. Before the beginning of each season the surrounding farmers meet and divide the projected food orders among themselves; each producer is then given a verbal guarantee of the order and price he can expect to receive.

A real reorganization of urban marketing through preordering and other innovations will require an altogether different type of farm technology and economy. Modern agriculture, with its large mono-cropped regions and short harvest seasons depends upon, and caters to, a multitude of processing and packaging techniques for storing and shipping food long distances. Millions of dollars are spent on government inspection, grading, culling, preserving, and bulk transportation so that a few large firms in California might provide half the country's fruit and vegetables. Such a system only seems rational if you think in large-scale terms—carloads of food, huge food-processing factories, and a penny profit on every pound. The problem is that people eat all year and demand a wide variety of foods in small quantities. So even more food dollars are spent to break down these large lots and to package, process, and preserve food for supermarket retailing operations. These operations consume an ever-larger portion of the food dollar and the nation's resources.[49] The end product of this massive industry and government effort is, in almost every way, inferior to fresh food right off the farm. The recurrent crop shortages, transportation problems, and absurdly high raw produce prices of recent years cast doubt on the wisdom of regional crop specialization, long distance shipping of foodstuffs, and the very industrial foundations of the modern food industry. Placed in the midst of these monocultures and huge food factories and supermarketing wonders, small diversified farms would be an aberration. But in the context of the preordering networks that are beginning to develop, such urban-fringe farms are not only desirable but necessary. A well-managed polyculture farm[50] produces a wide variety of animal and vegetable products in small quantities. Its rural, ecological, and social ramifications dovetail nicely with the aspirations and efforts of today's urban neighborhoods. Urban food organizing has been able to provide the new experimental "organic" farms with more than a dependable, predictable year-round outlet for its variety of produce.

Work brigades and other community groups have frequently gone to work on the farms, exchanging labor and ideas. A symbiotic economic union of these neighborhoods and rural settlements could provide an environment where a more rational and flexible *information technology* could begin to supplant the nutritionally destructive and energy-wasting *storage technology* of conventional food marketing.

A good case study of an attempt to operate America's food system without the aid of harmful chemical technology and without changing the basic marketing structure is the so-called organic food movement. About six years ago a number of food eaters, merchants, and farmers set out to change the destructive dietary and farming habits of Americans by supplying them with "organically grown" fresh produce and grains. In the beginning everything seemed to be going for the "clean-food" entrepreneurs. Consumer demand was incredible. Declaring organic food 1971's "glamour business," *Barron's Financial Weekly* predicted that by 1975, 40 percent of the nation's food supply would come from the "health food industry."[51] However, a few years later "nature's bounty" had begun to sour. Organic produce had a reputation for poor quality, high prices, and outright fraud. As demand dropped off, the "organic" food industry was left with the dubious distinction of making good, wholesome food a middle-class luxury.

The industry's problems were more than growing pains. By using the professional farmers, machinery, and handling techniques of the chemical food industry, it was doomed before it began. Without doing the necessary groundwork, organic growers produced an inferior product at twice the cost. Overnight wonders like California's New Age Organic Food Distribution Center ran aground simply because they were unable to devise the new production and distribution techniques demanded by organic foods. Many firms have been bought out by the very food industries to whom they purported to offer an alternative.

### III

From yet a third perspective, cooperatives can be part of a community-wide effort to develop and assert cultural identity separate from surrounding institutions and outside control. Often a prolonged strike, boycott, political protest, or just a wide disparity in culture or world view will prompt a group of people to begin to supply their own basic needs. Sometimes these impromptu efforts develop into community cooperatives. In Berkeley and Palo Alto, for example, "food conspiracies" were organized nearly five years ago as part of a tenants union strike. Other times these food networks were more short-lived. In the 1919 Seattle general strike, workers and farmers organized neighborhood milk depots and community kitchens that served over thirty thousand meals a day.[52] In May 1968, French students, workers, and peasants established a complex provisioning system amidst the barricaded Paris streets.[53] Rural federations of self-managed cooperative production units were well developed by agrarian anarchists during the Spanish Civil War, 1936 to 1939.[54] However long these "insurgent communities" may last, they broaden the potentials and motive power of any food organizing effort and should be recognized in any evaluation of cooperative institutions.

In the same light, it is impossible to understand the food cooperatives of the sixties without taking into account the cultural and political upheaval that prompted their organization. In this particular description, we have emphasized the practices and organizational forms the co-ops developed, rather than the theory or "movement" behind them. This emphasis should not obscure the fact that nearly all the co-op organizing was a conscious effort to preserve and develop the revolutionary and cultural energy of the period. On the other hand, few co-op workers see their "alternative" institutions as ends in themselves or as a substitute for the necessary transformation of industrialized society. Conditions have obviously changed from those of the sixties; the political spirit and cultural enthusiasm have waned. Yet the economic crises and raw material shortages of the seventies make this type of organizing even more desirable and effective.

One of the major failings of earlier cooperative movements was their lack of defined goals or general theory from which to evolve. A cooperative that survived as a business was almost always the one that had left its founding community and original purpose far behind. Who would guess that a giant supermarket complex like the Berkeley twin pines co-op was begun by the Finns, the most radical of all cooperative organizers. Today, because their communities have coalesced around a similar world view, many cooperative federations have adopted common principles or "operational guidelines." Both can be used as points of departure and as measures of progress independent of outside norms.

One of the best examples of this defined inner-community federation is the "cooperating community" in Seattle.[55] Other cities achieve the same effect through regular newsletters and "all–co-op" or "all-conspiracy" meetings.

If they are to avoid the fate of the organic food merchants, the cooperative food networks must develop broader, more innovative rural–urban alliances. Buying and selling food is simply not enough. Cooperative farms outside of Seattle and San Francisco are having trouble sustaining themselves even though they sell all their produce. Efforts at joint warehousing, transportation, and labor exchanges have met with only mixed success. Most regions have temporarily settled on a combination of preordering and "farmer's markets" (in cities that have them). Part of the answer is more efficient, "single exchange" marketing arrangements coupled with more sophisticated polycultural farms.[56] And part of the answer lies in not having specialized farms at all. The trend is toward rural communities that absorb more of the culture, industry, and people of the cities and urban groups who do more farming.

During their brief history, the new cooperative food network has made a significant break with American cooperative tradition. By taking a broad rural-urban approach and by making a radical departure from normal business practices, they have demonstrated that cooperative organizing need not be the futile diversion from fundamental social change that earlier cooperative movements have become. Already their accomplishments have confirmed the creative and practical potential of self-managed economic units and have provided a glimmer of the ecological rationality that must govern any viable food economy.

## Notes

1. Jerome Goldstein, *The Way to a Nation's Land Reform Is through Its Stomach*, presented at the First National Conference on Land Reform, 27 April 1973, San Francisco, California (Emmaus, Pa.: Rodale Press, 1973).

2. Sarah Shaver Hughes, "Agricultural Surpluses and American Foreign Policy 1952–1960" (Master's thesis, University of Wisconsin, 1964).

3. William Ronco, "The New Food Co-ops, Makings of a Movement?" *Working Papers* 1 (Spring 1973); Richard Margolies, "Coming Together the Cooperative Way: Its Origins, Development and Prospects," *New Leader*, 17 April 1972.

4. Paul H. Johnstone, "Old Ideals Versus New Ideals in Farm Life," in *Farmers in a Changing World: Yearbook of Agriculture, 1940* (Washington, D.C.: U.S. Government Printing Office, 1940); Jerome Goldstein, *The Way to a Nation's Land Reform Is Through Its Stomach*.

5. Linda Kravitz, *Who's Minding the Co-op*, Agribusiness Accountability Project, 1000 Wisconsin Ave., N.W. Washington, D.C., 1974.

6. Art Dansforth, Personal Correspondence, 1972.

7. Martin Buber, *Paths in Utopia* (Boston, Mass.: Beacon Press, 1949).

8. Ibid.

9. Fred A. Shannon, *The Farmer's Last Frontier: Agriculture, 1860–1897* (New York: Harper & Row, 1968).

10. Paul W. Gates, *The Farmer's Age: Agriculture, 1815–1860* (New York: Harper & Row, 1968).

11. Ibid.

12. William A. Williams, *Roots of Modern American Empire* (New York: Random House, 1969).

13. Rosa Luxemburg, *The Accumulation of Capital* (New York: Modern Reader Paperbacks, 1968); Carey McWilliams, *Factories In the Field* (Santa Barbara, California: Peregine Press, 1970); Richard Hofstadter, *The Age of Reform* (New York: Random House, 1955).

14. Thorstein Veblen, *Absentee Ownership*, "The Independent Farmer" (New York: Huebsch, Inc., 1923).

15. Karl Marx, *Capital—A Critique of Political Economy*, Vol. 1 (New York: International Publishers, 1867).

16. L. S. Tenny, "Historical and Interpretative View of Cooperation in the United States," *American Cooperation* (Washington, D.C.: The American Institute of Cooperation, 1925); Editorial. *Top Operator Magazine* (Philadelphia: Farm Journal Publications, May 1972).

17. F. A. Buechel, "Types of Cooperatives," in *American Cooperation* (Washington, D.C.: The American Institute of Cooperation, 1925).

18. Rainer Schikele, *Agrarian Revolution* (New York: Praeger, 1968); Ranaan Weitz, *From Peasant to Farmer* (New York: Columbia Univ. Press, 1971).

19. Shannon, *Farmer's Last Frontier*.

20. S. M. Lipset, *Agrarian Socialism: The Cooperative Commonwealth Federation in Saskatchewan* (Berkeley: Univ. of California Press, 1971).

21. Buechel, "Types of Cooperatives."

22. McWilliams, *Factories in the Field*.

23. Tenny, "Historical and Interpretative View"; Buechel, "Types of Cooperatives."

24. Ronco, "New Food Co-ops."

25. Linda Kravitz, *A Working Paper on Farmer Control of Cooperatives*,

presented 'at the National Conference on Land Reform, San Francisco, Calif., 27 April 1973 (Washington, D.C.: Agribusiness Accountability Project, 1000 Wisconsin Avenue, N.W., Washington, D.C. 20007).

26. James P. Warbasse, *Consumer Cooperation and the Society of the Future* (New York: Apollo Editions, 1972).

27. Margolies, "Coming Together the Cooperative Way"; Warbasse, *Consumer Cooperation*.

28. Buber, *Paths to Utopia*.

29. Margolies, "Coming Together the Cooperative Way."

30. Howard Adelman, "Youth, Bureaucracy, and Co-Operatives," unpublished paper, 1968.

31. A. C. Hoffman, "Reducing the Costs of Food Distribution," in *Agriculture, 1940* (Washington,

g Access to Markets," in *Who ecting the Organizational Struc-* Illinois at Urbana-Champaign, ion Service, Special Publication

dquist Kyle, "Who Will Control ," in *Who Will Control U.S.*

*The New Harbinger* 11 (1973), ganization, Co-op Publications,

efore the Select Committee on enate, Part 13A, "Nutrition and ernment Printing Office, 15, 17,

, *Food Pollution: The Violation* , Rinehart & Winston, 1972); ort from Ralph Nader's Center fety and the Chemical Harvest onggood, *The Poisons in Your*

ulture?

rketing, *Organization and Com-* y #7 (Washington, D.C.: Gov-

ic Research Service, "What's in tober 1971).

od Retailing.

ooperative Way."

American Student Cooperative 16–18 April 1971.

44. Peter Kropotkin, *Mutual Aid* (New York: New York Univ. Press, 1972).

45. Stephen A. Marglin, *What Do Bosses Do? The Origins and Functions of Hierarchy in Capitalist Production*, Discussion Paper No. 222 (Cambridge, Mass.: Harvard Institute of Economic Research, November 1971).

46. "The Movement: From the People to the People," The Plowshare Collective, Seattle, Washington (reprints available from Mill City Foods, 2552 Bloomington Avenue South, Minneapolis, Minn.).

47. P. Chaulieu, *Workers' Councils and the Economics of a Self-Managed Society* (London: Solidarity, 1972); "A Supplement on Self-Management," *New Morning*, February 1973, (P.O. Box 694, Berkeley, Calif. 94704).

48. Ed Place, "Pre-Order Systems and the Federated Cooperatives: Making the Better Way Work," *Journal of the New Harbinger* 4 (1973).

49. *Organization and Competition in Food Retailing*; "What's in the Package?," *Farm Index*.

50. A polyculture farm can be defined as a small-scale, labor-intensive farm raising a wide variety of crops and livestock with ecologically sophisticated techniques.

51. Margaret D. Pacey, "Nature's Bounty: Merchandisers of Health Foods Are Cashing in on It," *Barron's Financial Weekly*, 10 May 1971.

52. Jeremy Brecher, *Strike!* (San Francisco: Straight Arrow Books, 1972).

53. Daniel and Gabriel Cohn-Bendit, *Obsolete Communism: The Left-Wing Alternative* (New York: McGraw-Hill, 1968).

54. Sam Dolboff, ed., *The Anarchist Collectives*, "Workers' Self-management in the Spanish Revolution, 1936–39" (New York: Free Life Editions, 1974).

55. Operational Guidelines: Seattle Cooperative Community, 4030 22d Avenue, Seattle, Washington 98199, January 1973.

56. A polyculture farm can be defined as a small-scale, labor-intensive farm raising a wide variety of crops and livestock with ecologically sophisticated techniques. (See Chapters 16–18.)

## References

The Food Co-op Project, Loop College, 64 E. Lake Street, Chicago, Illinois 60601. This organization serves as a national clearing house for new food co-op groups. It publishes a newspaper, *The Food Co-op Nooz*, and a national biannual publication, *Food Co-op Directory*.

*The New Harbinger.* Co-op Publications, Box 1301, Ann Arbor, Mich. 48106. A bi-monthly publication of the North American Student Cooperative Organization.

"The How's and Why's of Good Co-ops." *Environmental Action Bulletin* 6, no. 6. Emmaus, Pennsylvania: Rodale Press, March 22, 1975. Contains a bibliography of recent books on food cooperatives.

NEFCO Handbook Collective. *The Food Co-op Handbook*. Boston: Houghton Mifflin, 1975. NEFCO is a federation of eastern co-ops and farm groups.

"The Natural Farmer." Newsletter of the Natural Organic Farmers Association, RFD #1, Plainfield, Vermont 05667. Published on occasional basis.

Vellel Tony, *Food Co-ops for Small Groups*. New York: Workman Publishing Company.

Wickstom, L. *The 1974 Food Conspiracy Cookbook*. San Francisco: 101 Productions, 834 Mission Street. An excellent history and description of Berkeley Neighborhood Food Conspiracies.

# 13

## ORGANIC FORCE

### Jerome Goldstein

The organic idea is basically a simple idea. It takes only a very few words to describe what organic means in the garden, or what organic means on the farm. Begin with the soil, get into the compost heap, natural cycles, the need to return garbage, sludge, and wastes back to the land, the hazards of pesticides and artificial fertilizers to the environment, and the personal health benefits that result from eating quality, nutritious food.

Yet more and more in the past few years, some grandiose concepts have been getting mixed in with the compost heap. It is still simple. It is still personal. But now, instead of only rating that heap for its ability to make a soil fertile, we talk about it in such terms as social practice and harmony with the environment. To many people, it's still just a pile of garbage and manure. To others, it's a vision of a society in harmony with the environment. The beauty of the organic idea as a social force is that it is firmly rooted in materials, methods, and efforts that most people usually refer to as idealistic, inefficient, or economically unfeasible. And as more and more people see it, the organic idea provides a model route from where we are now to where we would like to be in the future. "Organic force" has become meaningful in the marketplace and

Jerome Goldstein is the executive editor of Rodale Press, Inc., Emmaus, Pa.

on the farm, in the supermarket and in the classroom because it has become the code word for something even more significant than no pesticides, chemicals, or additives. It is becoming a linking symbol upon which a consumer can relate to a producer. It is a substitute for national-brand advertising via television, newspapers, or magazines; the word organic when truly defined cannot have a national brand name because its essence is its localization and personalization.

The organic concept has the ability to show government officials, industry executives, professors, and policy-makers everywhere how smallness can become economically viable—how money can be spent to subsidize projects other than those leading to mechanization, bigness, and so-called efficiency.

*The organic concept is a forerunner of how an economic base can be given to idealistic concepts.* Even now, organic force is developing models to show how ecology can be blended into the marketplace and into daily living. It offers us a game plan—a personal action plan—that takes us beyond wringing our hands, and preaching. This organic force may very well be our best reason to be optimistic at this time.

One of the clearest pictures of where we should be heading is contained in a document called the "Blueprint for Survival" developed by *The Ecologist*, a British publication edited by Edward Goldsmith.

To succeed, says "Blueprint," we must formulate "a new philosophy of life, whose goals can be achieved without destroying the environment, and a precise and comprehensive program for bringing about the sort of society in which it can be implemented."

In developing a strategy for change to a stable society, "Blueprint" deals with concepts that have been continually discussed and developed by organic gardeners and farmers in this country—minimum disruption of ecological processes; maximum conservation of materials and energy; a social system in which the individual can enjoy, rather than feel constricted by, those conditions; systematic substitution of the most dangerous aspects of present technology with ones that cause minimum disturbance to natural processes; decentralization of economy at all levels; and the formation of communities small enough to be reasonably self-regulating and self-supporting.

Based on the information we have received as editors of *Organic Gardening and Farming* magazine, we believe—as do the creators of "Blueprint for Survival"—that it is possible to change from an expansionist, overchemicalized, wasteful society to a stable, more natural, more personal, more organic society without loss of jobs, with less starvation and malnutrition than now plagues us, and without an increase in real expenditure. In discussing the real-life applications of these concepts, one immediately must think of how they relate to food production—land use, fertilizer use, pesticide use, corporate farming versus family farming, food distribution, marketing and, not incidentally, disposal of garbage and sewage sludge.

And these considerations lead right back to the backyard garden, the small farm, the compost heap, the organic foods, the neighborhood grocery store. All have become symbols of the new American Revolution.

### Social Significance of Organic Foods

A whole generation of Americans has grown up without any personal communication with the producers of their foods. The supermarket checkout clerk has been the closest human contact, as the food goes from shelf to freezer into shopping basket for transfer to pantry and freezer. The television screen gives the clearest picture of where the food comes from before it goes to the supermarket.

Organic foods have the ability to change this. The consumer can identify the farmer, and the farmer can identify the consumer. The human identity of each can surface and interrelate. The money spent for food—for its production and consumption—can become a real economic force for social and environmental objectives. Part of each dollar spent for organically grown foods should mean

1. less money for pesticides and artificial fertilizers—and less chemical pollutants in the environment;
2. more money for the farmers and farmworkers who make their homes on the land—thus more profits when land is used for producing food instead of for residential and industrial development;
3. more jobs with adequate compensation on farms growing crops

by labor-intensive organic methods—thus less forced migration to city ghettos;

4. more economic incentive to bring composted organic wastes from the city onto farmland where it builds up humus content;

5. More economic support to the *small* entrepreneur—the family farmer, the mama-and-papa grocery store, and the *local* brand name;

6. more demand for personal services and less for mass-distributed, environmentally hazardous products—that means more money for the person who is educated to advise on how to recycle wastes back to the land and less for the chemical fertilizer salesman;

7. more incentive to have local farmers supply the local market.

Today, the simple ideas have taken on a new aura of social significance. Organic methods in agriculture put an economic base under city planners' dreams of open spaces around urban areas. For city people to get high-quality, inexpensive foods, city people need country people on small farms. A good many of those farms should be close to the city. Every city in America faces a crisis with garbage disposal; local farms know that garbage can be useful once it's returned safely to the soil.

*When you buy organic foods, you should be aware that you are using your food dollar to encourage change in American agricultural methods.* For years, American farmers have been led to believe that food is no different from any other product. Churn it out in assembly-line fashion as fast and as cheaply and mechanically as you can. Just as we do with cars. Or envelopes. Or wigs. Or any other product.

But now you are taking your food dollar and you are dissenting. You are saying that all food is not equal in quality, regardless of how it is grown.

By your dissent, by your desire to buy organic foods, you are leading a consumer food revolution. When you buy organically grown foods produced by a family farmer who is not supposed to be able to make a living on the land, you become an organic force helping to reverse a trend that has driven people off the land and made farming the profession of an older generation. By buying organic foods at a mama-and-papa neighborhood store that is not supposed to be able to compete with supermarket chains, you are helping to change the makeup of America.

We are only beginning to understand how much force—how much political and economic clout—can be generated by the rapid expansion of the organic market. But that clout will only develop if the market develops for the reasons cited. If that market develops a corporate structure of its own, then the social and environmental benefits will diminish—even if the quality of the food harvested is maintained. [Unfortunately there are indications that this is starting to happen: (1) the presence of natural food distributors on the New York Stock Exchange, (2) the purchase of natural food distributors by major national companies, and (3) the mimicking of "natural" foods by the food industry. —Ed.]

What we are trying to show is that there is a rapidly growing trend for food grown with more manpower. Sure, we know that eggs produced by chickens allowed to run around will cost more than eggs produced by chickens that are penned four to a cage. But those organically produced eggs will get and deserve a higher price. Part of that higher price will provide decent wages for more farm-workers—right now those farmworkers of the future may be sweating it out in some urban tenement.

Does that mean organic foods are always higher priced and only for the upper middle class? We think not. Many environmentalists, including organic gardeners, have been worried that industry is trying to isolate them from labor and the poor. Statements that organic foods are only for the rich, that antipollution costs force layoffs and use funds needed to eliminate poverty are seen as an effort to set workers and the poor against environmentalists.

Groups like the Rural Advancement Fund (see Chapter 9) have shown the relevance of the organic idea to small farmers and rural poor, and the necessity to use organic foods as a way to relate the farmer to the consumer, and city problems with farm problems (see Chapter 11). Organic foods can help to solidify the coalition between the needs of sharecroppers and small farmers and concern for the right use of land.

## A Rebirth in the Ability to Communicate

It is most important to dwell on the food aspect of the organic force, since it is so tangible. Food is the great link between problems in the city and problems on the farm. Therefore, food becomes the great force for communication between the people who consume the foods and the people who grow the foods. Bigness in food production methods and marketing has come to mean anonymity—and a breakdown in communications. Smallness in food production and marketing has come to mean personal identity—and a rebirth in communications.

The great strength in organic force is how it has led to communication between individuals and groups who would not recognize each other were it not for their common interest in the quality of foods. Organic force comes from the black sharecropper in Georgia and from the volunteer in a Park Avenue office for consumer action, from an ecology center in Washington, D.C., from a dormitory group in Ann Arbor, from a senator's office on the Hill, from a center for community change in an urban ghetto, from health food customers and supermarket buyers, and from millions and millions of poor and rich alike. We are going through a time when dissatisfaction and unrest are widely evident. All of us, conscious of the phenomenon of organic force, should do everything we can to translate that dissatisfaction into something positive—something that will truly improve the quality of life while improving the quality of the environment. And a major part of the effort must be to show how widespread is this organic force.

### New Jobs, New Profits in Mama-and-Papa Stores, Farms, Services

About once a year, for the last three years, several of us have made a pilgrimage to the Department of Agriculture in Washington, D.C. to establish a more direct contact between the organic farming community and the USDA.

During one of these expeditions I decided to ask the USDA to name an "organic" man within the department—someone who could become intimately aware of the needs of an organic family farmer and try to relate what was going on within the department to those needs. To make my point, I told the USDA official that there were now more than 800,000 subscribers to *Organic Gardening and Farming* magazine and it was fair to assume that in that number there were between 10,000 and 50,000 farmers who could benefit from USDA recognition of their existence.

I did not get a positive response to my request, since the official pointed out that the USDA served all farmers. The official, however, was most impressed with the figure of 800,000. "How," he asked, "could *Organic Gardening* magazine have 800,000 subscribers and be thriving, and *Look* magazine be out of business?"

And the answer to that question, I maintain, is an indication of why the organic force can supply an economic base to far more than a publishing company and its employees in Emmaus, Pennsylvania.

Writing in the *New Scientist*, the Honorable Reginald Prentice, Britain's minister of overseas development from 1967 to 1969, speculates on the inadequacy of conventional approaches to get 280 million more jobs. He advocates a drastic change, to be accomplished in three major steps: a shift to labor-intensive projects; greater emphasis on the rural sector of the economy; and a shift of emphasis in the education systems so that they are more closely geared to employment opportunities. What we refer to as organic force shows up in each area mentioned by Mr. Prentice.

> Every project should be assessed primarily in terms of the effect on employment. This will mean a movement away from sophisticated methods towards intermediate technology. It will mean a larger number of smaller projects and fewer prestige projects. . . . Urban problems can only be tackled successfully against a background of rural development. . . . It also means a much more urgent need for land reform in many countries—otherwise there is a danger that only the big farmers and land owners will benefit from these trends and that they may use mechanized methods which lead to further unemployment. The educational system has to move rapidly towards vocational training, especially for middle level jobs, and especially so as to offer training that is dovetailed to the needs of the rural areas.

More and more, as I speak to people concerned with improving the environment, I find successful efforts that are both environmentally and economically sound occur when the individuals involved *think personal and small!* When a project (like recycling newspapers) involves one town as opposed to an entire state, or one type of product or service as opposed to an entire range of products or services, I repeatedly find reasons for optimism. But when the development is characterized by a computerlike conglomerate philosophy, I see reasons for pessimism.

To return to agriculture, what seems to be increasingly evident now is that organic farming methods with their emphases on polyculture farming, the return of organic wastes to the soil, the prohibition of pesticides and high-nitrate fertilizers, and the use of labor-intensive agriculture offer the environment and the farmer a most worthwhile and potentially profitable alternative.

Organic farming means that pesticides are not used; but pesticides are used in conventional farming to prevent crop damage even though most people consider them environmentally hazardous. What does an organic farmer use to control pest damage? Many rely on biological controls. Some entomologists who have done research in biological controls have started their own consulting firms to help organic farmers control pests without pesticides (see Chapter 18).

Here again, we can see costs from an environmentally hazardous pesticide transferred to an environmentally sound pest control *service*. Even if the costs are the same, the benefits to the environment are obvious. And a pest control service providing income and jobs for humans replaces a pest control product requiring energy and material from our ecosystem.

Organic farming methods also forge the link between city and farm, since, to farm organically, you need to return organic wastes to the land. It has been suggested that farmers receive payments based on the amounts of organic wastes allowed to be brought to their lands. This subsidy makes good sense. We are already spending great sums of money to burn our city wastes, so it seems eminently economical to direct those dollars and the wastes to the farms. A subsidy like this certainly makes more sense than damming rivers out West to provide water for corporations to grow grain to further glut the market.

Organic farming can also be a prime force in building up the humus content of our soils instead of our waterways. Organic farmers—whether or not they benefit directly from such "waste" subsidies—can get paid for the extra costs of putting composted wastes on their land by getting higher prices from the consumers who want their food grown by such methods. What's more, the standards of organic farming can help keep industrial contaminants from ever getting into the municipal waste stream.

Robert Rodale, editor of *Organic Gardening and Farming*, has pointed out that while the organic method is well-shaped as an idea, it remains to be fleshed out as a technology. What makes conventional farm technology so powerful is its input of science. Vast sums have been spent on scientific research for the chemical energy-consuming agribusiness technology. Without that scientific input, farming of large areas with little use of manpower would hardly be practical.

By contrast, organic farm technology has been able to feed only on the scraps and remnants of conventional science. Farming with natural materials, and without poisons, has been considered old-fashioned by most scientific institutions, and hardly any effort has been expended on perfecting ways for families to make small farms more productive within the ecological, organic framework. Although organic farmers can use some parts of conventional farm technology (modern tractors, combines, manure-spreading equipment, and vehicles, for example), the general thrust of agribusiness has been away from natural farming. Therefore, as conventional farming became more mechanized and chemicalized, farmers inclined toward natural methods found themselves at more of a disadvantage.

What would happen if significant scientific efforts were directed toward creating a more effective organic farm technology? What if the small farmer gradually found himself offered machines, fertilizers, plants, and techniques that would enable him to produce marketable amounts of fresh foods, at a reasonable cost and without chemicals? And what if techniques were combined with restrictions on tax and subsidy benefits now enjoyed by large-scale agribusiness farmers? The answer to those questions is that a farm revolution would happen, accelerated by the growing consumer demand for organically-grown food. If small-scale farming without chemicals suddenly were made even more competitive and satisfying to both farmer and consumer than agribusiness and its plastic food, then the population of rural areas would stabilize, if not increase.

Improved small-farm technology is feasible. No tremendous breakthroughs are needed, only the fleshing-out of concepts which have already been outlined. Organic farming, therefore, can be made much more practical and competitive than it now is, with only moderate help in the way of improved science.

## The Potentials of Recycling and Composting

The waste disposal business is the third largest area of domestic government expense—following road-building and education. Most of the money goes for collection and transportation to the dump.

Recycling wastes—which all agree is the route to take—needs development. It needs consultants and technology. It needs special transportation equipment to convey wastes from where they are pollutants to where they can be used. It needs a mass transit system designed for wastes. It needs heavy-duty grinding equipment, and sophisticated sorters and classifiers. It may involve separators based on cryogenics, or it may involve the same technology that now is used in oil refineries; research projects indicate that equipment developed in one industry can be adapted for use in the waste-recycling industry. It needs land around urban areas on which to apply composted wastes.

Eloquent statements on recycling organic wastes have come out of Washington, so we must assume that officials have something else in mind besides dumping, burning, or burying. Research and world-wide operating experience show that the compost process offers a real potential to return some wastes back to the land in an economical, safe, and environmentally sound way. A small percentage of all the compost can be marketed as an income-producing soil conditioner, but vast amounts must be used to improve our soils instead of destroying our waters.

Unfortunately, the EPA evaluates composting only as a commercial venture in two recent reports: "American Composting Concepts" and "Composting of Municipal Wastes in the United States" (See Chapter 16). At a time when so much talk is given to recycling, we must develop a national policy to return organic wastes to the land. The composting process can help achieve this goal without a lot of fancy buildings or equipment. But composting—like any recycling system, if it's going to succeed—needs a commitment from our elected and appointed leaders. Any improvement in waste treatment will cost us money, but composting is the only process that will get wastes back where they belong!

Here are a few examples of how treated organic wastes and waste-water are being recycled back to the land.

1. *Los Angeles County*: All of the digested sewage solids—about 100 tons of dried solids per day—are being composted in windrows for about sixteen days before being turned over to a private firm for marketing as a soil conditioner-fertilizer.

2. *San Francisco*: A San Francisco environmental management firm, headed by the highly regarded public health authority Frank Stead, recommends that San Francisco refuse be composted and used to raise the land level in the Sacramento–San Joaquin Delta

region. The delta land, below river level, has been steadily sinking, and the compost could solve the problem.

3. *Yakima, Washington*: Beef herds around Yakima consume some 538 tons of hay grown on land fertilized only with applications of sludge and effluent from the city's sewage plant digesters.

4. *University Park, Pennsylvania*: The experiment begun in 1968 has now become a landmark development for using sewage sludge and effluent to revegetate strip-mined land as well as to fertilize crops. William Sopper of Penn State University continues to report outstanding results as the project continues. The Muskegon, Michigan, system to put effluent back on the land is also a model for other cities to emulate.

5. *Madison, Wisconsin*: With increasing speed, grinding is being recommended and used at municipal landfill sites for garbage. Under the leadership of Ed Duszynski, Madison has successfully pulverized refuse—a first step in any composting program—and the pretreatment is now showing up on more and more sites.

6. *New Brunswick, New Jersey*: New Jersey has the largest population per acre of any state in the union. Professor Charles Reed of Rutgers University is especially aware of the significance of his research to incorporate biodegradable wastes in the soil by the plow-furrow-cover (PFC) technique. With so little farmland remaining in the state, PFC offers some hope to get city officials plagued with the need for disposal sites to work with farmers in an effective waste-recycling program.

Those are just a few examples of the progress throughout the country. Much research is concerned with overcoming the potentially hazardous buildup of heavy metals in soils after heavy waste applications. The big advantage of using wastes on land is that toxic substances can't be ignored as they now are when they float down the streams to the sea.

This could very well be the year the land won back its wastes, the year the United States decided to make wastes a national resource, thus contributing positively rather than negatively to our gross national product. More companies are entering the field; more legislation is beginning to put the bite on bad practices; more consultants in sanitary engineering are becoming comfortable with the land application concept; more professionals in agriculture and public health are beginning to see the connection between farming and wastes; and, most important of all, more Americans are pressuring for waste management which genuinely relates to environmental limits of resource and energy. The fact is that for the first time in U.S. history we're creating a land-waste relationship.

I think we should not try to belabor the point that the marketplace will, through our good wishes, our prayers, or our threats or seminars, produce an endless array of products *because* of the speeches and writings. I do think we should cling, with whatever optimism possible, to the idea that the same economic forces that brought us environmentally bad products will be the ones to get them out of the marketplace. And that means, one way or another, that it will be more profitable to produce those environmentally sound products.

The organic force that is surfacing in so many different areas of our society has the elements to revolutionize the marketplace for the benefit of us all. How great an impact it will make is the big unanswered question of the moment.

# 14

# POLYCULTURE FARMING
# IN THE CITIES

*Warren Pierce*

Human history more and more becomes a race
between education and catastrophe.
—H. G. WELLS

Although there is an increasing awareness of society's ecological
problems, more often than not the efforts of environmental groups
and schools have been directed toward "teaching" rather than
participating in the processes of our natural environment. In many
instances traditional courses in the sciences, especially biology, are
simply not able to present the complete concept of a dynamic
living environment with just field trips and classroom projects
presented as static events "out there." This chapter is meant to
provide some preliminary ideas and instructional material for
those who are looking for viable alternatives in the areas of natural
or environmental science.

"Organic" horticulture methods have the advantage that techniques of fertilizing plants and controlling pests are based primarily
on the use of natural processes rather than artificial or toxic

---

Warren Pierce works with the Santa Barbara Urban Farm Project.

chemicals.[1] In fact, the diverse organic vegetable garden with its soil and animal communities can be thought of as a cultivated ecosystem that is relatively simple (compared to natural ecosystems) and easily learned. This cultivated ecosystem is also more rapidly changing than wilderness situations, making it easier and more practical to learn the *dynamic* situations in nature. In other words, by maintaining a vegetable garden with its ecological potential rather than with artificial chemicals, we can come to understand the workings of ecological processes as well as the practical techniques of growing food.

For example, we not only become directly involved with the subjects of plant science, climate, biology, geology, etc., but also with various natural processes. These include direct involvement with (1) *recycling and soil biology*, by using rock minerals and composted wastes to fertilize the soil; (2) *ecological relationships*, by relying on natural enemies and diverse plant arrangements rather than poisons to control pests; (3) *the dynamics of a living community or ecosystem*, by recording and observing the growth of the garden during the seasons; and (4) *food and nutrition*, by tending food plants from seedling to meal table. Finally, by replacing chemical energy with human energy, an organic vegetable garden, especially a large one, becomes not only a dynamic and participatory tool, but also a highly cooperative social unit in which people of all ages can work together toward practical ends.

The value of the organic garden can be extended by including small livestock, insectaries, apiaries, worm cultures, grain crops, etc., plus small-scale alternative energy devices such as methane digesters and solar collectors. In this way, the organic garden becomes the heart of a more complex polyculture system that offers one even more opportunities to participate in integrated natural processes.

The first section of this chapter deals with urban self-sufficiency and examines briefly the concepts of self-sufficiency especially as it relates to the urban household and to alternative sources of energy. The reason for this is that most of us live in the cities and should be aware of our households in terms of energy and material inputs and wastes.

The second section describes earth cycles and energy flow, and can be the basis of an organic or polyculture garden.

The third section deals with an actual garden workshop and attempts to outline briefly the problems and resources needed. There is also a suggested mixed-husbandry method with alternatives.

## Urban Self-Sufficiency

Man's use of the wind, the sun, and other natural power sources is ancient indeed. However, the concept of integrating a collection of nonpolluting alternative energy sources to bring about self-sufficiency for large groups of people in a technological culture is relatively new and raises questions about many of our present attitudes toward energy and "convenience." For example, in a city or household, what types of energy-consuming devices are most frequently used? Why are they used? What is the source of their energy? Why are only a few sources of energy actually used when so many are potentially available? (See Table I.)

The question is how solar, wind, geothermal, and other unique sources of energy can be substituted for heat engines, appliances, and nuclear fission reactors. The following list suggests possible procedures and potential substitutes for presently used resources that we might consider.

I. Recycling
   A. Sewage
      1. Methods of recycling *water* for agricultural and domestic use.
      2. *Anaerobic digester operations* that break down municipal sewage (and organic wastes) to produce methane fuel gas and a liquid "sludge" fertilizer for a variety of purposes. The pumping and transportation of natural gas is reduced, as is the use of chemical fertilizers.
   B. Solid Wastes
      1. Procedures to assure conservation of *paper, glass, metal*, and *plastics* with or without the assumption that the city can provide adequate resources for local recycling.
      2. Methods of composting *organic wastes* (leaves, weeds, garbage, manure, etc.) from backyard, municipal, or large farm operations. Ways to encourage people and organizations to separate organic wastes. Techniques for distributing compost to home gardeners and farmers. Simple

TABLE I

SOME SOURCES OF ENERGY AND POWER USED
FREQUENTLY AND INFREQUENTLY IN OUR CULTURE

I. *Frequently Used Sources of Power and Energy*

| SOURCE | TYPE | PURPOSE |
|---|---|---|
| 1. *Sunlight*<br>photosynthesis<br>fossil fuel ...... { | natural gas ........<br>gasoline ...........<br>coal ............. | stoves, water heaters<br>combustion |
| 2. *Sunlight*<br>evaporation<br>precipitation<br>run-off ...... | water power ....... | electricity |
| 3. *Cosmic-Geologic*<br>mining<br>splitting atoms | nuclear fission ....... | |

II. *Infrequently Used Sources of Power and Energy*

| | SOURCE | TYPE | PURPOSE |
|---|---|---|---|
| 1. | *Sunlight* ....... | Greenhouses, solar water heaters, solar ovens, solar furnaces, desalinaters | Cultivation, heating, cooking, electricity |
| 2. | *Wind* ......... | Water pumps, ....... grinders, sailing ships, wind generators | Transportation, electricity, irrigation |
| 3. | *Methane* ....... | Stoves, water heaters, combustion engines .. | Heating, cooking, electricity |
| 4. | *Wood* ......... | Fireplaces, stoves .... | Heating, cooking |
| 5. | *Waves* ........ | Generates desalination | Electricity |
| 6. | *Geothermal power* | | |
| 7. | Alcohol ....... | Engines ............ | |
| 8. | Tides ............................. | | Electricity |
| 9. | Nuclear fusion ....................... | | |
| 10. | Magnetohydrodynamic ................ | | |
| 11. | Hydrogen ...... | Batteries, engines, fuel cells | |

methods of recycling organic wastes at homes (rabbits, worm cultures, composting, etc.).

II. Alternative Energy[2]

  A. Solar Power

    1. *Solar absorbers* with heat storage tanks for hot water or air, in place of gas-powered water or air heaters. Practicality of rooftop solar water heaters, with reference to practices in other parts of the world. Methods of constructing simple solar water heating devices.

    2. Feasibility of *solar ovens* projecting from the kitchen outside, to be used in conjunction with stoves burning methane gas generated by the city's waste disposal plant.

  B. Wind Power

    1. Design of *wind/water pumps* and methods of construction.

    2. Small and efficient *wind generators* mounted on rooftops to run several appliances plus a small generator for auxiliary power.

  C. *Tidal and Geothermal Power* for electricity

Finally, examples of recycling and alternative power sources can be extended to concepts of "efficiency" and energy consumption in agriculture (see Chapters 6 and 16). This can be done in several ways by learning the following: (1) methods of recycling municipal, agricultural, and animal wastes (especially feed lots) in order to reduce the use of chemical fertilizers; (2) ways of reducing the use of toxic pesticides by substituting biological control techniques, diverse cropping methods, and strains of domestic plants and animals resistant to disease and pests; and (3) the consequences of farms that are polycultural rather than monocultural and that generate much of their power from local sources of energy (wind, sun, and organic matter). One possibility would be to plan large agricultural and wilderness areas adjacent to or flowing throughout the city. Most food would then have to travel only a short distance to consumer markets; townspeople could readily hike into nearby park areas. Unique wildlife corridors extending through a variety of habitats could preserve many different plants and animals of different environmental zones. Adjacent to the wildlife corridors might be the city's recycling and energy centers, while business and residential areas could be integrated with dairies, fish farms, croplands, and orchards. Special transportation facilities for returning municipal sewage, compost, and other organic wastes to nearby farmlands, together with a system of bike- and footpaths could form most of the major connections in this urban-rural community.

There are, of course, a variety of other ways in which ecological processes can be described in terms of urban problems and planning. These then can be applied to the natural or cultivated situation.

## Earth Cycles and Flow of Energy

With personal experience the earth cycles and the flow of energy in ecosystems lend themselves well to the organic garden workshop. The processes of photosynthesis and respiration can be dealt with specifically in terms of the recycling of carbon dioxide and oxygen and generally as they relate to the energy metabolism of living things. The compost pile demonstrates the nitrogen and mineral cycles as well as details of soil biology and plant nutrition. The water cycle is related to basic principles of climate and weather. Finally, the distribution of food and energy in natural ecosystems can be applied to the garden with direct examples. With this basic knowledge we can begin to make observations in the garden as to the behavior and feeding habits of different insects and other animals, until food chains or even ecological "pyramids" (see Figure 1) can be constructed[3] for any particular garden ecosystem.

### PHOTOSYNTHESIS AND RESPIRATION

Living plants are composed largely of carbon, hydrogen, and oxygen. The carbon source for plants is carbon dioxide. Hydrogen comes from the splitting of water in a process called photosynthesis, which utilizes carbon dioxide, water, and light chlorophyll at a temperature suitable for making simple sugars, new water, some waste heat, and oxygen. Some of the oxygen is used by the plant for respiration; the excess becomes available to all other creatures of the earth that need oxygen. A general description of photosynthesis, respiration, and energy utilization by living things is summarized in Table II. The reactions of photosynthesis and

TABLE II

AN OUTLINE DESCRIPTION OF PHOTOSYNTHESIS, RESPIRATION, AND ENERGY UTILIZATION BY LIVING THINGS IN THE GARDEN

I. Photosynthesis and Respiration as Opposing Reactions

   A.  Photosynthesis (an energy-*producing* reaction)

$$CO_2 + H_2O \xrightarrow{\text{light, chlorophyll, suitable temperature}} Sugar + H_2O + O_2 + Heat$$

$\begin{bmatrix}\text{carbon} \\ \text{dioxide}\end{bmatrix}\begin{bmatrix}\text{water}\end{bmatrix}$

   B.  Respiration: All living cells respire 24 hours a day. There are two kinds of respiration: in one, energy is yielded with oxygen (aerobic) and in the other, energy is yielded without oxygen (anaerobic). In this discussion aerobic respiration is the main concern.* The ingredients for aerobic respiration are synthesized in photosynthesis—simple sugars and oxygen. The simple sugars may be immediately used by the plant for respiration, but the photosynthetic cell usually makes a great deal more than it needs. Some sugar is also converted to starch for storage. When animals eat plants they obtain simple sugars and starch for maintaining *their* respiration.

*Aerobic Respiration*

$$SUGAR + O_2 \xrightarrow{\text{suitable temperature}} ENERGY + CO_2 + H_2O + Heat$$

   Frequently with outside nutrients the sugar is converted into the building blocks for protein, fats, carbohydrates, nucleic acids, and vitamins.

*Alternate Course for Products of Photosynthesis*

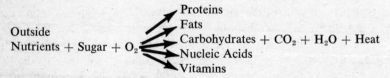

Outside Nutrients + Sugar + $O_2$ → Proteins / Fats / Carbohydrates + $CO_2$ + $H_2O$ + Heat / Nucleic Acids / Vitamins

---

   * In the aerobic process, sugar and oxygen in a living cell make enormous amounts of energy available to the cell while carbon dioxide, water, and heat are the end products.

Therefore, the simple sugar synthesized in photosynthesis may produce energy for building new materials or may produce the actual building materials themselves.

*Aerobic Respiration: Summarized*

Sugar + $O_2$
— Energy ——→ Proteins + $CO_2$ + $H_2O$ + Heat
— Building Materials (amino & fatty acids, glycerol, glucose, etc.)
→ Proteins
Fats
Carbohydrates
Nucleic Acids
Vitamins
+ $CO_2$ + $H_2O$ + Heat

The above equations illustrate only *summary* reactions. It should be clear that the biochemical workings of the cell are extremely complex, and that the sugar and oxygen do not actually react together.

II. Energy Utilization by Living Things

    A. Ultimately only two sources of energy are available to living things: solar energy and the chemical energy of organic material. Organisms that use sunlight are called phototrophs, those that use chemical energy are called chemotrophs.

        1. *Phototrophs* (energy from sunlight)
            algae, higher plants,
            some bacteria (green and purple)
        2. *Chemotrophs* (energy from chemicals)
            *Lithotrophs* (energy from inorganic chemicals)
            some bacteria
            *Organotrophs* (energy from organic chemicals)
            animals, fungi, most bacteria

    B. Other popular classifications

        1. *Autotrophs* (energy from nonliving sources, carbon from $CO_2$)
            *Photosynthetic* (energy from light)
            *Chemosynthetic* (energy from inorganic chemicals)
        2. *Heterotrophs* (energy and carbon from organic chemicals)

respiration can be abstracted and drawn schematically as the oxygen and carbon cycle (see Figure 2).

## THE NITROGEN AND MINERAL CYCLES

Other elements are essential to the life of the plant, and these are found in the soil in varying amounts that affect the growth, development, and future offspring of the plant. The major elements are nitrogen, phosphorus, potassium, calcium, magnesium, iron, sulfur, sodium, chloride, and copper. Trace elements (manganese, zinc, cobalt, iodine, boron, molybdenum, vanadium, silicon,

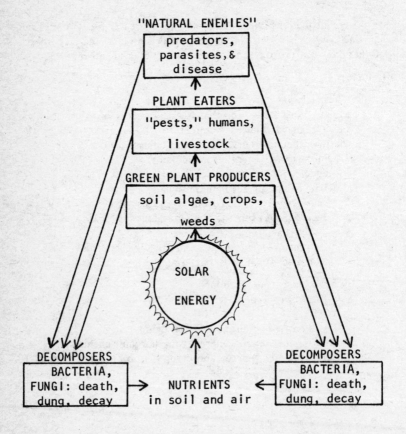

FIGURE 1. Basic food or "trophic" pyramid of a farm ecosystem.

fluorine, chromium, selenium, and tin) are found in minute quantities in living systems. Some are essential to the plant, while others may be essential only to those animals that feed on the plants.

In gardening and farming practices, important elements may become depleted in the soil and have to be replenished. Following is a list of good natural sources for these elements.

*Nitrogen:* Blood meal, manures, legumes, grass clippings, and wet garbage; necessary for protein, nucleic acids, and vitamins; makes healthy leafy plants.

*Potassium:* Citrus and banana peels, wood ashes, manures, wet garbage, kelp, granite rock; important cation; makes strong healthy stems.

*Phosphorus:* Rock phosphate, wet garbage, bone meal; essential for

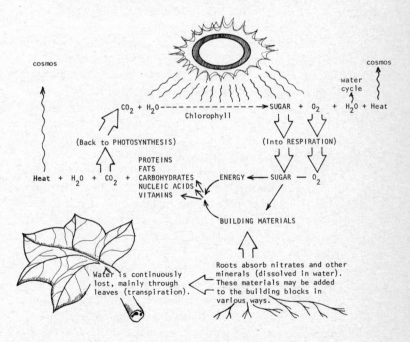

FIGURE 2. Carbon and oxygen cycles as illustrated by processes in the garden.

biochemical synthesis and energy transfer; produces healthy roots and fruit development.

*Calcium:* Egg shells, sea shells, limestone, dolomite, bone meal; corrects soil that is too acid; works in certain enzyme systems.

*Magnesium:* Dolomite, bone meal, all chlorophyll containing parts of plants; in chlorophyll and certain enzyme systems.

*Sulfur, Iron, Sodium, Chloride, Copper, and the Trace Elements:* Important in the functioning of various enzymes and in the exchange of various materials across barriers of the cell; may be the required element for a certain pigment. If one is worried about not having enough of these elements, alfalfa, kelp, and leaf litter are good sources.

When making and using compost for gardening and farming, a maintenance of sufficient soil nitrogen is the principal problem. If the compost is made from diverse materials and the nitrogen is sufficient, the other elements will most likely be sufficient also.

Nitrogen exists in several different forms in nature. In living organisms, decaying organisms, and fresh manures it exists mainly in the form of amino acids that make up proteins, and in nitrogenous bases of nucleic acids. Nitrogenous waste products of a variety of organisms are synthesized directly or indirectly from the amino acids and bases, and these include urea, ammonia, ammonium ion, nitrite, nitrate, and nitrogen gas. As nitrogen gas escapes to the atmosphere (the reservoir for the nitrogen cycle), it may become available to food chains again through nitrogen fixation performed by certain bacteria and blue-green algae (see Table III), and through industrial fixation (see Table IV). To date, denitrification is out of balance by about nine million metric tons due to the fertilizer industry. The effects of this on the earth's nitrogen cycle are still unclear, but nevertheless, the equilibrium has been disrupted.[4]

The soil contains nitrogen-fixing bacteria, and some live symbiotically with the more complex plant forms, namely legumes like alfalfa, peas, beans, vetch, clover, lupines, and peanuts. Many different kinds of legumes are cultivated and plowed into the soil in order to restore nitrogen used by a previous crop.

The flow of nitrogen is illustrated in Figure 3. From this diagram, it should be obvious that bacteria carry out most of the steps in the nitrogen cycle, and that the waste products of one organism

TABLE III

MICRO-ORGANISMS ASSOCIATED WITH THE NITROGEN CYCLE

| Representative Organisms | Type of Metabolism | Material Utilized | Descriptive Term of Process |
|---|---|---|---|
| *Pseudomonas* | autotroph, anaerobe | nitrites & nitrates | denitrification |
| *Thiobacillus* | autotroph, anaerobe | nitrites & nitrates | denitrification |
| *Azotobacter* | heterotroph, aerobe | nitrogen gas | nitrogen fixation |
| *Clostridium* | heterotroph, anaerobe | nitrogen gas | nitrogen fixation |
| *Nostoc* | photosynthetic, anaerobe | nitrogen gas | nitrogen fixation |
| *Rhodospirillum* | photosynthetic, anaerobe | nitrogen gas | nitrogen fixation |
| *Rhizobium* | heterotroph, aerobe (legume association) | nitrogen gas | nitrogen fixation |
| Many species of fungi, bacteria, & actinomycetes | heterotrophs, aerobes, & anaerobes | protein, amino acids, & nucleic acid | ammonification |
| *Nitrosomonas* | autotroph, aerobe | ammonia | nitrification |
| *Nitrosococcus* | autotroph, aerobe | ammonia | nitrification |
| *Nitrobacter* | autotroph, aerobe | ammonia | nitrification |

TABLE IV

COMPARATIVE FIXATION AND DENITRIFICATION
VALUES OF GLOBAL NITROGEN

| | |
|---|---|
| Industry (chemical fertilizer fixation) | 30 million metric tons |
| Biological Fixation | 62 million metric tons |
| | |
| Total Fixation | 92 million metric tons |
| Biological Denitrification | 83 million metric tons |
| | |
| | 9 million metric ton imbalance |

SOURCE: "The Biosphere," *Scientific American*, September 1970.

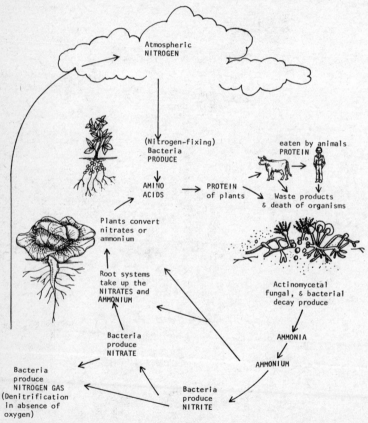

FIGURE 3. Simple diagram of the nitrogen cycle.

become the food and/or energy of the next. In addition, however, many species of fungi and actinomycetes play important roles in decomposition of organic materials. They put forth enzymes that break down cellulose, lignin, hair, nails, and many other compounds.

## COMPOSTING: SIMULATING NATURAL CYCLES

In making compost, man merely employs organisms that can live at high temperatures of 140°–160°F. At these temperatures, growth and reproduction are very rapid, so that material decomposes into available nitrates and plant nutrients at a more efficient rate than in nature. It might be looked upon as a "domesticated nitrogen cycle." One composting method requires only two weeks.[5]

In general, a compost pile must initially be four to five feet high in order to provide insulation and generate heat. A pile too high will compress too much and create a lack of oxygen. One-quarter to one-third of the pile should be legumes or manures (much less in the case of bird manures, which are high in nitrogen) or a combination of the two; two-thirds to three-fourths should be a mixture of leaves, garbage, grass clippings, and possibly kelp. These fractional differences bring about a desired C:N ratio (carbon to nitrogen) of about 30:1. This ratio ensures desired temperatures and low nitrogen loss. Also, with every eight inches of material in height, it is desirable to put one-quarter inch of soil to trap the ammonia vapors to the soil particles. This will lower nitrogen loss considerably from any method used.

Each time the pile is turned, large amounts of carbon dioxide and water will escape into the atmosphere, and so the pile will lose weight and volume. At the end of the composting time, 20 to 60 percent of the volume may be lost, and up to 50 percent of the weight. Since so much carbon is lost, a new C:N ratio will be established. This is normally around 12:1, which is suitable for agricultural purposes. Small amounts of rock fertilizers and wood ash may be added to the pile for extra minerals. Limestone, dolomite, and bone meal, which have a high $CaCO_3$ rating, should not be added to the quick compost methods since they may increase the ammonia production which will mean greater nitrogen loss and more flies attracted to the area.[6] These calcium sources are best applied to the soil, or to the long-term layered composts listed in Appendix IV of this chapter. Bird manure,[7] blood meal, cottonseed meal, and fish meal,[8] which are very high in nitrogen, should be sprinkled in lightly as a supplemental nitrogen source if the pile does not heat up. If no mammalian manure or legumes are available, approximately one-tenth of the pile should be composed of the above high-nitrogen sources and nine-tenths of leaves, garbage, and so forth. The various meals, all manures, and legumes contain enough nitrogen to activate vigorous growth and reproduction of compost organisms. When sufficient nitrogen is supplied, the pile should stay hot for many days, the exact time depending on the method. Maintenance of a hot compost pile is accomplished by keeping the pile moist to the consistency of a wrung-out sponge and by turning the pile to introduce fresh air. Table V shows how the length of composting time can be shortened by increasing the number of turnings.

In terms of ammoniacal nitrogen and a properly made compost, Table VI illustrates the methods that will give the highest values for ammonia (based on my personal experience with different composting methods). For use on young leafy vegetables the compost should have high ammonia value to promote a fairly rapid and uniform growth rate. Nitrate promotes growth too, but it is slower because it takes more energy on the part of the plant to convert the nitrate back to ammonia and then into protein. Old compost may have most of the original ammonia formed in the process converted to nitrates. This means that plants obtaining nitrogen from old compost may grow slowly, be eaten down by insects, become stunted, and even taste bad if they do reach harvest stage. The nitrates are also leached from the soil faster than ammonium salts, which is another reason why old compost may give poor results. When nitrates are present in large amounts in the soil (usually from adding too much chemical fertilizer), the plant many absorb too much and store them in various ways, possibly making the vegetables unhealthy for food use. You can make your own comparison with different composting methods.

For the highest ammonia values an anaerobic methane digester should be constructed (there can be no conversion of ammonia to nitrates without oxygen). Take a five-gallon bucket, fill it with garbage and manure, fill it to the brim with water placing a loose-

TABLE V

SOME COMPOSTING CONSIDERATIONS AND
LENGTHS OF COMPOST TIME

| *No. of turnings* | *Ready in* |
|---|---|
| zero | 4–6 months or longer |
| 1–2 | 2–3 months |
| 4–5 | 2 weeks if shredded |

| | |
|---|---|
| Pile will not heat up | 1. Not enough nitrogen to make the initial C:N ratio below 36:1. This means too much sawdust, wood-chips, paper, or straw which have very high C:N ratios because of their high cellulose and lignin content. Add more of a good nitrogen source. |
| Strong smell of ammonia | 2. Too much water initially applied. This suffocates the aerobic organisms and the cool-living anaerobes take over, producing ammonia and putrid smells. This is remedied by frequent turning, or layering dirt into the pile and making it a long-term pile.<br>3. C:N ratio is below 30:1 in the initial stage. Too much nitrogen was added in the beginning. This can be remedied by adding old leaves, straw, sawdust, or shredded paper in small amounts.<br>4. Too much limestone or other element high in calcium carbonate added. Difficult to remedy, but add acid leaf litter and wet garbage. Next time add the calcium to the soil. (Applies to 2-week method only.) |
| Indications of finished compost | 5. Ammonia smell is absent.<br>6. The temperature has completely cooled down.<br>7. The compost is crumbly, dark, and sweet-smelling.<br>8. At least three species of arthropods are present. Examples are both the sow and pill bug, ground beetles (carabids), and centipedes. |
| Indications of semi-finished compost that can best finish up in the soil | 9. Slight smell of ammonia.<br>10. The temperature has started to decline but steam still comes off. (About the 12th day of a well-made 2-week compost.)<br>11. Possibly one or two species of "sow bugs," beetles, etc. (about the 4th month of a 6-month pile). |

TABLE VI

AMMONIACAL NITROGEN VALUES OF
VARIOUS COMPOSTING METHODS

*With soil layers*
| | | |
|---|---|---|
| 1. | No turnings, used after 4 months | Highest NH₄ (ammonium) |
| 2. | No turnings, used after 6 months | |
| 3. | One turning, used after 3 months | |
| 4. | Two turnings, used after 2 months | |
| 5. | Four to five turnings, used after 2 weeks | |

*No soil layers*
| | | |
|---|---|---|
| 6. | Four to five turnings, used after 2 weeks | Lowest NH₄ (ammonium) |

fitting lid on top, and wait six months. The dark, bubbly tea that forms can be applied in various dilutions to various plants. The leftover can be buried or thrown into a compost pile. Simple methane digesters can also be built from easily available materials such as inner tubes,[9] discarded containers and reinforced concrete, or oil drums.[10] Digesters produce methane gas for cooking use in addition to a fertilizer tea.

## WATER CYCLE

Water is the major substance of all living cells, being the source of hydrogen in the chemistry of life. In studying the water cycle one should realize that the water one drinks has been recycled continuously for millions of years; however, the amount is finite, as only so much water is distilled by the sun through surface evaporation from the waters and soils of the earth. Animals sweat or perspire, emitting water from their bodies, and plants transpire most of their water through pores in the leaves, thus returning a little more water into the cycle. Figure 4 illustrates the basic water cycle. From this can be seen how details of the water cycle relate back to weather and climate.

## ENERGY FLOW

Biological materials like oxygen, carbon, nitrogen, water, and minerals eventually find their way into the compost pile or even the decomposing litter of a forest floor. Organic molecules and

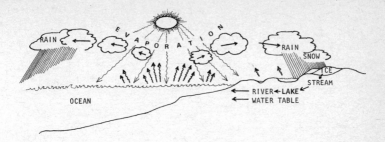

FIGURE 4. Basic water cycle and its relation to weather patterns.

oxygen are taken into the compost pile and large amounts of carbon dioxide, water vapor, and heat, as well as a little ammonia, are given off into the atmosphere. Temporarily left behind are substances like lignin and cellulose that do not break down readily, usable nitrates and other minerals, plus a stable bacterial, fungal, and actinomycetal population mixed with worms, insects, centipedes, and isopods. These substances, collectively called compost, are in time claimed by succeeding generations of plants and animals, the atmosphere, the hydrosphere, and the lithosphere. Thus cycles are continuously going on for these materials.

Energy flows in very intricate patterns in a community of organisms. About 10 to 20 percent of the original energy is retained with each transfer. In the hypothetical energy flow diagram (Figure 5), an enormous amount of energy from sunlight strikes the surface of the earth and the plants; but only a fraction (mainly blue and red wavelengths) is actually fixed by chlorophyll in the photosynthetic process. In this example, only one-hundredth of the light energy striking the surface is converted into chemical energy (or fixed in photosynthesis). For the entire earth, probably 1/4000 of the solar energy would go into photosynthesis because of large unproductive areas of the world like the poles, mountains, middle-ocean areas, and deserts. The great mass of solar energy goes into reflection, heating the earth, and causing evaporation, wind, waves, and other currents.

In Figure 5, the plants convert the 2,200 units of sunlight fixed by chlorophyll into 1,000 units of net plant production (energy stored in starch, cellulose, and other organic molecules); this conversion is highly efficient in that it captures 45 percent of the

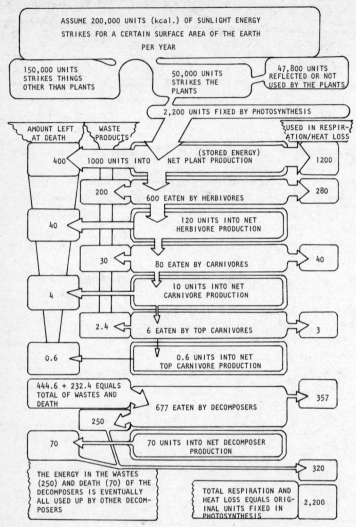

ASSUME 200,000 UNITS (kcal.) OF SUNLIGHT ENERGY
STRIKES FOR A CERTAIN SURFACE AREA OF THE EARTH
PER YEAR

150,000 UNITS
STRIKES THINGS
OTHER THAN PLANTS

50,000 UNITS
STRIKES THE
PLANTS

47,800 UNITS
REFLECTED OR NOT
USED BY THE PLANTS

2,200 UNITS FIXED BY PHOTOSYNTHESIS

AMOUNT LEFT
AT DEATH

WASTE
PRODUCTS

USED IN RESPIR-
ATION/HEAT LOSS

400    1000 UNITS INTO    (STORED ENERGY)
NET PLANT PRODUCTION    1200

200    600 EATEN BY HERBIVORES    280

40    120 UNITS INTO NET
HERBIVORE PRODUCTION

30    80 EATEN BY CARNIVORES    40

4    10 UNITS INTO NET
CARNIVORE PRODUCTION

2.4    6 EATEN BY TOP CARNIVORES    3

0.6    0.6 UNITS INTO NET
TOP CARNIVORE PRODUCTION

444.6 + 232.4 EQUALS
TOTAL OF WASTES AND
DEATH    677 EATEN BY DECOMPOSERS    357

250

70    70 UNITS INTO NET DECOMPOSER
PRODUCTION

320

THE ENERGY IN THE WASTES
(250) AND DEATH (70) OF THE
DECOMPOSERS IS EVENTUALLY
ALL USED UP BY OTHER DECOM-
POSERS

TOTAL RESPIRATION AND
HEAT LOSS EQUALS ORIG-
INAL UNITS FIXED IN
PHOTOSYNTHESIS    2,200

FIGURE 5. Hypothetical energy flow diagram of a natural ecosystem.

energy input. Plants store some of this material until they die.
Some disappears as leaf litter and dead branches or roots to the
decomposers (bacteria, fungi, actinomycetes); and the rest is
grazed upon by the herbivores (various insects, rodents, deer,

cattle, bears, men, pigs, and fish, depending on the type of community). In time, all the energy of these particular plants will dissipate to the universe, possibly to be recycled in some other cosmological year. The transfer of energy to the decomposers or the herbivores operates at a rate of efficiency of about 10 to 20 percent. The transfer to carnivores that feed on the herbivores is again about 10 to 20 percent. The transfer to top carnivores that feed on other carnivores is again about 10 to 20 percent efficient, as is each successive transfer to another carnivore or to a decomposer.[11]

In order to gain net production, each trophic level other than that of plants must digest and respire, and this is where the major energy loss occurs. Some of the digested food passes and does not become absorbed; the decomposers work on this portion. The energy from respiration goes into the making of digestive enzymes and the work of chewing, digestion, chasing after prey, and general body maintenance, including the maintenance of body temperature.

In conclusion, tremendous amounts of irretrievable energy are lost every day in the form of heat to the universe. Only a small fraction comes from biological systems even though the transfers operate at somewhat low efficiency levels. The 2,200 units of energy fixed in photosynthesis (see Figure 5) are all eventually consumed by respiration and heat loss in the chemical reactions so that the total fixed energy equals the total energy used in the flow diagram.

### Garden Workshop

#### MATERIALS

1. Adequate tool storage area.
2. Transplanting tools: hand shovels, spatulas, seed flats, pots, strainers of $\frac{1}{8}$-inch and $\frac{1}{16}$-inch size, wheelbarrows, and protected area for raising seedlings (lath house or greenhouse).
3. Transplanting ingredients: compost, leaf mulch, topsoil, sand, and bone meal.
4. Regular working tools: pointed and flat-nosed shovels, turning forks, manure forks, rakes, regular and planter's hoes, four-pronged and single-pronged cultivators, brooms, and large $\frac{1}{2}$-inch strainers for compost.

5. Seeds.
6. A small farm machine for running a plow, rototiller, and sickle-bar mower if over one acre is planned.
7. For large compost operations a skiploader, shredder, and sifter may be necessary; however, no gas-powered motors are really needed if one has a large manual workforce.

## CLIMATE, DRAINAGE, SOIL, AND GARDEN LAYOUT

An adequate irrigation system and good sun exposure are of prime importance. Almost any slope or soil can be modified to produce, but bricklike soil and steep slopes may be too much of a challenge to the beginner. Also, ground containing cement footings and overrun with Bermuda grass or other tough perennial weeds can be discouraging. After acquiring the land, determine pathways and angles of the sun for a one-year period. Study the drainage patterns by letting water run over the property. The county agricultural department already has information on the soil type, minerals, and pH. Fifty-year rainfall and temperature readings should also be easily obtainable. A min.-max. thermometer and a rain gauge are advisable because the readings may explain a lot of the successes and failures of the project. Do not ignore wind, salt spray, and nearby shrubs and trees as possible agitators of the garden.

The numerous urban and rural sources of garbage, manures, and garden wastes is the next consideration. The best contacts are made at dairies, horse stables, and at the entrances to the city dumps, because the gardener or hauler has to pay a dumping fee and would probably drive a mile or so out of his way to dump compost materials at your garden. You initially may provide materials from home to get the first pile started, but later piles should have other sources. Consider also:

1. Minimum 400 square feet for leafy vegetable plot.
2. Sixty-four square feet for compost bins, 32 square feet for bins, and 32 square feet for loading space; each bin should be 4' x 4' x 4'.
3. Minimum 200 square feet for squash, corn, and bean patch; or the same for tomatoes, peppers, eggplant, squash, beans, and melon.
4. Cement-covered areas for seed flats, potted herbs, chicken and rabbit hutches, and compost bins if area is limited. Compost bins on cement rather than dirt can be kept clean and orderly with

less fly problem. However, a dirt bottom makes it easier for worms and beneficial arthropods to gain access.

5. Easy access to compost pile by truck or cart.
6. Garbage wastes sorted and stockpiled in closed containers until a new compost can be started.
7. Fenced garden areas in an inconspicuous place as a protection against vandalism and to keep city dogs out (wild rabbits, squirrels, domestic herds, and birds have to be considered as potential pests for a small garden as they can destroy the whole garden in one night, organic or not).

Start with a small plot, even if the total area is large. If a large area is available, use it for fruit trees, berries, legumes, and grains. Small farm machinery might be used here, but save a small plot for hand-tool methods. You will then have two types of gardening

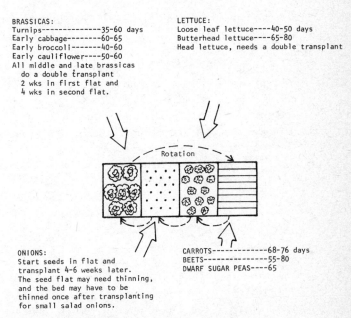

SUGGESTED LIST FOR CHICKEN TRACTOR
(1 of 10 beds illustrated)

BRASSICAS:
Turnips--------------35-60 days
Early cabbage--------60-65
Early broccoli-------40-60
Early cauliflower----50-60
All middle and late brassicas
   do a double transplant
   2 wks in first flat and
   4 wks in second flat.

LETTUCE:
Loose leaf lettuce----40-50 days
Butterhead lettuce----65-80
Head lettuce, needs a double transplant

Rotation

ONIONS:
Start seeds in flat and
transplant 4-6 weeks later.
The seed flat may need thinning,
and the bed may have to be
thinned once after transplanting
for small salad onions.

CARROTS-------------68-76 days
BEETS---------------55-80
DWARF SUGAR PEAS----65

FIGURE 6. Rotation scheme for one raised bed. Vegetables are grown closely together in a carefully cultivated rectangular area 3–4 feet wide and 15–20 feet long.

available, both of which will demand different sensitivity for the earth. No written work can substitute for that experience.

A surprising amount can be grown on a 40' x 25' plot. This plot can be worked by a simple "chicken-tractor" method (see below and Figure 7), by the two-week or long-term compost method, or by some combination of the two. In the layout, ten raised beds (one of which is shown in Figure 6) are prepared. Beds are divided into four different vegetable-type sections: (1) Carrots, beets, or Chinese peas; (2) one of the brassicas; (3) onions; and (4) lettuce. The four sections are rotated each time a new planting occurs (Figure 7). All footpaths are planted in rye grass and legumes (alfalfa, vetch). These areas will provide extra compost, mulch material, or feed supplement for the chickens (see below). Two or three borders are planted with flowers like borage, sunflower, Jerusalem artichokes, scarlet runner beans, chrysanthemums, or various pinks. The flower "hedges" are places for beneficial insects to live in. Some tomatoes, peppers, or herbs can be mixed with these borders. Some Brussels sprouts make good insectary plants which attract and house pests, especially aphids, which in turn serve as a food source for beneficial insect predators and parasites.

If the "chicken-tractor" mixed-husbandry method is used (Figure 7), chickens six weeks old or older (about eight of them) must be raised for the "tractor." Barelegged brown egg-layers are the best since they get along well with each other, scratch up the earth well, and produce fine eggs. The tractor should be roughly the length of the bed and maybe a foot wider to allow the chickens to forage in one of the paths beside the bed (see Figure 7). If they are fed grains, garbage, grass, and alfalfa, the chickens will give an egg a day in good weather, they will clear out all weeds, slugs, and insects, and they will work their manure and some garbage scraps into the soil fairly uniformly. If sorghum is in the feed they may leave this behind; it will sprout after the chickens have moved on to the next bed. Move the chicken pen to the next plot every two to four weeks, depending on how rich the soil should be. Immediately after the move, the ground should be cultivated. Fluff up the soil with a turning fork, cultivator, and a rake so that it is two to three inches higher than the normal ground level (called "raised beds"), and thus the roots can obtain more

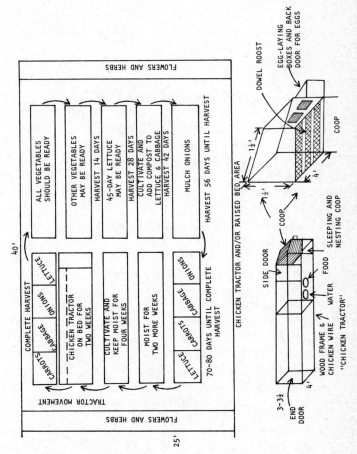

FIGURE 7. Suggested planting and rotation scheme for the "chicken-tractor" mixed-husbandry method of cultivation with ten beds under rotation.

oxygen.[12] Leave the bed open and keep it moist for about four weeks. At the end of this period, add 50 to 100 pounds of compost to the bed. Then carrots may be seeded, or else onions, lettuce, and brassicas transplanted into the bed. Approximately three to four weeks before the chicken tractor is to be moved on to a certain bed, seed that area with oats, wheat, or rye. Then when the chickens are moved to that place they will have the equivalent of spring grass, which is probably the best food of all. Do not expect great results from this method at first, because the timing and bed preparation may have to be adjusted for your area. However, you will learn a great deal from it.

Other possibilities are to fence off two or more plots, letting the chickens run in one plot for an entire growing period, and then rotate them into another plot and spade up the area they left for a new vegetable plot during the following growing period. This method may work best where winters are too severe for planting.

If one does not wish to use the above mixed-husbandry method, the same layout with a more intense planting schedule can be used, preferably with a long-term, two-to-four-month compost. In this case, one does not have to wait four weeks to seed or transplant. Spread approximately a one- to two-inch layer of compost over the bed and work into the top four inches of soil; you may add a light application of bone meal or dolomite if the soil needs this calcium requirement. Plants may be set in immediately. About midway to maturity, brassicas may need a second dressing of compost worked carefully into the soil. Every plant should be cultivated at least once during its growth phase. Onions should receive the most compost. Lettuce and brassicas should not be mulched because of the pests that live in the mulch; if these two crops are transplanted close enough at the beginning, the entire bed will close up with no soil showing, thus creating a natural mulch effect. Brassicas have the highest calcium requirement and therefore usually need a dusting of bone meal with the initial compost that is added. Carrots receive the lowest amount of compost and usually no calcium.

The raised-bed plot for leafy vegetables and food crops should be watered overhead. For the summer crops like corn, beans, and squash use another area with irrigation furrows.

If the growing season is short, composting programs can still be carried out during the cold periods. Quick-maturing cold crops can be planted first and hopefully harvested at the end of the spring, while hardy, late-maturing summer crops may be planted during the last month of spring so that by the end of the summer you may have pumpkins, corn, tomatoes, and the like.

## APPENDIX 1
### Seed Flat Preparation

BOTTOM PREPARATION:

1. Cover bottom cracks and holes with a few old, rotten leaves so materials in the seed flat will not fall through.
2. Sprinkle a very fine layer of bone meal on the leaves.
3. Sprinkle a ¼-inch layer of compost strained through a ⅛-inch screen and then dampen.

TOP PREPARATION:

1. Prepare a mix of ⅓ river sand strained through a $1/_{16}$- to ¼-inch screen; ⅓ topsoil strained through a ⅛- to ¼-inch screen; and ⅓ oak or similar leaf mulch strained through a ⅛- to ¼-inch screen. The size of the strainer depends on how fine one wants the mix.
2. Mix the above thoroughly, by hand or with a shovel; sprinkle water while mixing until damp only (rub mix between hands to break up lumps).
3. Fill seed flat with mix, compressing it a little, to the last ⅛ to ¼ inch.
4. Sprinkle in seeds and cover with leftover mix to the top of the seed flat.
5. Water with fine spray and keep seed flat moist and in warm place; the soil should settle about ¼ inch.

## APPENDIX 2
### Compost and Extras for Raised Beds

|  | Lbs./sq. yd. of compost | Lbs./sq. yd. of lime | May be started in flats | Other |
|---|---|---|---|---|
| Onions | 20–25 | 0 | yes | Mulching is recommended. |
| Celery | 20–25 | 0 | yes* | Needs a lot of water. |

| Potatoes | 20–25 | 0 | no | Hill up soil around plants or mulch. |
| Brassicas | 15–20 | ½ | yes* | Hill up soil around stems. |
| Beans | 10 | ¼ | no | |
| Lettuce | 20 | ¼ | yes | |
| Radishes | 10–15 | ¼ | no | Hill up soil around plants. |
| Parsley | 10–12 | 0 | yes | |
| Carrots | 7 | 0 | no | Use old seed flat mix for filling in seed drills. Thin and hill in. |
| Beets | 10 | 0 | no | Do same as for carrots. |
| Chard | 10 | 0 | yes | |

* Germinate on new moon.
Transplant and spread out the seedlings into new seed flats on full moon.
Transplant second flat to the raised bed on next full moon.

## APPENDIX 3
### Two-Week Compost Method

1. Needed: organic material (derived from living organisms) to make a pile 4 to 5 feet high, 4 to 5 feet wide, and 4 or more feet long. Much of the organic material should be chopped up into 2-inch or smaller pieces. A shredder is best. These dimensions and factors are necessary if the compost is to become hot and finished in two weeks.

2. Composition: The material should be composed of approximately ⅓ animal manure or legumes ($^1/_{10}$ if bird manure), and ⅔ leaf litter, grass clippings, algae, garbage, or other suitable plant material. Only small amounts of straw, branches, sawdust, and other woody, hard-to-digest, low-nitrogen materials should be used (use in six-month composting method or as chicken litter). One can supplement the pile with blood and fish meals, rock fertilizers, and wood ash. A wheelbarrow of dirt layered in will save on nitrogen loss.

3. Once an adequate pile is made, it will heat up in one to three days. It will begin cooling about thirteen to sixteen days later. Because the organisms that decompose this material consume much of the oxygen within three days, the heap has to be turned every third day. The temperatures will be around 140°–150° for twelve to fifteen days, and then begin to cool off. When it has cooled, it is ready to be used for fertilizer. If the compost is coarse, strain it through a ½-inch screen.

4. The main advantage of the two-week compost is the speed. It

demonstrates adequately the flow of energy and the recycling of materials in biosystems.

5. The main advantage of compost is the humus end product. Humus holds water and minerals in the soil; it releases minerals at a uniform or natural rate so that the plant will not accumulate toxic levels of certain salts; it allows air to reach the roots, promotes good texture and healthy soil organisms, and helps maintain a slightly acid soil. A 1- to 2-inch layer of humus dug into the top 4 inches of soil is adequate for most plants. Fly larvae, dangerous parasites, and pathogenic bacteria are killed when turned into the hot interior of the pile, and therefore the two-week method is encouraged in tropical areas where night soil is used. The pile initially will attract flies from the neighborhood, especially if too much nitrogen material or water has been added, but most of the eggs and maggots will die in the pile, which is a service to your neighbors. As an added fly control, check the corners and edges of the bin at the floor level for dark red fly pupae. These can be swept up and thrown into the hot interior of the compost.

## APPENDIX 4
### Long-Term Compost

The six-month or Indore composting method consists of layering. Begin by digging out a pit 4 to 6 inches deep, 8 feet wide, and as long as you want the pile to be (this is to help introduce the worms from the bottom). Pile the dirt close by. Begin the first layer with leaves, straw, grass, and/or weeds. This should be about 6 inches high. The next 2 inches should be legumes, fresh lawn cuttings, wet garbage, high nitrogen meals, and/or fresh manure (this layer contains the necessary high nitrogen substances). The soil goes on next as a fine ¼-inch layer, in order to trap escaping ammonia gas (a source of nitrogen) as ammonium ions. The last layer is a fine dusting of bone meal, dolomite, or limestone as a buffer to acidity. Rock minerals and wood ash may be added at this point also. Repeat these four layers again and again until a pile 4 to 5 feet high is obtained. Now cover with a final 1-inch layer of soil and a 2-inch layer of straw or other mulch material; make sure the top is flat so that rain water can seep in. During dry months one should sprinkle the pile for an hour or so.

One may use the compost in two to four months, depending on the season and on the number of turnings. There are several possible programs for turning: (1) the pile can be turned every three weeks and used; (2) it can be turned once at six weeks and used six weeks later; or (3) it can not be turned at all and used in two to four months. If time is not a factor it is recommended that the pile not be turned, because then most of the ammonia will be trapped in the soil. Although it hastens the rate of decay, every time the pile is turned some ammonia will escape, thus lowering the nitrogen quantity of the compost.

## Notes

1. A. Howard, *The Soil and Health: A Study of Organic Agriculture* (New York: Schocken Books, 1947); C. Pendergast, *Introduction to Organic Gardening* (Los Angeles: Nash Publishing, 1971); Philip Edinger, ed., *Sunset Guide to Organic Gardening* (Menlo Park, Calif.: Lane Magazine and Book Co., 1971); Robert Rodale, ed., *The Basic Book of Organic Gardening* (New York: Ballantine Books, 1971); Hamilton Tyler, *Organic Gardening Without Poisons* (New York: Van Nostrand, 1971).

2. United Nations, Conference Proceedings, *New Sources of Energy*, vol. 5, *Solar Energy II* (New York: U.N., 1961); Brace Research Institute (1965–1973), various pamphlets on alternative energy experiments: *Simple Solar Still, How To Build a Solar Water Heater, How To Make a Solar Still (plastic covered), Solar Steam Cooker, Plans for a Glass and Concrete Solar Still* (Brace Research Institute, MacDonald College of McGill Univ., Ste. Anne de Bellevue 800, Quebec, Canada); Volunteers in Technical Assistance, *Village Technology Handbook: Solar Water Heater* (VITA, 3706 Rhode Island Ave., Mt. Rainier, Md. 20822), pp. 321–322.

See also Richard J. Williams, *Solar Energy: Technology and Application* (Ann Arbor, Mich.: Science Publications, Inc.); D. S. Halacy, *Solar Science Projects* (Englewood Cliffs, N.J.: Scholastic Press); R. Merrill et al., eds., *Energy Primer* (Menlo Park, California: Portola Institute, 1975).

3. E. P. Odum, *Fundamentals of Ecology* (Philadelphia: W. B. Saunders Co., 1971); W. B. Clapham, *Natural Ecosystems* (New York: Macmillan, 1972); R. H. Whittaker, *Communities and Ecosystems* (New York: Macmillan, 1975); John Phillipson, *Ecological Energetics* (London: Arnold, 1966).

4. *The Biosphere: Scientific American*, September 1970 (San Francisco: W. H. Freeman & Co., 1970).

5. "Make Compost in 14 Days" (Emmaus, Pa.: Rodale Press).

6. University of California, "Reclamation of Municipal Refuse by Composting," Technical Bulletin No. 9, Sanitation Engineering Research Laboratory, University of California, Berkeley, 1953.

7. If bird manure is used, it is not recommended for the two-week method, as it tends to clump and most of the decay goes anaerobic.

8. The various meals should be weighed for their mercury and pesticide residues.

9. John Fry, Richard Merrill, and Yedida Merrill, *Methane Digesters for Fuel Gas and Fertilizer*, Newsletter #3 (Pescadero, Calif.: New Alchemy Institute, 1973).

10. Ram Bux Singh, *Bio-Gas Plant: Generating Methane from Organic Wastes*. (Ajitmal, Etawah, India: Gobar Gas Research Station, 1968); Center for Maximum Potential Building Systems, *Horizontal Drum Biogas Plant* (Austin, Texas: CMPB, 1974).

11. Odum, *Fundamentals of Ecology*; Clapham, *Natural Ecosystems*; Whittaker, *Communities and Ecosystems*; Phillipson, *Ecological Energetics*.

12. For an interesting guide to this form of intensive organic vegetable horticulture, one that describes in detail practical instructions see John Jeavons, *How to Grow More Vegetables* (Ecology Action of the Mid-peninsula, 2225 El Camino Real, Palo Alto, Calif., 94306, 1973).

## References

### SERIES

"Brooklyn Botanic Garden Record: Plants and Gardens." Brooklyn, N.Y.: Brooklyn Botanic Garden. Illustrated horticultural handbooks on various subjects (e.g., #20, *Soils*; #23, *Mulches*; #24, *Propagation*; #27, *Handbook on Herbs*; #34, *Biological Control of Plant Pests*).

"Pictured Key Nature Series." Dubuque, Iowa: W. C. Brown Co. Various illustrated keys for identifying plants and animals of the garden and farm. (*Beetles*, Jaques, 1951; *Butterflies*, Ehrlich, 1961; *Grasses*, Pohl, 1968; *Immature Insects*, Chu, 1949; *Insects*, Jaques, 1947; *Weeds*, Jaques, 1959; *Economic Plants*, Jaques, 1958).

Rodale Press, Inc., Emmaus, Pa. Publishers of many books on natural gardening and farming techniques. These include *Basic Organic Gardening Course*, staff and editors, 1972; *Poisoned Power*, Gofman & Tamplin, 1971; *Encyclopedia of Organic Gardening*, J. I. Rodale, ed., 1959; *Teaching Science with Garbage*, Schatz, 1971; *Teaching Science with Soil*, Schatz, 1972; *The Complete Book of Composting*, J. I. Rodale, ed., 1969; *Organic Way to Plant Protection*, J. I. Rodale, ed., 1966.

"Rural Studies Series." London: Blandford Press, 167 High Hilborn. Various books for rural education. (#1, *Soils*; #2, *Plant Life*; #3, *Animal Husbandry*; #4, *Smaller Livestock*; #5, *Insect Life*; #7, *Science in the Garden*; #8, *Growing and Studying Trees*).

### GENERAL

Chevron Chemical Company. "A Child's Garden." San Francisco: Chevron Chemical Company, 1972.

Frieden, Earl. "The Chemical Elements of Life." *Scientific American*, July 1972, pp. 52–60.

Hardy, Jack, and Foxman, S. *Gardening for Schools and Students*. London: Allman & Son, 1958.

Janick, Jules, et al., eds. *Plant Agriculture: Readings from Scientific American*. San Francisco: W. H. Freeman & Co., 1970.

Kormondy, Edward J. *Concepts of Ecology*. Englewood Cliffs, N.J.: Prentice-Hall, 1969.

Loveday, Evelyn. *Complete Book of Home Storage of Vegetables and Fruits*. Charlotte, Vt.: Garden Way Publ. Co., 1972.

Pfeiffer, Ehrenfried. *Weeds and What They Tell*. Stroudberg, Pa.: Biodynamic Gardening and Farming Assoc., 1960.

Philbrick, Helen, and Gregg, Richard. *Companion Plants and How To Use Them*. Old Greenwich, Conn.: Devin-Adair Co., 1966.

Philbrick, Helen, and Philbrick, John. *Gardening for Health and Nutrition*. Blauvelt, N.Y.: Rudolf Steiner Pub., 1971.

Shewell-Cooper. *The Complete Vegetable Grower*. London: Faber & Faber, 1955.

Stratford, B. *First Steps in School Gardening*. London: Mellifont Press, 1963.

Sunset editorial staff. *Vegetable Gardening*. A Sunset Book. Menlo Park, Calif.: Lane Magazine & Book Co., 1961.

## WEATHER AND CLIMATE

Flohn, Herman. *Climate and Weather*. New York: World Univ. Library, McGraw-Hill, 1969.

Lehr, Paul; Burnell, R. Will; and Zim, Herbert. *Weather*. A Golden Science Guide. New York: Golden Press, 1965.

Neuberger, Hans and Cabir, John. *Principles of Climatology*. New York: Holt, Rinehart & Winston, 1969.

Sutcliffe, R. C. *Weather and Climate*. New York: Signet Books. 1966.

U.S. Dept. Commerce. *Pilots' Weather Handbook*. Civil Aeronautics Administration. C.A.A. Technical Manual No. 104, 1955.

U.S. Dept. of Commerce. U.S. Weather Bureau. Climatological Data. Monthly Report. Also: U.S. Dept. of Agriculture. Weekly Weather & Crop Bulletin. Washington, D.C.: Agricultural Climatology Service Office.

Watts, Alan. *Instant Weather Forecasting*. New York: Dodd, Mead & Co., 1968.

## PLANTS AND THEIR NUTRITION

Galston, Arthur W. *The Life of the Green Plant*. Englewood Cliffs, N.J.: Prentice-Hall, 1961.

Ishizuka, Yoshiaki. *Nutrient Deficiencies of Crops.* Taipei, Taiwan: Food and Fertilizer Technology Center, 116 Nuai Ning St., 1971.

Kessler, George M. *Fruits, Vegetables and Flowers: Physiology and Structure in Relation to Economic Use and Market Quality.* Minneapolis, Minn.: Burgess Publ. Co., 1954.

Leeper, G. W. *Six Trace Elements in Soils.* Carlton Victoria, Aust.: Melbourne Univ. Press, 1970.

Reid, D. *Botany for the Gardener.* New York: Taplinger Publ. Co., 1966.

Sprague, H. *Hunger Signs in Crops.* New York: David McKay Co., 1960.

Wellman, Frederick L. *Plant Diseases: An Introduction for the Layman.* American Museum of Natural History. Garden City, N.Y.: The Natural History Press, 1971.

## SOILS

Bridges, E. M. *World Soils.* New York: Cambridge Univ. Press, 1970.

Burrows, William. *Textbook of Microbiology*, 16th ed. Philadelphia: W. B. Saunders Co., 1954.

Jackson, Richard H., and Raw, Frank. *Life in Soil.* New York: St. Martin's Press, 1966.

La Motte Research Dept. *The La Motte Soil Handbook.* Chesteraun, Md.: La Motte Chemical Products Co., 1970.

Russell, E. J. *Lessons on Soil.* New York: Cambridge Univ. Press, 1950.

## COMPOSTING

Bruce, M. E. *Common Sense Compost Making.* London: Faber & Faber, 1967.

Golueke, Clarence. *Composting: A Study of the Process and its Principles.* Emmaus, Pa.: Rodale Press, 1972.

Gotaas, H. B. *Composting, Sanitary Disposal and Reclamation of Organic Wastes.* World Health Organization, Monograph Series No. 31, Geneva, 1956.

Poincelot, R. D. *The Biochemistry and Methodology of Composting*, Bulletin 727. New Haven, Conn.: Connecticut Agricultural Experimental Station, 1972.

Veselind, D. A. "A Laboratory Exercise in Composting." *Compost Science,* Sept.–Oct. 1973.

## INSECT MANAGEMENT

Borror, Donald J., and White, Richard E. *A Field Guide to the Insects of America North of Mexico.* Boston: Houghton Mifflin, 1970.

Fichter, George. *Insect Pests.* A Golden Nature Guide. New York: Golden Press, 1966.

Olkowski, Helga. *Common Sense Pest Control*. Berkeley, Calif.: Consumers Cooperative of Berkeley, 4805 Central Ave., Richmond, Calif. 94804, 1971.

Van den Bosch, Robert, and Messenger, P. S. *Biological Control*. Scranton, Pa.: Intext Educational Publishers, 1973.

# PART V

# Food, Energy, and the New Rural Renaissance

# 15

# A MODEST PROPOSAL:
# SCIENCE FOR THE PEOPLE

*John Todd*

I HAVE been assured by a very knowing *American* of my Acquaintance in *London*; that a young healthy Child, well nursed, is, at a Year old, a most delicious, nourishing, and wholesome Food. . . . I GRANT this Food will be somewhat dear, and therefore very *proper for landlords*; who, as they have already devoured most of the Parents, seem to have the best Title to the Children.
—JONATHAN SWIFT, 1729
*A Modest Proposal*

A single overview is increasingly dominating human affairs while diversity and indigenous approaches are being set aside with the flourishing of modern science and technology. If the present trend continues, the world community will be shaped into a series of highly planned megalopolises that are regulated by an advanced technology and fed by a mechanized and chemically sanitized agriculture.[1] This future course is countered largely by the tenacity of

John Todd is a director and cofounder of the New Alchemy Institute, Woods Hole, Mass.

many peoples throughout the world, including many indigenous peoples, marginal and peasant farmers, traditional craftsmen, and new generations seeking alternatives to the modern industrial state. However, national and international agencies and business enterprises are vigorously attempting to "raise the standard of living" of most of these peoples and to incorporate them into the framework of the dominant societies. The rapid influx of populations into urban areas indicates that these attempts are successful in at least one respect—namely, that the numbers and impact of those who live apart from the mainstream of society are constantly being reduced. The world is rapidly becoming more homogeneous, and therein may lie one of the most serious problems confronting modern societies.

Most current solutions to the immense problems facing us utilize the latest techniques of systems engineering and involve resource and social management on a previously unattainable scale.[2] Increasingly, governments and international agencies are coping with the future by planning and acting on a world-wide basis. FAO's (Food and Agricultural Organization of United Nations) ambitious plan known as the Indicative World Plan for Agricultural Development exemplifies this approach and will strongly influence, if not dictate, agricultural development in many of the poorer nations over the next quarter century.[3] There are a number of dangers built into top-down management at national or supranational levels as progressively fewer people are going to be making more and more of the recommendations. This could lead to a lessening of the representation of people and points of view involved in shaping society, particularly if systems-specialists take it upon themselves to select the inputs and come up with the answers to future planning. Unfortunately, there is no guarantee that the methods currently in vogue will do any more than identify the crises which are piling up, and if important social or environmental variables are omitted, then these plans may actually aggravate our problems.

It is my contention that we are in danger of losing an important amount of social variability in the human community at the same time that we are losing the required amount of biological variability in our life-support bases. If we continue on our present path, at the present rate, then our chances of maintaining healthy communities and environments will be reduced dramatically before the year 2000, perhaps beyond a point where society as we know it will be capable of functioning.

A few years ago a group of scientists and humanists began a search for ways in which science and the individual could come to the aid of people and the stressed planet. We all shared the uneasy feeling that modern science and technology have created a false confidence in our techniques and abilities to solve problems. We were also disturbed that most futurology seemed to jeopardize the continued survival of man by displaying a real ignorance of biology. It was clear from the outset that social and biological diversity needed to be protected and, if at all possible, extended.

We felt that a plan for the future should create alternatives and help counter the trend toward uniformity. It should provide immediately applicable solutions for small farmers, homesteaders, native peoples everywhere, and those seeking ecologically sane lives, enabling them to extend their uniqueness and vitality. Our ideas could also have a beneficial impact on a wider scale if some of the concepts were incorporated into society at large. This modest and very tentative proposal suggests a direction that society might well consider.

At the foundation of the proposal is the creation of a biotechnology which by its very nature would

1. function most effectively at the lowest levels of society;
2. be comprehensible to and utilizable by the poorest of peoples;
3. be based upon ecological as well as economic realities, leading to the development of local economies;
4. permit the evolution of small decentralist communities which in turn might act as beacons for a wiser future for much of the world's population; and
5. be created at local levels and require relatively small amounts of financial support. This would enable poorer regions or nations to embark upon the creation of indigenous biotechnologies.

## Unnatural Selection: Loss of Diversity

It is necessary, before describing a way of reviving diversity at all levels, to evaluate how its loss threatens the future of man. Suppose some wise alien from another planet were commissioned to investigate earth. He would no doubt be dismayed at the outset by the tendency of the dominant societies, whether "communist" or "capitalist," to be constantly selecting the most efficient or profitable ways of doing things. Our visitor would ascertain clearly that our narrow approaches are reducing our options and that people are being conditioned and habituated to the options that remain. To him it would represent an evolutionary trap, and after his survey of energy use and agriculture was completed, he would confidently predict a major catastrophe. There would be no need to go on to industry, the university, or government, despite the fact that much ecological insanity resides in them also.

Examples of unnatural selection are everywhere.

For hundreds of years prior to the industrial revolution a wide variety of energy resources were used by man. Besides animal power and human toil there was a subtle integration of resources such as wind and water power, and a variety of fuels including peat, wood, coal, dung, vegetable starches, and animal fats.[4] This approach of integration through diversity in providing the energy for society has been replaced by an almost exclusive reliance on fossil fuels and nuclear power. Energy sources are often linked together into huge transmission grids which provide electric power over large sections of the country. The industrial revolution took place only where there was a large-scale shift to fossil fuels as an energy source. The costs resulting may yet overshadow its benefits. The production of air pollutants and highly dangerous radioactive wastes continues to increase rapidly, and no downward trend is immediately in sight, despite an increased environmental awareness. Modern society, by reducing the variety of its basic energy sources while increasing its per capita energy needs, is now vulnerable to disruption on an unprecedented scale. It would be foolhardy to disregard the very real possibility of a small group of people destroying our power transmission systems. Tragically, there are no widely disseminated backup sources of power available to help the majority of people in a nation hooked on massive amounts of electricity. Our society was not as precariously based as this in 1776, or even 1929.

On this country's farmlands changes have taken place over the last fifty years that have not yet had their full impact on the nation. The majority of the population has been displaced from relatively self-sufficient farms by large monoculture farm industries. That many of the displaced farm people are on welfare or adding to the ghetto's problems is not usually considered by agricultural planners. Unfortunately, the trend is world-wide as former colonial regimes and the present economic involvement by powerful industrial nations have created a climate of uncontrolled urbanization in Third World countries. There is a contemporary theory that contends that the industrial powers have contributed directly to the conditions that led to their dangerously high population levels.[5]

Proselytizers on behalf of modern agribusiness rarely consider the key role of numerous and diverse small farms as a social buffer during periods of emergency or social breakdown. This oversight could well be the result of a lack of civilian research into the needs of a major industrial nation under the stress of severe crises, despite the fact that a disaster could occur.[6] A depression like that which befell the country in 1929 could well take place; but if one should happen in the 1970s the social consequences would be much more severe. In 1929, a large percentage of Americans had friends or relatives on farms that could operate on a self-sufficient basis during lean periods. Today the situation is alarmingly different, as the rural buffer is largely gone and far fewer people have access to the land. The problem is compounded by the fact that today's farms have little resemblance to those of forty years ago; the modern farm is in no way independent, and like other businesses requires large amounts of capital, machinery, and chemicals to maintain its operations.

The replacement of rural populations and cultures by agribusiness operated primarily on the basis of short-term incentives rather than as legacies for future generations, is resulting in a tremendous loss of biological and social diversity in the countryside. When the land and landscapes become just another commodity, society as a whole suffers. It would not be so serious if the loss of a viable countryside were all that was threatened by modern agriculture, but a closer look at present agricultural methods suggests that many of them are causing a severe loss of biological variability, so vital to any sound and lasting agriculture.

### The Green Revolution: Unnatural Selection

Over the past several decades the agricultural sciences have created a number of major advances in food raising, and the widely acclaimed green revolution has come to symbolize the power of applied science and technology working on behalf of all people. Our confidence has been renewed that mushrooming populations can be fed if only Western agriculture can be spread rapidly enough throughout the world.[7] But the green revolution has not been shaped by an ecological ethic, and its keenest enthusiasts are usually manufacturers of chemicals and agricultural implements backed by government officials, rather than farmers and agricultural researchers who are generally aware of the immense complexity of stable agricultural systems. A brief examination of the ads in a wide variety of American journals and magazines would lead one to believe that the agricultural revolution is actually a chemical revolution, and perhaps it basically is.

A number of biologists and agricultural authorities are cautious about the future, as they foresee environmental decimation which will offset the agricultural gains before the turn of the century.[8] Among some of them, there is a disquieting feeling that we are witnessing the agricultural equivalent of the launching of the *Titanic*, only this time there are several billion passengers.

The modernization of agriculture has resulted in the large-scale use of chemical fertilizers upon which many of the new high-yielding strains of grains depend. Coupled with this is a basic emphasis on single cash crops which are grown on increasingly larger tracts of land. The dependency on fertilizers for successful crops has created depressed soil faunas and an alarming increase in nitrates in the ground waters of some areas. The nitrate levels are often above the safety limits set by the U.S. Public Health Service for infants' drinking water.[9]

Accompanying the widespread use of chemical fertilizers has been the rapid increase of biocides to control pests and weeds. These, in turn, have reduced the number of species of soil animals in many farm fields, with subsequent reductions in the quality of humus, which is essential to the sustained health of soils.[10] Unfortunately, these changes are occurring just as we are beginning to discover how much the soil fauna, particularly the earthworms, contribute to plant growth and health.[11] The use of biocides has triggered a vicious cycle: soils decline in quality, which in turn makes crops more vulnerable to attack by pests or disease organisms. This creates a need for increasingly large amounts of pesticides and fungicides for agricultural production to be sustained.

The full impact of biocides has yet to come. It is as if ecology and agriculture represent a modern Janus in their antithetic stances. While a team of ecologists has recently announced that the full impact of DDT often does not show up in long-lived birds, predatory animals, and humans for twenty-five years after application,[12] agricultural planners confidently predict a 600-percent increase in the use of pesticides in Third World countries over the next few years.[13] By the year 2000 the developing nations, as the beneficiaries of an uncontrolled experiment, will have reason to resent the blessings of modern technology.

The most notable achievement of the green revolution has been the creation of new high-yield strains of rice, wheat, and corn.[14] World agriculture has in the space of a few years been made more efficient, and in the short run, more productive because of these supergrains, particularly the Mexican semidwarf varieties of wheat. They represent a triumph of the modern plant breeder's art, but are in no way a panacea to the world food shortage. The grain revolution has an Achilles' heel; the new varieties, grown on increasingly vast acreages, are causing the rapid extinction of older varieties and a decline in diversity of the germ plasm in nature. The genetic variability which initially enabled the new types to be created is threatened, and the very foundation of the new agriculture is being eroded. In Turkey and Ethiopia thousands of local wheats have become extinct over the last several decades, and the phenomenon is widespread.[15] It is possible that the genetic variability of wheats could be irreplaceably lost. Erna Bennett of FAO has stated that "the world is beleaguered as far as its genetic resources are concerned."[16] Some of the most influential agricultural experts are deeply aware of the problem and are attempting to create the necessary "gene banks" before it is too late. It has been suggested that the race to save our genetic resources may be hampered by another biological fact of life, namely that seed storage may not be enough since "reserves" of the original microclimates and ecosystems may also be required if the viability of the local strains is to be maintained.[17]

The trend away from cultivating local varieties to a few higher-yielding forms is placing much of the world's population out on a limb. If the new varieties are attacked by pathogens the consequences could be world-wide rather than local, and plant breeders may not be able to create new strains before it is too late. Such events are not without precedent. An earlier counterpart of the green revolution occurred in Ireland in the eighteenth century, with the introduction of the Irish potato from the western hemisphere.[18] Production of food dramatically increased, and by 1835 a population explosion had taken place as a result of the land's increased carrying capacity. During the 1840s a new fungal plant disease appeared, destroying several potato crops, and one-quarter of the Irish people died of starvation.[19] The recent devastation of coffee plants in Brazil is partly the result of their narrow genetic base and their consequent vulnerability to leaf rust disease.[20] The 1970 corn leaf blight in the U.S. was caused by a fungus which attacks plants that carry the T gene for male sterility, and 70–90 percent of the corn hybrids carry this gene.[21] Despite heavy applications of fungicides, corn blight spread with heavy crop losses.

Clearly, a modern agriculture frantically struggling to right the wrongs of its single vision is not ecologically sane, no matter how productive, efficient, or economically sound it may seem.

There are other hidden perils associated with the modernization of agriculture,[22] but the loss of genetic diversity is perhaps the most obvious example of general changes taking place at every level of society. Since a scientific or technological advance on one level (e.g., the supergrains) may be pushing us closer to disaster, on another, it is time to look carefully at the roots of the alternatives before these avenues have disappeared behind us.

### Psychic Diversity and the Human Experience: A Narrowing Path?

The environmental dilemma is mirrored by comparable changes in people themselves. Unnatural selection is causing a loss of diversity in the human sphere, and this loss may lead toward social instability. The roles of most individuals are becoming ever more reduced as they relinquish the various tasks of living and governing to myriads of machines and specialists. Unlike our ancestors we have little direct control over the creation of our power and energy, food, clothing, or shelter. Claude Lévi-Strauss has shown how far this narrowing of roles has progressed, particularly with regard to our direct experience of the world around us. People fly faster, travel farther, and partake of more of the world, and yet in doing so, the world, sampled widely but without depth, becomes more elusive and farther from their grasp.[23]

It is highly probable, although difficult to prove, that the simplification and impoverishment of the lives of most of us lie close to the roots of much of the chaos threatening modern society. Erich Fromm has suggested that violence particularly is related to boredom:[24] It seems highly likely that boredom is one result of impoverishment or retreat from function.

Retreat from function is a negative trend since it removes the individual from the totality of his world. Restoring and extending genuine interaction with the life processes is the only lasting way to reverse this course, and this should begin at the basic functional levels of society, within the life-spaces of the individual or the small group. Fraser Darling, in his perceptive studies of remote Scottish peoples, showed how self-sufficiency was a positive force in their lives.[25] The most independent communities were far more diverse and socially vital than single-industry towns heavily dependent on a lifeline to the outside. He also came to realize that they coped far better in their dealings with the world at large. Equally important, the independent communities cared for their environment and were less prone to despoil it for short-term monetary gain. Another study of two California farm communities (Arvin–Denuba) revealed a comparable story.[26]

Modern science and its technologies have shaped industrially based societies that dominate the world today. These societies have an almost unlimited capacity to manipulate and destroy nature and men. In the long run they will not prove adaptive: as our options narrow, the specter of a future which is inhumane and in violation of nature looms larger. To reverse this trend, a moral, intellectual, and scientific renaissance will be required. Fortunately the basis for an adaptive view of society in nature is beginning to emerge, and an attendant science and philosophy exists in embryonic form today.

### New Alchemy and a Reconstructive Science: An Alternative Future

The direction of contemporary science is powerfully influenced by its patrons: the military and large corporations with their governmental cohorts. If a major scientific project or discipline does not hold out some promise of profit or military supremacy, it is not usually supported. The driving wheel of science in industrial societies is not a dispassionate seeking of knowledge. Science rarely addresses itself to the needs of human beings at the level of the individual or small group. With a sprinkling of notable exceptions, particularly in medicine, modern science and its technologies affect the majority of mankind in a negative or oppressive way, if at all. Science ignores, rather than addresses itself to, the richness and range of human potential. Knowledge is being replaced by hardware, not so much because hardware is superior to knowledge, but because it is more profitable. Unfortunately, technology as we know it cannot be expected to correct its own ills. These must be replaced with wisdom and practices that are fundamentally restorative rather than destructive.

An alternative science must seek to act on behalf of all people by searching for techniques and options that will restore the earth and create a new sense of community along ecological lines. Many talented people are working in the cities on urban problems, trying to make the cities livable and human, but very few are interested in making the countryside and farmlands livable by providing viable alternatives to the present rural destruction. Tools and techniques for individuals or small groups, however poor, must be sought to enable rural dwellers to work toward recapturing and extending their biological and social diversity. This new science must also link social and scientific purpose with the aim of creating a reconstructive knowledge that will function at the basic levels of society. If it did address itself to social and environmental microcosms, any group of people would be able to create its own indigenous biotechnic systems, gain more control over its own lives, and become more self-sufficient.

The ideal is to find ways of living that will help alleviate oppression at all levels, against the earth as well as against people. Ecology and personal liberation together have the potential to create environments within which people can gain increasing control over the processes which sustain them. This philosophy, call it "New Alchemy," in seeking modes of stewardship, attempts to fuse ethics with a scientific commitment to microcosms, because in caring for the immediate, a dynamic may be born that will ultimately lead to a saner tomorrow.

### Centers for New World Research

The New Alchemy Institute has established a few small independent centers in a variety of climates and environments, including the tropics. In this way we hope to induce a high degree of diversity into research and approaches to land stewardship. However, there does run within the organization a common thread, namely a holistic view of the task ahead. No research is undertaken

Windmill partially constructed. (PHOTO BY ALAN PEARLMAN, M.D.)

in a vacuum. Energy is linked to food production, food production to the larger questions of environment and communitas. Where possible, wastes, power, gardens, aquaculture, housing, and surrounding ecosystems are studied simultaneously. In the foreseeable future all elements of the systems will be linked in a variety of ways so that the most viable living environment can evolve. Thus a holistic view becomes possible at the level of the social microcosm.

The New Alchemy farm on Cape Cod in Massachusetts typifies our research approach to the rural problems of tomorrow. The fundamental strategy has been to integrate an array of low-cost yet sophisticated and efficient biological and solar energy systems. This has created a productive and self-contained microcosm—stewardship responsive to local conditions.

With respect to our *preliminary* model at Cape Cod, windmills, solar heaters, intensive vegetable gardens, field crops, and fish cultures are linked together in mutually beneficial ways. Brief descriptions follow.

*The Wind Generator*: A wind generator is a streamlined windmill that generates enough power to run an electric generator. It was once popular in rural areas during the 20s and 30s, before the advent of rural electrification. Our wind generator, which cost very little, was assembled primarily from scrap auto parts.[27] However, it is by no means perfected, and a great deal still needs to be learned about producing electricity inexpensively from the wind. Recent designs and new gearing systems, discoveries of solid-state power converters, and efficient storage batteries and air-foil blade designs, coupled with a dwindling supply and increased cost of fossil and nuclear fuels are making wind generators increasingly practical as an alternative energy source.

*Fish Ponds*: Below the windmills, at the entrance to the gardens, are two small solar-heated aquaculture minifarms. One is covered with a dome having a clear plastic skin and curved surface to trap the sun's heat and store it in a "tropical" pond fashion.[28] The other covered pond of more conventional design uses a solar heater for additional warming of the water. Both ponds are maintained around 80°F throughout the late spring and summer months. Within the 25-foot-diameter pools, *Tilapia*, a tropical fish of high food value, is raised. These fish derive their feed primarily from massive algae blooms whose growth is stimulated in waters warmed

Sailwing windmill.            (PHOTO BY ALAN PEARLMAN, M.D.)

by the sun and enriched by small amounts of animal manure. Edible-size fish have been cultured in as brief a time as ten weeks. (See Figures 1 and 2.)

Other food sources will come from research involving the production of high-protein insect-food in polluted waters. In order to accomplish this, insects with an aquatic larval stage are being reared in large numbers in the tiny ponds. These provide an ecological food source for *Tilapia* and other fish.[29] The insects currently being cultured are midges or Chironomids, tiny nonbiting mosquitolike insects which commonly swarm on summer evenings. The larval stage, normally found in the bottom muck in ponds, is cultured on burlap mats suspended in the fertilized ponds. The problems of food production and water purification are interconnected at the point of fertilization. At present, in order to obtain high yields, animal manures are used as fertilizer. The ponds are, in fact, polluted to increase production. While growing, the larval midges help to purify the ponds. They accomplish this by feeding on microscopic organisms whose populations are increased by the manure, and perhaps also by direct assimilation of nutrients in the enriched waters. At this stage the insect-rearing ponds use only manure, but there are plans to shift some of the culture over to

(PHOTO BY ALAN PEARLMAN, M.D.)

Dome backyard fish farm.

human sewage, thereby linking sewage purification with the rearing of insects for fish culture.

The sun and the wind are coupled in the backyard fish system to optimize productivity. It is a self-sufficient approach to the rearing of aquatic foods and there is little in the way of capital involved. It requires only labor and a large array of ecologically derived ideas, many of which have yet to be completely elucidated.

*Household Purification System*: Human sewage is being partially purified in one practical experiment. A small glass-sided A-frame structure is used to elevate temperatures over a series of pools that purify household sewage and wastes, through the culture of aquatic plants, live-bearing fishes, and insects of a variety of species. The produce from the household waste purification system is fed in turn to a flock of chickens. The wastes, partially purified by the living organisms, are subsequently used for irrigating the lawn and tree crops. Sewage, ordinarily an expensive and awkward problem for society, becomes a beneficial source of energy when dealt with on a small scale. New animal feeds are found and local soils enriched.

*Intensive Vegetable Gardens*: The birthplace of much of our agricultural research is in the gardens below the ponds. Several experiments intended to help find ways of culturing plants and animals without using expensive and harmful biocides have been initiated. One large project, is a systematic search for varieties of vegetables that may have some built-in genetic resistance to insect pests. Most modern plant breeders have assumed that pesticides are an inevitable tool in agriculture; consequently, knowledge is scant concerning vegetable varieties with an intrinsic ability to resist pest attacks. In another research project the efficacy of inter-planting vegetables with herbs and flowers that have a suspected ability to trap or repel pests is being tested, along with techniques for performing reliable yet simple experiments in highly productive

(PHOTO BY ALAN PEARLMAN, M.D.)

Backyard fish farm, set up for summer culture without solar heater.

N.A.I.'s 5,000 Gallon Fish Rearing Complex
Adapted To Northern Climates

FIGURE 1. New Alchemy's backyard fish farm, with solar heater, reflectors, and night covers for cool-season culture.

vegetable gardens.[30] Each of the experimental gardens, regardless of the research taking place, is treated as a miniature ecosystem, and many of the biological processes are monitored to determine aspects of diversity and "stability" in each of the systems.

*Integrating Gardens–Fish Ponds*: Ideas for future research projects are being tested in the gardens. For example, one experimental plot is being used to look into the value of using nutrient-laden water from the small fish ponds for irrigating crops. Some fish species, when cultured in high densities, apparently secrete a fatty substance that tends to reduce evaporation. Consequently, pond water containing moisture-conserving substances as well as nutrients may be highly useful for irrigating crops, especially under arid conditions. Early laboratory trials with lettuce and parsley indicated that the water from tanks containing fish has a "hermetic" quality that conserves moisture around the roots of the plants. Field trials conducted in 1973 demonstrated the agricultural value of using aquaculture wastes. Lettuce yields were increased up to 112 percent over controls.

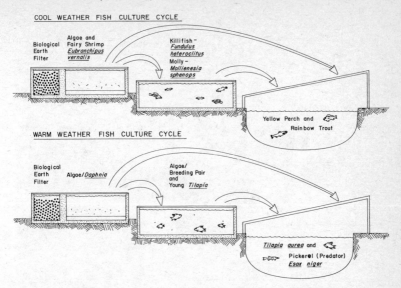

FIGURE 2. Biological relationships in the backyard fish farm.

Already we can begin to envisage closely linked aquatic and terrestrial food systems suited to regions where water is seasonal and limited. Vegetation for food and for shelter from the sun could be nurtured from water stored and enriched in aquaculture ponds. Many of the earth's arid regions may one day sustain small communities within microenvironments that are biologically complete without the need to import large amounts of food, energy, and capital.

*Ecologically Derived Structures*: The investigations of a small group of people at a single New Alchemy center are coming together most completely in a project initiated in 1973. A direct involvement in process has drawn us toward the idea of creating living structures that are ecologically derived and reflect all that we have learned. Our initial approach to such housing is to have the structures evolve directly out of the ongoing aquaculture, waste, greenhouse, and solar and wind energy research. On a microscopic scale such a strategy seems to make good sense, as the threads of each person's investigations are spun together to create a structure that mimics nature and perhaps will enable us to live in, rather than apart from, her. These structures will be self-regulating and

(PHOTO BY ALAN PEARLMAN, M.D.)

Intensive vegetable gardens.

eventually will provide inhabitants with shelter and a wide variety of aquatic plant and animal feeds as well as vegetables and fruits.[31] Such systems have the potential to provide the majority of food needs as well as housing for their inhabitants. Only the essential grains would need to come from outside.

*The Ark*: Our first structure, just started, is called the "ark." It is a solar-heated greenhouse and aquaculture complex adapted to the rigorous climates of the northeast. If suitable internal climates can be maintained, we will eventually attach living quarters to the structure. The prototype will include a sunken greenhouse, an attached aquaculture pond, and a diversity of light and heat conservation and distribution components (see Figures 3 and 4). It will be an integrated self-regulating system requiring the sun, power for water circulation, waste materials, and labor to sustain its productivity. The electricity to drive the circulation pump will be provided by a windmill. The heat storage-climate regulation component will be a 13,500-gallon aquaculture pond. Solar heat will be trapped directly by the covered pond and by water circulating through the solar heater. The attached greenhouse will be built below the frost line and will derive its heat from the earth, direct

(PHOTO BY ALAN PEARLMAN, M.D.)

Gathering wheat at New Alchemy.

sunlight, and from the warmed pond water passing through pipes in the growing beds within the structure.

The intensive fish-farming component will be comparable to those already pioneered by New Alchemists. Several crops of *Tilapia* fish will be cultured through the warm months and a single crop of perch and trout during the cooler seasons. The aquaculture system may prove productive enough to underwrite the construction and maintenance costs of similar food-growing complexes in the future.

The greenhouse will be used to raise high-value vegetables and greens fertilized by wastes from the aquaculture pond. If our solar-heated ark should prove successful, then ecologically derived low-energy agricultures may thrive in northern climates.

So far, while building our models, we have learned that incredibly little is known about devising and caring for small-scale systems for communities that are both ecologically complete and restorative of environments. The contemporary colossal sense of scale, combined with the fragmentation of knowledge by the scientific establishment, has effectively blocked the development of an alternative for the future that is humble and yet ecologically wise.

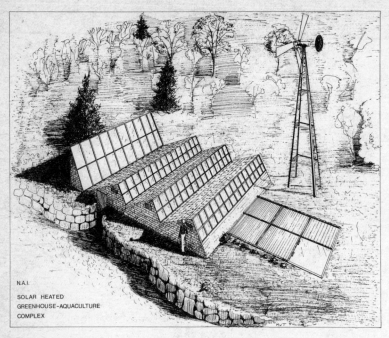

N.A.I.

SOLAR HEATED
GREENHOUSE-AQUACULTURE
COMPLEX

FIGURE 3. The Ark: New Alchemy's proposed solar-heated greenhouse-aquaculture complex.

There are as many mysteries to be explored in the workings of the
wind, the sun, and the soil on a tiny plot of ground as exist in the
grandiose schemes of modern science. The totality of the human
experience becomes available to each of us as we begin to learn to
function at the level of the microcosm.

### Beyond Ourselves: A People's Science

A few people working at a handful of centers cannot alone
affect the course of human events. The elitism underlying contemporary science must be eliminated and a reconstructive science
created. Knowledge should become the province of many, including all those struggling to become pioneers for the twenty-first
century. If responsibility and diversity are to be established at the
level of the individual, then individuals with a wide array of back-

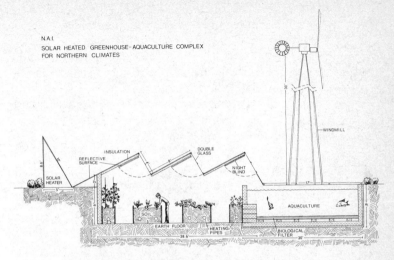

FIGURE 4. Cutaway of the Ark showing the interior.

grounds and experiences should take part in the discovery of the knowledge and techniques required for the transformation ahead. A lay science, addressing itself to problems at basic levels of society, could restore diversity to the human sphere and establish an involvement for many in the subtle workings of the world around them.

Already a number of lay scientists are working with us investigating the backyard fish farm concept, experimenting with the raising of *Tilapia* under intensive culture conditions.

Other lay researchers are involved in experiments to determine the value of ecologically designed food gardens. One of the experiments is a systematic search for varieties of vegetables that may have some genetic resistance to insect pests. Another is a search to determine the techniques of interplanting vegetables with herbs and flowers with a suspected ability to trap or repel pests and to nurture natural control agents of those pests.

Only with the help of hundreds of earth scientists could this kind of information be acquired on a country-wide scale, in a relatively short period of time. The research on resistance and ecological design must take into account soils, environments, and climates from a diversity of regions in order to comprehend the forces underlying a balanced and restorative agriculture. Such a study has not been attempted by orthodox research organizations, nor is it likely to be attempted.

It is too soon for us to have developed much experience in guiding a lay science that will create its own independent dynamic. If we are at all successful, individuals and groups will within a few years branch out and explore the questions that seem most relevant to them and their own lives. Indigenous centers for learning through direct involvement in the process of reconstruction will spring up, providing an alternative to the colonization and fragmentation of knowledge by the universities.

Our initial approach was to compile two working manuals covering the research that is now part of the peoples' research program.[32] One of the manuals deals with agricultural research and the other is a guide to the fish farm project. The agricultural manual attempts to show the garden as an experimental system and leads the potential investigator through problems often faced by ecologists. After working with the manual in an experimental garden the problems and concepts gradually become comprehensible. The aquaculture investigator's manual uses a somewhat different method, following more of a "cook book" approach, with a step-by-step guide through the fish culture experiment.

There is a strong tendency in the academic world of modern science to publish more and more (usually about less and less) to be considered a successful scientist. To peruse most research publications is an almost absurd experience. The contents are dreary, fragmented, and usually border on irrelevancy. What a far cry from the scientific writings of men like Charles Darwin, who would not publish his theory of evolution without years of intensive labor. Today, a Darwin would probably be sacked from even the most progressive college. Despite almost a lifetime of illness, his intense intellectual activity ultimately resulted in the publication of books on evolution and natural selection, earthworms, the formation of coral reefs, and the behavior of humans and animals which remain of real value to this day.

Although the criticism of scientific publishing is usually valid, it still seems clear that publishing will have to remain a corner-need for publications that are readily understandable, relevant, and directly applicable to the needs of the new pioneers. They should reflect education in the broadest sense, in which individuals, society, and the biosphere are seen in holistic and meaningful terms.

Already it is apparent that an alternative science is evolving on a world-wide scale, and will continue to grow. There are common threads weaving the tapestry that underlies the lives of the new pioneers and scientists; among these are a strong sense of the human scale, a desire to comprehend the forces of communitas, and a passion for ecology and its teachings, which imply ethics and awakened sensibility and morality. These are forces in their own right, and though pitted against the shadow of technological man destroying man and nature, and a science operating in a moral vacuum, they may still represent the beginning of a hopeful path along which we may one day travel.

## Notes

1. R. B. Fuller, *Operating Manual for Spaceship Earth* (New York: Pocket Books, 1970). See also J. B. Billard, "The Revolution in American Agriculture," *National Geographic* 137, no. 2 (February 1970). Both works represent the prevailing views in global engineering, city design, and agriculture.

2. J. Platt, "What We Must Do," *Science* 162 (1969): 1115.

3. A. H. Boerman, "World Agricultural Plan," *Scientific American* 223, no. 2 (1970).

4. Murray Bookchin, "Ecology and Revolutionary Thought," *Anarchos* 1 (1968). This essay is also in "Post-Scarcity Anarchism"—a much more accessible source.

5. Barry Commoner, *The Humanist*, November–December 1970.

6. Platt, "What We Must Do."

7. Boerman, "World Agricultural Plan"; L. R. Brown, *Seeds of Change* (New York: Praeger, 1970).

8. Barry Commoner, "Soil and Fresh Water: Damaged Global Fabric," *Environment* 12, no. 3 (1970); W. C. Paddock, "How Green Is the Green Revolution?" *BioScience* 20, no. 16 (1970); John H. Todd, Editorial, The New Alchemy Institute Bulletin 1 (1970), Box 432, Woods Hole, Mass. 02543.

9. Commoner, "Soil and Fresh Water."

10. C. R. Malone, A. G. Winnett, and K. Helrich, "Insecticide-Induced Responses in an Old Field Ecosystem," *Bulletin Environ. Contam. Toxicol.* 2, no. 2 (1967).

11. R. Rodale, ed., *The Challenge of Earthworm Research* (Emmaus, Pa.: Soil and Health Foundation, 1961).

12. H. L. Harnson et al., "Systems Studies of DDT Transport," *Science* 170 (1970): 503.

13. Paddock, "How Green Is the Green Revolution?"; President's Science Advisory Committee, *The World Food Problem* 1 (1967). See also *Chem. Eng. News* 49, no. 2 (1971).

14. L. P. Reitz, "New Wheats and Social Progress," *Science* 169 (1970): 952.

15. G. Chedd, "Hidden Perils of the Green Revolution," *New Scientist* 48 (1970): 724.

16. Ibid.

17. John E. Bardach, personal communication.

18. *World Food Problem.*

19. Redcliffe N. Salaman, *The Influence of the Potato on the Course of Irish History*, Tenth Findlay Memorial Lecture, University College, Dublin, 27 October 1943 (Dublin: Brown & Nolan).

20. Chedd, "Hidden Perils."

21. N. Gruchow, "Corn Blight Threatens Crop," *Science* 169 (1970): 961.

22. N. Pilpel, "Crumb Formation in the Soil," *New Scientist* 48 (1970): 732.

23. Claude Lévi-Strauss, *The Savage Mind* (Chicago: Univ. of Chicago Press, 1966).

24. Erich Fromm, *The Revolution of Hope: Toward a Humanized Technology* (New York: Harper & Row, 1968).

25. F. F. Darling, "The Ecological Approach to the Social Sciences," *Amer. Sci.* 39, no. 2 (1951).

26. W. Goldschmidt, *As You Sow* (Glencoe, Ill.: Free Press, 1947).

27. E. Barnhart, "A Windmill for Generating Electricity," *The Journal of the New Alchemists* 1 (1973): 12–15.

28. W. McLarney, "An Introduction to Aquaculture on the Organic Farm and Homestead," *Organic Gardening and Farming*, August 1971, pp. 71–76; J. H. Todd and W. O. McLarney, "The Backyard Fish Farm," *Organic Gardening and Farming*, January 1972, pp. 99–109. W. McLarney, *The Backyard Fish Farm Working Manual*, Readers Research Project No. 1, New Alchemy Institute (Emmaus, Pa.: Rodale Press, 1973). See also *Journal of the New Alchemists* 2 (1974) New Alchemy Institute, Woods Hole, for more recent work with endemic aquaculture systems in cool-temperate climates.

29. W. McLarney, S. Henderson, and M. Sherman, "The Culture of Chironomids," *Journal of the New Alchemists* 2 (September 1974).

30. J. Todd and R. Merrill, "Insect Resistance in Vegetable Crops," *Organic Gardening and Farming*, March 1972; R. Merrill, "Companion Planting and Ecological Design in the Organic Garden," *Organic Gardening and Farming*, April 1972; R. Merrill, *Designing Experiments for the Organic Garden: A Research Manual*, Readers Research Project, New Alchemy Institute, Woods Hole, Mass. 1973.

31. J. Todd, R. Angevine, and E. Barnhart, *The Ark: An Autonomous Fish Culture—Greenhouse Complex Powered by the Wind and the Sun and Suited to Northern Climates* (Woods Hole, Mass.: New Alchemy Institute, 1973).

32. McLarney, *Backyard Fish Farm Working Manual*; Merrill, *Designing Experiments for the Organic Garden.*

# 16

# TOWARD A SELF-SUSTAINING AGRICULTURE

## Richard Merrill

> Take not too much of a land, weare not out all
> the fatness, but leave in it some heart.
> —PLINY THE ELDER
> A.D. 23–79

> Farming isn't a way of life, it's a way to make
> a living.
> —EARL BUTZ
> U.S. SECRETARY OF AGRICULTURE

The fact that a culture can produce more food on less land with less human toil has been cited by people of all persuasions as *the* example of human "progress." Until recently there has been little reason to challenge this belief. As long as agriculture produced a surplus of food for its people and for foreign trade, the farm technologies which were used and the economic incentives encouraging them were justified.

The fallacy here is the assumption that the *only* purpose of

Richard Merrill is a director of the New Alchemy Institute and coeditor of *The Energy Primer* (Menlo Park, Calif., Portola Institute). He currently teaches courses in alternative food and energy systems in California.

agriculture is to produce food. Over the years all sorts of propaganda have locked us into this illusion, and we tend to forget that agriculture is dynamic and that its historic role has been to *maintain* productive land in order to sustain its people. In addition, a thriving rural culture has always played a vital part in providing food and fiber and absorbing dispossessed people during wars and economic depressions.[1] In a healthy society agriculture not only provides food; it also provides a reliable buffer during social crises and a culture of land stewards for posterity.

But like most rapid revolutions, the green revolution has created more long-term problems than short-term "solutions." At first glance the new farm technology has been praised because it has allowed millions to leave their lands in favor of an urban paradise free from rural toil. But in the haste to free the majority of people from the need to be farmers, an enormous dilemma has been created. The tools of liberation—chemicals, machinery, monocultures, hybrid crop strains, etc.—have alleviated scarcity and work (of one sort), but they have precipitated a number of mounting economic and ecological problems which not only compound themselves, but which threaten the sustaining potential of our farmlands. Since World War II we have so altered our rural environment and have become so totally dependent upon a single chemical strategy for food production that we face a future in which a major human concern will be the increasing hazards of the fuels and chemicals needed to continue the food supply. We brag of being a nation in which food is relatively cheap and agriculture is efficient; yet we ignore the fact that most measures of food prices and farm efficiencies fail to account for the endangerment of many valuable resources such as soil fertility, water, wildlife, public health, and a viable rural economy. When we really stop to consider the full impact of the agricultural tools that have replaced people, it is clear that "modern" agriculture is rapidly causing more problems than it is solving.

There are other problems associated with the green revolution besides environmental hazards and the destruction of a healthy rural base. Today's farms require massive inputs of fossil fuel energy to maintain them in a stable state. In fact, during the last few decades we have simply been exchanging finite reserves of fossil fuels for our food and fiber supply. Obviously this trade-off cannot continue indefinitely; if agriculture is energy-intensive, then fuel shortages must inevitably lead to food shortages. So in the very near future, we will have no choice but to adopt agricultural techniques that utilize *renewable* energy supplies. These include recycling of organic wastes to supplement synthetic fertilizers, using renewable forms of energy (solar, wind, and organic fuels) to help supply rural power needs; and applying ecologically diverse cropping patterns and integrated pest control programs to reduce the use of pesticides. Without a broad approach to these alternatives, the future of modern agriculture could very well become self-defeating rather than self-sustaining.

More and more, the full consequences of the green revolution present a number of unresolved questions concerning the relationship between modern agriculture and the quality of life. In the past many of these questions have been considered rhetorical or academic. Today they suggest forcefully that we have not been adjusting our priorities to the speed of current events. For example:

1. What is the *total* impact of modern agriculture on our indispensable natural resources?
2. What are the long-term effects of pesticides and synthetic fertilizers on public health and the continued ability of farmlands to produce quality food?
3. Is the displacement of our rural culture by high-energy technology an inevitable or even desirable consequence of social "progress"? If not, how can we change economic policy and public sentiment to encourage the success of independent farmers who are best able to steward the rural environment for future generations?
4. What are the consequences of an agricultural system that is totally dependent upon nonrenewable supplies of fossil fuel energy? Is such an agriculture itself nonrenewable? Do further increases in food production justify additional uses of fossil fuel resources?
5. Will the increasing costs of fossil fuels mean the total monopoly of agriculture by corporations, industry, and their petroleum technology? If so, what effect will this have on the price, availability, and quality of food?
6. Why does agricultural research continue to focus attention on developing farming methods geared to machines and fossil fuels rather than people and renewable energy inputs? Why do ecologically sophisticated techniques of agriculture continue to be

considered "inefficient" and "backward" by the U.S. Department of Agriculture and most of the scientific establishment?
7. To what degree can the polluting, high-energy techniques of agriculture be replaced by the renewable and self-sustaining energy of natural resources and biological processes?

## Modern Agriculture: A Wasteland Technology

By its very nature, agriculture makes a heavy impact on the environment. Ever since neolithic tribes began to cultivate endemic wild plants, agriculture has involved a tradition of people manipulating their surroundings in order to grow plants and to husband animals for food. The results have often been unfavorable. Throughout history, people have accelerated natural erosion by rapid deforestation, poor soil management, and the cultivation of plants intended for export at the expense of local economies.[2] In the United States land destruction has been a matter of record since seventeenth-century tobacco and cotton farming in the South. Wind erosion and the Midwest dust bowls of the early 1900s are now infamous history. Current data from the national land-use inventory show that 64 percent of U.S. croplands are in need of soil conservation.[3]

Today, however, the traditional hazards of agriculture have been overwhelmed by an arsenal of sophisticated technologies that place the environmental impact of agriculture into new dimensions.

### PESTICIDES: OVERKILL AND DIMINISHING RETURNS

Pesticides are poisons that kill pests: insects (insecticides), weeds (herbicides), and plant diseases (fungicides). Prior to 1940 pesticides were made from inorganic materials (mostly heavy metals, arsenic, and sulfur) or plants. Then, just before World War II, it was discovered that DDT, a synthetic organic compound first made in 1874, had remarkable insecticide properties. Wartime conditions increased the demand for DDT, and after the war U.S. production soared[4] from 9.5 million pounds in 1944 to 179 million pounds in 1963.[5] Other requirements led to the development of more toxic insecticides, including variants of DDT and certain nerve poisons such as the organophosphates (developed as by-products of nerve gas research during World War II) and carbamates.

The popularity of herbicides was also inaugurated by a revolutionary chemical developed during World War II. The compound, 2,4-D, was especially appealing because it acted like a plant hormone and selectively poisoned broadleaf plants but not grasses. This was an important discovery for the grain farmer; it was also important for the military, which used 2,4-D to destroy millions of acres of farmlands and forests in South Vietnam.[6] Today, 2,4-D continues to be the most widely produced herbicide in the United States, and the most widely used by American farmers.

In many ways World War II can be thought of as the instigator of an agricultural revolution. Since the introduction of DDT, 2,4-D, and organic phosphates, U.S. production of synthetic organic pesticides has increased from 33,000 tons of DDT and 1,000 tons of 2,4-D in 1945 to 552,000 tons of over 100 different pesticides in 1968.[7] Today there are over 100 industrial firms producing about 1,000 pesticide chemicals variously combined in over 50,000 registered commercial pesticides. See Table I.

Most of the public debate concerning pesticides has centered on problems associated with their uses in agriculture. However, only about 50 percent of the pesticides used in the United States are applied to farms (see Table II): the remainder is used by government, industry, and urban dwellers. In fact, suburban lawns and gardens probably receive the heaviest concentrations of pesticides of any land area in the United States.[8]

Nevertheless, pesticides are used extensively on U.S. farmlands, although exact figures are unavailable. Usually records of production, sales, imports, and exports give some indication of farm usage, but these are imprecise. The best estimates are for the years 1964 and 1966,[9] and even these data are based on a limited survey of farmers. Still, there are some interesting facts from these surveys.

1. Over half of all pesticides used in agriculture are applied to three crops: cotton, tobacco, and corn (see Table III).
2. Thirty-seven percent of U.S. farmers use herbicides; 29 percent use insecticides; and 4 percent use fungicides. These figures have undoubtedly increased since 1966, especially in regard to herbicides and organophosphates.
3. A greater proportion of large farms use pesticides than small farms (see Figure 1).

TABLE I

PRODUCTION OF PESTICIDES IN THE UNITED STATES
(IN THOUSANDS OF TONS)

| | Herbicides† | | Insecticides | | | Total |
| | Fungicides* | 2,4-D‡ | Other Organic Herbicides | DDT | Other Organic Insecticides | Inorganic Insecticides | (1,000 tons) |
|---|---|---|---|---|---|---|---|
| 1959 | 123.9 | 28.4 | 21.6 | 78.4 | 80.7 | 9.7 | 342.7 |
| 1960 | 147.5 | 35.1 | 16.1 | 82.1 | 93.4 | 8.3 | 382.5 |
| 1961 | 132.4 | 40.1 | 20.4 | 85.7 | 109.6 | 9.2 | 397.4 |
| 1962 | 98.7 | 40.0 | 35.4 | 83.5 | 119.2 | 7.3 | 384.1 |
| 1963 | 97.2 | 45.4 | 41.9 | 89.5 | 119.5 | 5.6 | 399.1 |
| 1964 | 98.2 | 54.0 | 59.2 | 61.9 | 128.1 | 8.1 | 409.5 |
| 1965 | 109.2 | 63.3 | 68.2 | 70.4 | 174.8 | 5.6 | 491.5 |
| 1966 | 120.4 | — | — | 70.7 | 205.3 | 5.1 | — |
| 1967 | 112.3 | 80.4 | 124.3 | 51.7 | 196.2 | 3.9 | 568.8 |
| 1968 | 120.9 | 86.7 | 147.8 | 69.7 | 214.9 | 6.2 | 646.2 |
| 1969 | 120.8 | 52.0 | 144.7 | 61.6 | 209.4 | 5.2 | 593.7 |
| 1970 | 115.4 | — | 201.9 | 29.7 | 177.3 | 3.0 | 527.3 |

SOURCES: Data for inorganic pesticides are from USDA, *Agricultural Statistics*; data for organic pesticides are from U.S. Tariff Commission, *Synthetic Organic Chemicals*.

* includes copper sulfate and organic compounds; excludes sulfur
† organic compounds only
‡ includes esters and salts

4. Farmers in the Southeast and Delta states use over 40 percent of all insecticides. The corn belt accounts for nearly a third of the herbicides used.[10]

These facts suggest that farm use of pesticides is concentrated on a few crops in specific regions and on large farms. However, many pesticides have been used so much that they must now be considered an integral part of our biological systems—present in our flesh, drifting in the air, flowing with the rivers, and falling with the rain.[11] After more than a generation of unrestrained use it is obvious that pesticides have produced, and are still producing, serious side effects whose full consequences have yet to appear. The major and most disturbing of them include the following:

1. *Pesticides are persistent:* Most pesticides are developed to withstand degradation by climate and microbes. Some are more persistent than others. Organophosphates break down in days or months while organochlorides can remain in the environment as

TABLE II
SELECTED PESTICIDES USED BY U.S. FARMERS, 1966

| Type of Pesticide | Total Use in U.S. 1,000 tons[a] | | % Used by U.S. Farmers |
|---|---|---|---|
| FUNGICIDES:[b] | | 63 | 27% |
| HERBICIDES:[c] | | 114 | 55% |
|   2,4-D; 2,4,5-T | 43 | | 48% |
|   others | 71 | | 59% |
| INSECTICIDES[d] | | 165 | 57% |
|   DDT | 25 | | 54% |
|   Aldrin-Toxaphene[e] | 39 | | 68% |
|   Other | 101 | | 54% |
| TOTAL PESTICIDE | | 341 | 51% |

SOURCE: USDA, Economic Research Service, *Quantities of Pesticides.*
[a] Calculated by subtracting exports from production.
[b] Excludes sulfur and pentachlorophenol.
[c] Includes plant hormones, defoliants, and desiccants.
[d] Includes fumigants, rodenticides, miticides.
[e] Includes aldrin, chlordane, dieldrin, endrin, heptachlor, and toxaphene.

TABLE III FARM USE OF DIFFERENT PESTICIDES ON MAJOR CROPS, 1966

| | Million Acres Planted (1966) | Pesticides Used (1,000 tons) | | | | | % Total Pesticides | Concentration (lbs./acre) |
|---|---|---|---|---|---|---|---|---|
| | | Fungicides | Herbicides | Insecticides | Misc.† | Subtotal | | |
| Cotton | 10.3 | .2 | 3.3 | 32.4 | 14.2 | 50.1 | 26.7% | 4.9 |
| Corn | 66.3 | — | 23.0 | 11.8 | .5 | 35.3 | 18.8% | .5 |
| Tobacco | 1.0 | — | — | 1.9 | 13.4 | 15.3 | 8.2% | 15.3 |
| Other Fruits & Nuts | 2.6 | 2.1 | 1.4 | 3.3 | 8.7 | 5.5 | 8.2% | 6.0 |
| Other Crops* | 661.5 | 2.3 | 10.1 | .8 | .1 | 13.3 | 7.1% | — |
| Peanuts | 1.5 | .5 | 1.4 | 2.8 | 7.0 | 11.7 | 6.2% | 7.8 |
| Apples | .7 | 4.2 | .2 | 4.2 | 1.1 | 9.7 | 5.2% | 13.9 |
| Other Vegetables | 3.7 | 2.1 | 1.7 | 4.1 | .4 | 8.3 | 4.4% | 2.2 |
| Soybeans | 37.4 | — | 5.2 | 1.6 | .1 | 6.9 | 3.7% | .2 |
| Irish Potatoes | 1.5 | 1.8 | 1.1 | 1.5 | .4 | 4.8 | 2.6% | 3.2 |
| Wheat | 54.5 | — | 4.1 | .4 | .1 | 4.6 | 2.5% | .1 |
| Citrus | 1.2 | 2.0 | .2 | 1.4 | 1.1 | 4.7 | 2.5% | 3.9 |
| Alfalfa | 29.0 | — | .6 | 1.8 | .3 | 2.7 | 1.4% | .1 |
| Sorghum | 16.4 | — | 2.0 | .4 | .1 | 2.5 | 1.3% | .2 |
| Rice | 2.0 | — | 1.4 | .2 | — | 1.6 | .8% | .8 |
| Sugar Beets | 1.2 | — | .5 | .1 | .1 | .7 | .4% | .6 |
| TOTAL | 890.8 | 15.2 | 56.2 | 68.7 | 47.6 | 187.7 | 100.0% | .07 |
| | | | | | | | | .26‡ |

SOURCE: USDA, Economic Research Service, *Quantities of Pesticides*.

* includes misc. field crops, pasture, mixed grains, summer fallow, plus those crops not listed individually.

† includes rodenticides, fumigants, plant hormones, miticides, repellents, defoliants, desiccants.

‡ excludes "other crops."

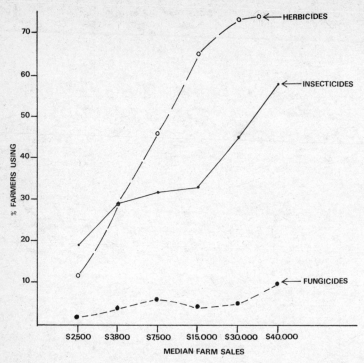

FIGURE 1. Proportion of U.S. farmers using pesticides in 1966 according to size of farm operation. (From USDA, Economic Research Service, *Extent of Farm Pesticide Use.*)

poisons for years. One product of DDT—DDE—may be the most common and widely distributed synthetic chemical on earth.[12]

Herbicides are degraded quickly by soil microbes (i.e., soils rich with organic matter); but the by-products themselves are often toxic.[13] Chemical reports will often recommend that farmers reduce their soil organic matter to make herbicides more effective! The general decrease in organic matter from U.S. farm soils suggests that herbicides will become more persistent in rural areas, especially in view of their increasing use.

2. *Pesticides affect public health:* There is a whole spectrum of opinion here. Some contend that no important human disease can be associated with pesticides.[14] Then there are the "objective" government reports that conclude that there is no danger of pesticides to public health but that calls for administrative controls and

more research. The Mrak report[15] concluded that no one seems to have a clear picture of pesticide hazards in America and predicted that the number of accidents actually exceeds those reported. Others have tried to make correlations between modern-day increases in cancer, poliomyelitis, heart disease, hepatitis, and other diseases and the increasing use of pesticides.[16]

Whatever one's opinion, some definite facts have been established:

a. Certain organophosphates, herbicides, and other pesticides can cause cancer, birth defects, and genetic mutations in animals.[17]
b. Organochlorides do accumulate in fatty tissues of livestock and humans.[18]
c. Organophosphates account for the majority of *known*[19] pesticide deaths in this country, particularly among farmworkers.[20] Thus the trend to replace DDT with organophosphates has largely been a trade-off in health hazards.
d. There also seems to be a relationship between low protein diet and susceptibility to pesticide toxicity.[21] This means that pesticides may have a greater effect on poor people in rural areas. Most likely the total impact of pesticides on human health will not be realized for years to come.

3. *Pesticides kill wildlife and jeopardize natural ecosystems:* Pesticides are poisons . . . they kill other things besides "pests." They are also mobile, chemically stable, and have an affinity for biological systems. It is the combination of these characteristics that can cause unpredictable damage to wildlife; hundreds of papers have been written on the subject.[22]

Not well publicized, but of particular importance to agriculture is the effect that pesticides are having on honeybees.

Large-scale monoculture, necessary for economic production . . . provide no continual source of pollen and nectar necessary to maintain strong colonies [of honeybees]. . . . Use of herbicides . . . further reduces bee forage. The use of pesticides highly toxic to bees either weakens or destroys many colonies. . . . This presents an impending dilemma, with a reduction of profitable beekeeping and native pollinators on one hand and an increased need for bees for crop pollination on the other.[23]

Even pesticides of low toxicity are known to reduce pollen- and nectar-gathering activity of bees.[24]

Most studies of pesticides and wildlife have focused on individual species of animals, notably birds and mammals. Much more important, and far less understood, are the effects of pesticides on the integrity of natural ecosystems. In living communities the activities of each organism impinge on and interact with the activities of other organisms sharing the same general area. So if a pesticide affects only a few individual species this still has strong implications for the living community as a whole. Persistent pesticides are likely to cause the most damage, but even rapidly degrading ones may have lasting effects. One study on a grassland ecosystem concluded that "although the insecticide [Sevin] remained toxic in the environment for only a few days, long-term side effects on . . . arthropod density and diversity, and mammalian reproduction were demonstrated."[25]

These and other studies suggest that the most meaningful way to assess the long-term effects of pesticides on wildlife is to study the ecosystems in which they live.

As an example, consider the impact of pesticides on the most complex and vital land ecosystem of all, topsoil. The problem of determining effects is confounded by the incredible array of life forms living in the soil, and by the many physical factors that influence the persistence of pesticides in the soil. More than 500 papers now exist demonstrating that pesticides may or may not have an effect on various species or groups of soil flora and fauna.[26] However, as noted above, the combined reaction of an ecosystem is a far better indicator of the effects of pesticides than are the responses of its individual members. The vigor of crops, that is, their ability to resist diseases and pests, is dependent upon the complementary metabolisms of many interacting soil microbes that provide a wide variety of major and trace nutrients to the plants.[27] Since pesticides tend to reduce the number of species of soil microbes[28] it is likely that they also tend to upset the balanced nutritional relationship between soil and plant. This suggests that the resistance of crops to pests may be the best measure of the health of a soil ecosystem as it responds to pesticides.

Besides the effects of pesticide fallout on public health and natural ecosystems, there is the fact that, for ecological reasons, the single strategy of chemical control is rapidly becoming an economic disaster for agriculture. In most cases pesticide use actually increases numbers of target pests, fosters new pests, and creates demands for new or more toxic pesticides. There are several rasons for this.

1. In many ways, pesticides free pests from control by their natural enemies (predators and parasites). First, because pesticides are concentrated at the end of food chains . . . being passed from plant, to pest to pest predator . . . the predator tends to suffer greater mortality in the long run than the pest. Second, most natural enemies have longer generation times, are less abundant, and are slower to recover from the effects of poisons than pests. Third, natural enemies contract larger doses of pesticides than pests since they forage over greater areas in search for food. Fourth, herbicides often remove weeds that provide nourishment and refuge for beneficial insects.

2. Pesticides are not capable of "controlling" pests in the ecological and most meaningful sense of the word, that is, over long periods of time. In order to do so they would have to kill a fraction of the pest population that is always proportionate to its density.

3. In general, pesticides reduce biological diversity, and this is likely to produce less stable cropland ecosystems. This point is discussed later in the chapter.

4. Pests often become more resistant to pesticides more rapidly than their natural enemies, which have longer generation times and less opportunity to develop resistances through genetic adaptation. Of more than 225 species of arthropods in which resistant straints have been documented, only four are natural enemies of pests.[29]

5. Repeated pesticide treatments often produce outbreaks of secondary pests—those normally kept in low numbers by natural enemies and competitors. A classic example involves the spider mites (*Tetranychidae*). A minor pest a quarter of a century ago, today it is the most serious insect pest affecting world-wide agriculture.[30]

These factors have precipitated a series of ecological backlashes which in turn have produced a definite pattern of pesticide failure and economic distress. For the farmer the pattern usually starts with a desire for higher yields, and the intensive use of pesticides on a regular schedule. Next, pest resistances compel the farmer to use larger doses or more toxic pesticides. In the meantime, the number of old pests increase and new kinds of pests arise. This forces the farmer to use still more pesticides at higher cost. Finally, as the farmer exhausts his arsenal of pesticides, as pest problems become more severe, and as spiraling costs of pesticides cut into his profits, the farmer simply ceases to become a farmer. The economically depressed rural areas of the Rio Grande River Valley, Lower Mississippi Valley, Imperial and San Joaquin Valleys of California, Canete Valley of Peru, and many other once-prosperous farm communities attest to the fact that the farmer has become a pawn of economic and corporate forces that have placed him on a costly pesticide treadmill. Claims by the USDA that the use of insecticides can produce a net return to the farmer for every dollar invested[31] are shortsighted and grossly misleading. For over a generation the agribusiness establishment has deluded farmers, through economic incentives[32] and propaganda directed by the USDA and the petrochemical industry, to accept pesticides with little regard for their long-term economic and ecological impact. The pattern becomes all the more absurd when we consider the many ways in which a farmer can greatly reduce the use of pesticides without sacrificing yields. (See Chapter 18.)

### SYNTHETIC FERTILIZERS: SALTING THE EARTH

Prior to the mid-nineteenth century, virtually all fertilizers were natural organic materials—plant and animal wastes, manures, etc. Then in 1840 a German chemist, Justus von Liebig, brought together his findings in a book that changed the course of Western agriculture and that laid the foundation for the modern chemical fertilizer industry.[33] Simply put, von Liebig's thesis was that plants could be rapidly nourished by mineral salts in solution instead of the slowly available by-products of decaying wastes in the soil. Enormous yields could be produced by simply mixing salts (nitrogen, potassium, phosphorus) into the moist soils of farmlands. The first chemical fertilizer was a phosphorous compound made by mixing sulfuric acid with bone materials.[34] The process was patented in England in 1842, and by the end of the U.S. Civil War, tens of thousands of tons of "chemical manure" were being produced each year by the British fertilizer industry.

Similar advances with nitrogen and potassium compounds provided the impetus for the complete substitution of synthetic fertilizers for natural fertilizers. By 1954 organic materials accounted for less than 3 percent of the total fertilizers in U.S. agriculture. There have been other trends in the adoption of chemical fertilizers.

1. The *rate* of use has increased steadily since 1850, and rapidly since the beginning of World War II. Most forecasts predict a continued increase in use through the 1980s.
2. Since 1959 nitrogen has been the primary nutrient fed to U.S. crops (Figure 2). Use has increased from 0.5 million tons in 1945 to 6.6 million tons in 1970. By 1980, U.S. agriculture will be using over 11 million tons or about 10 percent of the total world consumption.[35]
3. New developments in production technology have increased the purity of chemical fertilizers and hence the concentrations at which they are used (Figure 2). For example, nitrogen is being used less and less in mixed forms (combined with other nutrients) and more and more in concentrated forms like anhydrous ammonia (Figure 3).
4. In 1970 about 70 percent of all chemical fertilizers used were applied to four crops: corn (95 percent of the corn acres received

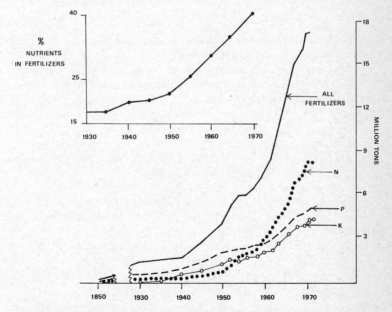

FIGURE 2. Trends in the use of chemical fertilizers in the United States. Top graph shows the increasing concentration of primary nutrients (NPK) used in fertilizers. Bottom graph shows the total use of all fertilizers and primary nutrients. (Adapted from USDA, Economic Research Service, *Fertilizer Situation*.)

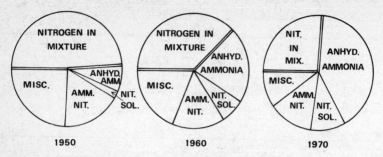

FIGURE 3. Changing proportions of mixed and concentrated nitrogen fertilizer used in U.S. agriculture. Concentrated forms include anhydrous ammonia, nitrogen solutions, ammonium nitrate, and other miscellaneous types such as aqua ammonia, urea, and ammonium sulfate. (Adapted from Tennessee Valley Authority, National Fertilizer Development Center, *Fertilizer Trends*.)

fertilizer); cotton (72 percent); wheat (59 percent); and soybeans (29 percent). For most crops the proportion of acres receiving applications of chemical fertilizers increases each year.[36]

5. Cheapness and ease of shipment have made synthetic fertilizers most attractive. Still, prices appeared to level off around 1969. Because fertilizer production is so heavily geared to fossil fuel inputs, prices will certainly rise sharply in the coming years.[37] For example, the cost of natural gas as a raw material (source of hydrogen) and fuel (to fix atmospheric nitrogen) accounts for 60 percent of the manufacturing cost of ammonia which now supplies about 90 percent of all fertilizer nitrogen. The costs of phosphorous and potassium fertilizer will also increase since they require fossil fuel energy for manufacturing and for the production and mining of raw materials (e.g., phosphoric and sulfuric acids, phosphate and potash rocks).

6. Chemical fertilizers are likely to be used more and more on low-fertilizer using crops, since many of the high fertilizer-using crops such as corn are approaching the maximum profitable rate of application.[38]

Because they are highly soluble salts, chemical fertilizers can overstimulate natural cycles. Before large-scale manufacturing of fertilizers a balance existed between nitrogen removed from the atmosphere by natural fixation and nitrogen returned to the atmosphere by natural denitrification. Today, due to the extensive use of nitrogen fertilizers, there may be an accumulation of nearly 10 million tons per year of fixed nitrogen compounds in the biosphere.[39]

One consequence of this buildup of fertilizer salts has been the overenrichment of local water reserves and the destruction of aquatic animals by eutrophication.[40] Another consequence has been the accumulation of toxic forms of nitrogen in water supplies and crops. Nitrogen toxicity occurs in two major ways: (1) Nitrites ($NO_2$) may react chemically with blood hemoglobin and impair the circulation of oxygen in the blood (methemoglobin), or cause vitamin deficiencies. Nitrite concentration is common in some fodder crops where it poisons livestock, or in vegetables of the brassica or spinach family where it constitutes a real health hazard, especially in baby foods.[41] (2) Nitrites may react with amines in the body to form nitrosamines and related nitrosamides. These compounds are known to induce cancer.[42] Concern that increased use of nitrogen fertilizer may be linked with the growing incidence of cancer in modern society has been expressed by even the medical establishment.[43]

Nitrites are found in large amounts in the soil, in crops, and in ground water where nitrogen fertilizers are used extensively. Apologists for the chemical industry claim that organic wastes, industrial and domestic effluent, and natural processes are more responsible for nitrate and nitrite accumulation in rural areas than fertilizers. But several studies have shown that nitrogen fertilizers can indeed percolate through soils and accumulate in local water supplies.[44] The long-term effects of this contamination are virtually unknown, although one observation by the USDA is ominous:

> The rate of water recharge from deep percolation is so slow that the possible nitrate pollution of aquifers . . . will take decades. However, once nitrate gets into the aquifer, decades will be required to replace the water with low nitrate water. . . . By the time the trend was established, a dangerous situation could be in the making that could not be corrected in a time shorter than it took to create.[45]

Finally there is the whole controversial issue dealing with the effects of chemical fertilizers on soil fertility (as defined by the activities of soil microbes and soil animals) and the quality of crops (as defined by their nutritional value and their ability to resist diseases and pests). These questions represent fundamental gaps in traditional agricultural research, and will not be dealt with here. Suffice it to say that heavy applications of chemical fertilizers may produce plants which are susceptible to attack by insect pests.[46] There appear to be at least three processes involved. First, chemical fertilizers place metabolic stresses on plants, which increase production of aromatic compounds that attract pests;[47] second, chemical fertilizers cause plants to take up water and produce succulent growth favored by pests for shelter and food; and third, the exclusive use of NPK chemical fertilizers reduces incentives for recycling organic material and trace minerals as part of a fertilizer program. Humus and trace minerals provide building blocks for plant enzymes that are important to a plant's defense mechanisms.

## GENETIC EROSION AND MONOCULTURES

For millennia people have domesticated wild species of plants and animals for food, selecting out strains that were palatable and easy to grow. About seventy-five years ago genetic engineers began to develop controlled breeding programs and to select crop varieties that were resistant to some of the notorious diseases and pests which have plagued societies throughout history. These efforts continue today,[48] but they have taken a back seat to the development of a few high-yielding uniform crops that meet the demands of mechanical harvesters and a competitive market economy. There have been three general methods to produce uniform strains: (1) reproduction from a single plant by vegetative cutting (e.g., potato); (2) reproduction by seeds from self-pollinating crops (e.g., lettuce, tomato, wheat, beans); and (3) reproduction by seeds from controlled cross-pollination and inbreeding (e.g., hybrid corn, hybrid cucumber, hybrid onion).

On the surface, results have been spectacular. Yields of most major crops have soared,[49] and machines are now able to harvest "efficiently" and provide the finicky consumer and processor with an abundance of eye-appealing produce. But for several reasons, the new genetic strategy has placed modern agriculture in perhaps its most vulnerable position. For one thing, by forsaking biological quality for yield and appearance, the genetic base for most major

TABLE IV
DOMINATION BY A FEW CROP VARIETIES IN U.S. AGRICULTURE

| Crop | Number of Popular Commercial or Certified Varieties | Major Varieties | |
|---|---|---|---|
| | | No. | % of Crop Acreage |
| Bean, dry | 25 | 2 | 60 |
| Bean, snap | 70 | 3 | 76 |
| Cotton | 50 | 3 | 53 |
| Corn* | 197 | 6 | 71 |
| Peanut | 15 | | |
| " | " | 3 | 70 |
| Peas | 50 | 2 | 96 |
| Potato | 82 | 4 | 72 |
| Rice | 14 | 4 | 65 |
| Soybeans | 62 | 6 | 56 |
| Sugar beet | 16 | 2 | 42 |
| Sweet potato | 48 | | |
| " | " | 3 | 84 |
| Wheat | 269 | 9 | 50 |

SOURCE: Adapted from the National Academy of Sciences, *Genetic Vulnerability of Major Crops.*
* Released public inbreds only, expressed as percentage of seed requirements.

crops has become dangerously narrow. As farming practices rely more and more on a few productive varieties (see Table IV), the numerous strains once grown in local communities and regions are abandoned. Changing land-use patterns further reduce the diversity and distribution of germ plasm by destroying habitats of endemic wild plants. This combined loss of genetic diversity reduces the gene pool from which plant breeders can choose to breed future varieties resistant to future diseases. Such a genetic reserve is important because the evolutionary contest between disease microbes and cultivated stocks is a continuous exchange of mutual adaptations; short-lived microbes mutate and recombine to new diseases, while longer-lived crops struggle to adapt resistance.[50] Likewise, the development of resistant varieties of crops is a continuous process and needs a diverse genetic base from which to operate. Unfortunately, a large proportion of the genes from old varieties has now been discarded; the new varieties represent only a fraction of the gene pool once planted.[51] This loss, although not well publicized and understood by the public, has very serious consequences for the availability of future food supplies.

Moreover most high-yield crops encourage the use of pesticides and synthetic fertilizers; in fact, they have actually been developed in concert with agricultural chemicals. Many of the new varieties appear to be prone to pest attack. This means that farmers are forced to use pesticides to protect their high yields, as well as herbicides to stretch production. In addition these varieties are productive only *because* they are responsive to heavy doses of chemical fertilizers; they are ineffective without this input. Thus, the abundance of high-yield crops always must be weighed against the hazards of the chemicals needed to make them productive.

Finally, planting large areas with the same kind of crop encourages the spread of diseases and pest outbreaks. When the monocultures are extended over broad geographic areas, as they are today in the United States, the potential for crop epidemics is compounded. This is precisely the condition that precipitated the great Irish potato famine of the 1840s and the U.S. corn leaf blight in the early 1970s, when over 15 percent of the total U.S. corn crop was destroyed. These and other examples show that crop monocultures and genetic uniformity actually invite crop diseases, increased pesticide use, and the potential for higher food costs and food shortages.

Unfortunately the market economy rather than common sense determines whether a new crop variety is used. The farmer requires uniform crops for tending and mechanical harvesting. The middlemen require uniformity for processing and mass merchandising. The competitive market permits no alternatives, in spite of the fact that our dependence on a narrow genetic base for our food supply destroys genetic reserves and encourages polluting inputs and crop epidemics. This fundamental dilemma, brought to focus by the 1970 U.S. corn leaf blight, was described in a report issued by the National Academy of Sciences that came to the following conclusions: (a) most major crops are impressively uniform genetically, and impressively vulnerable; (b) this uniformity derives from powerful economic and legislative forces; and (c) increasing vulnerability to epidemics is not likely to automatically generate self-correcting tendencies in the marketplace.[52]

## FARM ENERGY, RISING FOOD COSTS AND CHANGING DIETS

A characteristic of chemical farming is the close relationship between pesticides and chemical fertilizers. These two technologies have developed together and are interdependent. As noted earlier, the use of agricultural chemicals, together with crops genetically geared to their use, have forced the farmer into a treadmill of chemical routines and resources (see Figure 4).

Still, the chemical treadmill is only part of a regime of gas and oil technology that now fuels the fields and cares for crops. Virtually all of the agricultural tools used today depend on fossil fuel energy in one form or another (see Table V). In view of growing fuel shortages (whether real or political), there seems to be little doubt that our present euphoria about farm production is a passing fancy, and that future fuel shortages will have increasing effects on the production, consumption, and price of food.

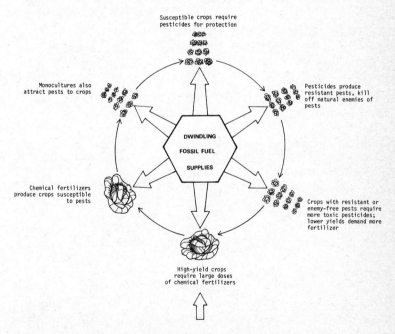

FIGURE 4. The treadmill of chemical dependence, fostered by the high-energy technologies of modern agriculture, and nurtured by a series of ecological backlashes.

TABLE V  FOSSIL FUEL REQUIREMENTS FOR DIFFERENT ASPECTS OF AGRICULTURE

| Category | Energy Source (in millions of units) | | | | | | Total |
| | Natural Gas Therms | Electricity Kwh | Diesel Fuel Gal. | Gasoline Gal. | LP Gas Propane Butane Gal. | Aviation Fuel Gal. | EQUIVALENT million bbls crude oil |
|---|---|---|---|---|---|---|---|
| Field crops | 364.784 | 464.681 | 96.400 | 19.477 | 2.381 | — | 9.34 |
| Vegetables | 165.999 | 358.193 | 38.792 | 25.031 | 4.441 | — | 4.62 |
| Fruits and nuts | 127.168 | 410.773 | 26.158 | 12.602 | 3.296 | — | 3.39 |
| Livestock | 107.111 | 1,460.966 | 46.443 | 7.813 | 12.261 | — | 4.19 |
| Irrigation | 40.618 | 7,177.441 | 6.531 | .487 | 4.521 | — | 5.16 |
| Fertilizers | 305.748 | 579.362 | 6.738 | 3.529 | 1.114 | — | 5.87 |
| Frost protection | — | 40.501 | 60.003 | 6.854 | .904 | — | 1.63 |
| Greenhouses | 102.700 | 83.427 | — | — | — | — | 1.82 |
| Agr. aircraft | — | — | 1.072 | 1.607 | — | 8.994 | 0.25 |
| Vehicles (farm use) | — | — | 10.447 | 117.798 | — | — | 2.77 |
| Others | — | — | — | — | 23.711 | — | 0.39 |
| TOTAL | 1,214.128 | 10,575.344 | 292.584 | 195.198 | 52.629 | 8.994 | |
| EQUIVALENT (Million bbls crude oil) | 20.93 | 6.21 | 7.06 | 4.17 | 0.86 | 0.19 | 39.43 |

SOURCE: Cervinks et al., *Energy Requirements for Agriculture in California*, California Dept. of Food and Agriculture, Sacramento, California.
NOTE: Data are for California, 1972.

Agriculture, of course, is just the starting point of a large food industry that includes production, processing, transportation, and marketing, plus domestic storage and cooking. In 1963 these food-related activities consumed about 12 percent of the total U.S. energy budget, or the equivalent of about 240 gallons of gasoline per person (see Table VI). Assuming that one person eats about one million kilocalories of food energy per year, or about 29 gallons of gasoline, it seems that our notions about food as an energy supplier are largely an illusion.

There is some evidence that agriculture itself has become an energy sink. Pimentel has shown that with regard to U.S. corn production, the ratio of energy in yields to energy in production inputs ("production efficiency") has started to decline in recent years.[53] (See Figure 5.) This decline has a profound effect on other food industries since corn supplies livestock feed as well as oil and food.

The production efficiencies of other raw foods are listed in Table VII. Although there is considerable variation in energy intensiveness among different crops, on the average most seem to

---

TABLE VI

DISTRIBUTION OF TOTAL PER CAPITA ENERGY REQUIREMENTS FOR FOOD CONSUMPTION IN THE UNITED STATES (1963)

| Source | Million BTUs | Heat Equivalent Gal. Gasoline | % of Total |
|---|---|---|---|
| Agriculture | 5.8 | 43.0 | 18 |
| Food Processing | 10.6 | 78.5 | 33 |
| Transportation | 0.9 | 6.7 | 3 |
| Wholesale & Retail Trade | 5.2 | 38.5 | 16 |
| Domestic (Storing, Preparing, Transporting) | 9.9 | 73.3 | 30 |
| TOTAL | 32.4 million BTUs per person per year | 240 gallons of gasoline per person per year | 100% |

SOURCE: Adapted from Eric Hirst, *Energy Use for Food in the United States*, Natural Science Foundation Environmental Program, Oak Ridge, Tennessee.

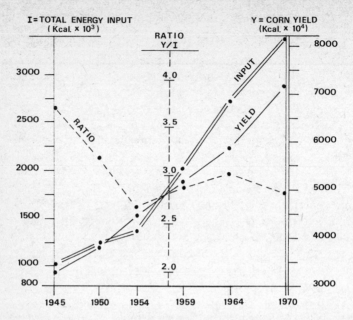

FIGURE 5. Trends in the energy efficiency of U.S. corn production (1 Kcal = 3.97 BTU). Energy inputs include labor, machinery, gasoline, NPK fertilizer production, seeds, irrigation, pesticides, crop drying, electricity, and transportation. Note that on scale, figures for yields are ten times greater than those for energy inputs. (Adapted from Pimentel et al., "Food Production and the Energy Crisis.")

use about as much energy for production as they provide for sustenance.[54]

Viewed in terms of an energy budget, then, modern agriculture does not seem so efficient. In fact, it may be less efficient than more "primitive" forms of agriculture.[55] So despite the high yields of modern farm technology, there does not appear to be an obvious net return of energy to society. In effect, the benefits of solar energy fixed in our foods are offset by the subsidy of fossil fuel energy needed to produce the foods.

Obviously there are strong implications in the fact that the principal raw material of modern agriculture is a dwindling, non-renewable resource. For one thing, there is the relationship between the inevitable rise in fossil fuel prices and the availability of food, especially products like processed foods and animal protein that require high energy inputs.[56] In fact, meat may become so expen-

TABLE VII

ENERGY EFFICIENCIES OF DIFFERENT CROPS PRODUCED
IN CALIFORNIA, 1972

| Commodity | Crop Energy Value Gal. Gasoline/Ton | Primary Energy Inputs* Gal. Gasoline/Ton | Ratio: Crop Energy/ Input Energy |
|---|---|---|---|
| *Field Crops* | | | |
| Barley | 101.9 | 15.4 | 6.6 |
| Corn | 107.5 | 46.5 | 2.3 |
| Rice | 106.0 | 41.5 | 2.6 |
| Sorghum | 97.0 | 38.3 | 2.5 |
| Wheat | 97.2 | 18.1 | 5.4 |
| *Raw Vegetables* | 12.3 | 15.9 | .77 |
| Range | (5.0–35.9) | (8.4–65.9) | |
| *Raw Fruits* | 15.7 | 28.8 | .54 |
| Range | (11.1–19.6) | (12.9–53.1) | |
| *AVERAGE OF ALL* | | | |
| *RAW FOODS* | 29.9 | 21.9 | 1.36 |
| | | | |
| *Canned Vegetables* | 17.1 | 67.1 | .25 |
| Range | (6.1–28.0) | (36.7–97.3) | |
| *Canned Fruits* | 12.3 | 48.7 | .25 |
| Range | (8.8–14.9) | (35.9–57.9) | |
| *Frozen Vegetables* | 14.8 | 68.6 | .22 |
| Range | (6.4–29.8) | (52.2–92.0) | |
| *Dried Fruits & Nuts* | 142.4 | 222.8 | .63 |
| Range | (93.6–170.9) | (133.4–322.4) | |
| *AVERAGE OF ALL* | | | |
| *PROCESSED RAW FOODS* | 45.5 | 97.2 | .47 |

SOURCE: Adopted from Cervinka et al., *Energy Requirements*.
* Inputs include machines (planting, culture, harvest, transport, storage, processing); fertilizers (production, transport, handling, application); irrigation; farm vehicles; frost protection; and airplanes.

sive in the future that it will probably be replaced by plant protein in the diets of many people. But plant protein may also be hard to get as the United States carries out its basic 1970s foreign policy and uses domestic grains and legumes to reduce balance of payments and to barter for oil and natural gas. Inevitably fuel shortages will lower the quality of food for most people, especially the poor. On the positive side, the changes should precipitate a renewed interest in urban "Victory Gardens" and microfarms, and provide impetus to develop food cooperatives and local food economies from the inner city to suburbia. (See Chapter 12.) Also the strain of fuel shortages on food production may finally stimulate a broad-based approach to the development of ecologically sound, energy-saving approaches to agriculture. These might include such things as incentives for using cropping patterns that limit the scale of mechanization; the utilization of renewable energy resources (solar, wind, organic fuels) to supplement rural power needs (see Chapter 20), and increased emphasis on integrated pest control, organic waste recycling, and other low-chemical[57] farming practices.

### AGRICULTURAL RESEARCH: RESTRICTING THE OPTIONS

Unfortunately, very little has been done to develop techniques for people to make farms more practical and productive within a low-energy framework. Traditional agricultural research continues to develop, promote, and extend chemical, energy-consuming technologies for the sake of pragmatism and economic "efficiency." Even the scientific establishment has criticized the USDA and state research institutions for supporting "pedestrian and inefficient work," for being guided by policies "repressive to the vitality of science," and "detrimental to the interests of agriculture," and for neglecting basic research.[58] This last criticism is especially important, since it points to the inability of conventional agricultural research to investigate and promote alternative methods of farming which stem from *basic* ecological principles.

> The antitheoretical bias of agricultural science is reinforced by the search for marketable products (mostly chemical) as the central strategy for improvement of agriculture . . . and by a narrow acceptance of the present structure of agriculture as a given condition which restricts options. For example, the consideration of mixed plantings is inhibited by the present design of farm machinery. Therefore research into the ecology of a mixed sowing only makes sense as part of a broader program that must include an engineering effort to redesign the machines.[59]

In other words, much agricultural research is heavily biased and restricted by a narrow set of technological assumptions (see also Chapter 7). In contrast, a true agricultural *science* is not guided exclusively by economic restrictions, but also by biological realities; it examines the potentials of food production from an ecological point of view, from a self-sustaining point of view, realizing that applied ecology is nothing more than long-term economics.

### Ecosystem Farming: A Self-Sustaining Technology

Agricultural systems are essentially artificial communities of domesticated plants and animals. In order to understand the use of ecological tools in agriculture, we might first consider briefly the ecological characteristics of natural plant and animal communities. These characteristics can then be used as models for an alternate agriculture.

#### MODELS IN NATURAL ECOSYSTEMS

Although not always obvious, wildlife communities have a biological integrity. Not only do particular groups of creatures usually live in a particular habitat, but the habitat is modified and new habitats are created by the living community itself. The two dynamics evolve to form a self-sufficient habitat or "ecosystem." We have, for example, a grassland ecosystem, a forest ecosystem, a pond ecosystem . . . an agricultural ecosystem.

The sun provides the energy for the running of ecosystems. This energy is stored by green plants and passed on through a food chain of plant-eaters and flesh-eaters. Wastes and dead bodies become food for microbes which decompose complex organic matter into simple materials that can be reused by plants. Available food energy is gradually lost as work and metabolic heat at each link along the food chain. *In ecosystems, then, matter cycles and energy flows.*

For the efficient conversion of matter and energy back into the life cycle, ecosystems are held together by very diverse but specific kinds of plant-animal relationships. Since energy and matter are lost as they pass along the food chain, plants are more abundant than plant-eaters and these, in turn, are more numerous than flesh-eaters. Since most animals have a variable diet, food chains intermingle in an ecosystem and form a complex food "web." Generally speaking, the more complex the food web, the less likely it is that a natural disturbance or outbreak will alter the integrity of the ecosystem or cause individual members to become extinct. Hence, *in natural ecosystems, there is a relationship between biological diversity and internal stability.*

Other important characteristics of natural ecosystems have been described by Pimentel:

1. Animal populations in natural biotic communities are relatively stable; outbreaks . . . are generally rare.
2. Most species are rare in relative numbers.
3. The majority of animals feed on living matter as opposed to non-living. . . . Although many animals are associated with dead matter, most of these animals . . . are feeding on [microbes] . . . present in the decaying matter.
4. Population outbreaks frequently occur with newly introduced animal species. The plant hosts on which the newly introduced animal is feeding often lack resistance to it.
5. Resistance factors which limit the feeding of animals on host plants are common in nature. [These might include] spines . . . toxins . . . growth inhibitors, etc.[60]

## THE FARM AS AN ECOSYSTEM

Farmlands, on the other hand, are artificial ecosystems; they reflect important differences from the workings of natural ecosystems.

1. Agro-ecosystems are *open communities* of limited duration. Because of cultivation and harvest, there is little opportunity for plant nutrients to be recycled. Natural ecosystems, however, are nearly *closed communities* since plants feed on the decayed bits of their recycled predecessors.
2. Cultivated plants and animals are particularly susceptible to attack from pests and diseases; most natural defenses have been bred out of them in favor of productivity and palatability.
3. Demands for agricultural efficiency are really demands for biological simplicity and uniformity. Hence, agro-ecosystems contain only a few species of plants and animals, which are substituted for the more complex network of the wildlife community. Strong interactions often develop among these few species and their associated competitors and predators. The system is simple, unstable, and easily disturbed. This is why pest populations are often larger in monocultures than in mixed-species stands.[61] This "simplicity"

of agro-ecosystems takes on several forms: (a) *Crop simplicity*: Fields are usually planted to a single kind of crop, occasionally two (intercropping), very rarely several (mixed stands). Herbicides eliminate weeds so that only one kind of plant prevails. (b) *Genetic simplicity*: Crops and livestock are usually of one high-yielding variety or inbred line of hybrid stock that often, in the name of production, forsakes disease resistance and adaptability. (c) *Structural simplicity*: The farm landscape is structurally simple, without hedgerows, trees, weeds, and other refuges of beneficial animals to interfere with efficiency. (d) *Ecological simplicity*: the farm ecosystem is simple with respect to the number of relationships among links in the food chain. For example, in the case of crop simplicity, or when livestock are removed to feeding lots so that manure cannot be recycled locally, or when natural enemies of pests are eliminated by indiscriminate pesticides, etc., the farm is made unstable in an ecological sense and requires large inputs of fossil-fuel energy to replace the biological energy that usually maintains balance and stability.

Generally speaking, farming practices of fertilizing soils and controlling pests fall in a spectrum between the chemical and ecological extremes. Today, practically all farming is at the chemical extreme. But, as I have tried to show, there is a growing imperative for ecological alternatives in farming which foster a more stable and closed agro-ecosystem. We might consider briefly some of these alternatives.

### ORGANIC WASTE RECYCLING AND CONVERSION

In 1971, the cities spent over $3.5 billion to collect and dispose of solid wastes. Next to schools and roads this was the costliest of all domestic public services. In that same year, the United States generated about 880 million tons of organic waste solids (see Table VIII). Less than 0.1 percent of this was sold commercially as compost, dried manure, and processed sewage sludge. An equally small amount was used by farmers as fertilizer and soil conditioner (see Table IX).

Unfortunately, most organic waste is not returned to the land but is either burned or dumped into oceans, rivers, and landfills. This unwillingness to close a basic ecological cycle occurs at both ends of the human food chain. On farmlands, the centralization of

TABLE VIII

DRY, ASH-FREE ORGANIC WASTES PRODUCED IN U.S., 1971
(IN MILLIONS OF TONS)

| Source | Waste Generated | Readily Collectable |
|---|---|---|
| AGRICULTURE | | |
| Crops & food waste | 390 | 22.6 |
| Manure | 200 | 26.0 |
| URBAN | | |
| Refuse | 129 | 71.0 |
| Municipal sewage solids | 12 | 1.5 |
| INDUSTRIAL WASTES | 44 | 5.2 |
| LOGGING & WOOD MANUFACTURING | 55 | 5.0 |
| MISCELLANEOUS | 50 | 5.0 |
| TOTAL | 880 | 136.3 |

SOURCE: L. L. Anderson, *Energy Potential from Organic Wastes: A Review of the Quantities and Sources*, Bureau of Mines, Information Circular 8549 (Washington, D.C. U.S. Department of the Interior, 1972).

livestock production has produced a waste "problem" instead of a fertilizer resource, while the expediencies of a few salt fertilizers have replaced the traditional patterns of crop rotation, green manuring, and lea farming. In the cities, the return of organic wastes to farmlands is hindered by the attitude that municipal composting is economically inefficient and *therefore* not a viable alternative to simple dumping and burning.[62] True, at this time the composting process is expensive and there is little demand for the benefits of compost; operating costs are about the same price as incineration.[63] Also the bulk of organic wastes makes them impractical to the farmer when compared with salt fertilizers that can be applied easily by machines or in irrigation water.

On the other hand, the shipment of millions of tons of agricultural produce from the country to the city and then into waterways or up into the air is a classic example of how rural and urban problems have been separated from their common ecology. The recycling of organic wastes on farmlands has the potential for creating new jobs, reversing the rapid loss of humus in farm soils, and offering a low-energy alternative to agriculture (see Table X).

TABLE IX
NATURAL ORGANIC MATERIALS USED (IN TONS) AS
FERTILIZERS IN U.S. AGRICULTURE

|  | | *1967* | *1968* | *1969* |
|---|---|---|---|---|
| 1. | DRIED BLOOD | 2,154 | 2,173 | 2,251 |
| 2. | CASTOR POMACE | 2,924 | 2,584 | 1,626 |
| 3. | COMPOST | 35,171 | 41,284 | 19,472 |
| 4. | COTTONSEED MEAL | 3,920 | 2,522 | 5,416 |
| 5. | DRIED MANURE | 366,013 | 363,418 | 350,557 |
| 6. | SEWAGE SLUDGE | 124,681 | 138,355 | 131,062 |
| 7. | TANKAGE | 9,407 | 6,606 | 6,732 |
| 8. | OTHER | 9,709 | 14,005 | 18,601 |
|  | | 553,979 | 570,947 | 535,717 |

SOURCE: U.S. Dept. of Agriculture, *Consumption of Commercial Fertilizers in the United States* (Washington, D.C.: Statistical Reporting Service, USDA, 1968, 1969).

Most important, it provides the first positive step toward the much-needed integration of cities and farms and the mutual solution of urban and rural problems.

TABLE X
ENERGY BUDGET OF ORGANIC AND CHEMICAL
FERTILIZER APPLICATIONS

| *Organic Fertilizer* (10 tons of cow manure per acre) | *Chemical Fertilizer* (N: 112 lbs; P: 31 lbs; K: 60 lbs per acre) | |
|---|---|---|
| HAULING AND SPREADING (Kcal/acre) | PRODUCTION (Kcal/acre) | APPLICATION (Kcal/acre) |
| 398,475 (11 gal. gasoline) | 1,415,200 (39 gal. gasoline) | 36,325 (1 gal. gasoline) |

SOURCE: Adapted from Pimentel et al., "Food Production."
NOTE: Substituting cow manure for chemical fertilizers could save a potential 1.1 million Kcal/acre.

Organic matter is not only a valuable fertilizer and soil conditioner, it can also be converted into energy sources. For example, the U.S. produces about 136 million tons of *easily available* organic solids each year (see Table VIII). This is equivalent to 170 million barrels of oil (3 percent of 1971 U.S. consumption) or 1.36 trillion cubic feet of methane gas (6 percent of the 1971 U.S. consumption). Put another way, it is equal to 150 percent of the energy used to run all U.S. farm tractors.

At present there are few economic incentives to begin such a tradition of recycling and converting organic wastes. Broad-based changes will no doubt depend on the inevitable high cost of fuels and synthetic fertilizers, sufficient research rationale, the acceptance of municipal composting and sludging, and the decentralization of livestock production. Meanwhile, there are always local efforts.

### DECENTRALIZED STOCK BREEDING—RESISTANCE AND DIVERSITY

If genetic uniformity makes crops vulnerable to pests, then genetic diversity is the best insurance against outbreaks. Following are several approaches:

1. The preservation of local plant life and wilderness ecosystems as genetic reservoirs[64]
2. The selection and storage of crops with diverse gene pools
3. The selection and rotation of varieties adapted to regional conditions and resistant to local as well as pandemic pests. This would establish buffer areas against widespread crop epidemics, and provide more options for the success of local food production
4. The reintegration of livestock with plant crops and the selection of animal varieties that can fend for themselves in reproduction, protect themselves from weather and disease, and develop their own patterns of group behavior[65]

Obviously these approaches are based on a dramatic decentralization of current plant-breeding programs and a sharp turn in research priorities from quantity to biological quality. Because the requirements for yield and crop uniformity are so ingrained in our food economy, it is difficult to imagine incentives for change, short of actual epidemics or radical economic reconstruction. Meanwhile, at the local level, community efforts in new rural areas and grass-roots research can promote the genetic diversity of crops by protecting indigenous plant life, by seeking out adaptable crops from independent seed and livestock companies (especially out-crossing and non-hybrid types), and by selecting for local varieties and strains whenever possible. Also, cooperative gene banks may have as much value for endemic agricultures in the future as cooperative distribution groups have had in the past.

### BIOLOGICAL CONTROLS AND DIVERSE FARMLANDS

As pointed out by Rudd,[66] any analysis of pesticide failure suggests three qualities to look for in alternative methods of pest control: (1) capability of keeping pests at a harmless density; (2) ability to prevent pests from developing resistances; and (3) ability to work with and not against the controls provided by natural enemies. Several methods fulfill these requirements in one way or another. Sex-scent attractant, reproductive hormones, sterile-male radiation, and traps have all been used successfully. But by themselves these techniques are based on a strategy of *discouraging the pest*. A more permanent and stabilizing strategy in terms of closing the agro-ecosystem is based on *encouraging the natural enemies of pests*—their predators, parasites, and dis-eases.

Pest enemies can be encouraged in two ways: they can be reared in large numbers under controlled conditions (insectaries) and released at strategic times into crop areas; or, crop patterns and the local environment can be modified so as to favor the life histories of beneficial insects already in the fields. Changes might include the cultivation of companion crops, the maintenance of uncultivated areas, or the establishment of permanent refuge habi-tats. There are several ways to improve the ecological stability of croplands; only a few specific relationships will be mentioned here for illustration.

1. *Flower crops*: The nectar and pollen of many flowers provide food for beneficial adult insects.[67] In orchards, for example, wild flowers can nurture populations of parasitic wasps and thereby reduce certain pests.[68] Research in Russia has shown that when the weed *Phacelia* was planted in orchards, a parasite of the tree's scale pest thrived in the orchard by subsisting on the nectar of the weed. When the population of the pests increased to dangerous

levels, the parasites were sufficient in number to control the pests, thus avoiding the unpredictable lag period that normally occurs between the appearance of a pest and its natural enemy. Another Russian study showed that when small plots of umbellifers were planted near vegetable fields in a ratio of one flower plant to 400 crop plants, up to 94 percent of the cabbage cutworms were parasitized. Flowers of crop plants such as brassicas, legumes, and sunflowers can also serve as alternative food sources for beneficial insects.

2. *Repellent crops*: Most insects are selective as to the kinds of plants they eat. It is generally held that insects are attracted to the odors of "secondary" substances in plants rather than to the food value of the plants themselves.[69] Experiments have shown that odors given off by aromatic plants interplanted with crops can interfere with the feeding behavior of pests by masking the attracting odor of the crops.[70] This means that certain kinds of plant diversity per se may have a profound effect on pests over and above that conferred by natural enemies. Repellent crops so far described include various pungent vegetables (*Solanum, Allium*) and aromatic herbs (*Labiatae, Compositae*, and *Umbelliferae*).

3. *Trap crops*: Some plants can be used to attract pests away from the main crop. With careful monitoring, these "trap" crops can also serve as insectaries for natural enemies. For example, when alfalfa strips were interplanted with fields of cotton, the Lygus bug (a serious pest in California) migrated away from the cotton and into the alfalfa.[71] With their concentrations of Lygus bugs, the alfalfa plots then provided a food source for several predatory insects in the area. Trap crops of alfalfa may also have applications in walnut and citrus orchards and bean fields. In the coastal climate of California, members of the cabbage family (*Brassicae*), which attract large numbers of aphids, can function as overwintering insectaries for parasitic wasps. When aphids attack other crops in the spring, wasp populations, having fed on aphids during the lean winter, are large enough to respond quickly and control the aphids.

4. *Hedgerows and shelter belts*: For centuries hedgerows have been planted between field crops to slow down winds and thus reduce wind erosion and improve microclimates. The presence of uncultivated land near cultivated fields also has a profound effect on the distribution and abundance of insects associated with crops. Wild plant stands can provide alternative food and refuge for pests and their natural enemies alike. In fact, almost every advantage offered

to the one is, to some degree, available to the other. Hence, it is not always clear whether uncultivated land is beneficial or harmful to pest control. However, in England, where much farming is done near wild vegetation, pest problems are generally less severe than in the United States, where monoculture farming persists.[72]

Many other kinds of plant relationships can be cultivated to advantage. Some "component" species probably serve more than one beneficial function. Repellent herbs, for example, also produce food-rich flower heads, as do many trap crops. Garden models (see Figure 6) exist for a variety of mixed cropping schemes,[73] and these undoubtedly could be tested and applied on a larger scale. Interest, however, will probably remain focused on monocultures until the "costs" of pesticides and poor farm management exceed the "costs" of ecological designs.

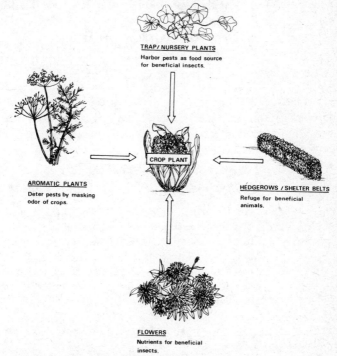

FIGURE 6. Some possible components and plant interactions of a diverse cropping system. Based on a garden model of "companion planting" arrangements.

Speaking generally, interplanting the farm landscape with trees, hedgerows, and other perennial stands, together with rotations,[74] strip cropping, and mixed stands will promote stability and effective natural pest control. But diverse landscapes and mixed farming methods per se will not create stability.[75] Sometimes diversity decreases pest damage, other times it may increase it. The web of possible plant-animal relationships is immeasurably complex, and each situation and crop ecosystem is unique. In other words, the right kind of diversity must be established, and we can only know that by practical experience in local areas and ecological studies of the agricultural environment.

There is yet another level of ecological complexity in agriculture that has tremendous potential on a decentralized scale, but which is as yet to be fully explored. This is the idea of the polyculture farm. The concept is borrowed from practices of the rice-vegetable-fish-livestock economies of southern and eastern Asia. Adapted to current information about ecological principles and a holistic science, modern polyculture farms would link several artificial ecosystems in a balanced and relatively self-sufficient complex of renewable energy systems, mixed crops, aquaculture, plus livestock and insect husbandry (see Figure 7). At present, several grass-roots groups in America (see Chapter 15) and Europe are investigating various ways to integrate renewable food and energy systems into endemic polyculture schemes.[76]

### Back to the Land—Forth to the Land and Postindustrial Agriculture

> And so I believe that the Back to the Land idea is a long-term goal; no one now living will live to see it fully developed. It will be a long, slow movement . . . not, I hope, toward an Earthly paradise, urban or rural, but toward a new nativity of our people in the real world and in the scheme of things.
>
> —Wendell Berry
> *The Long Way Back to the Land*

Popular notions about agriculture in the future often depict great monoculture deserts, rows of high-rise livestock cages, and antiseptic greenhouse complexes being nurtured by chemical robots

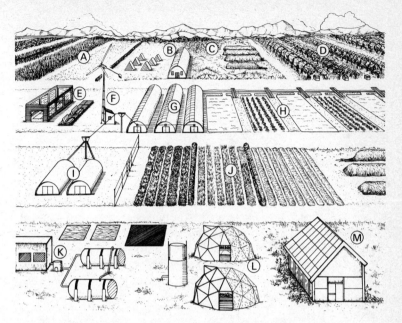

FIGURE 7. An idealized polyculture farm and research center.

A = Grains, green manure, and forage crops.

B = Solar grain-drying structures.

C = Windrow compost piles on fallow section.

D = Diverse, multi-storied orchards with undergrowth of refuge plants and green manure crops.

E = Fish-food stocks; aquaculture insectary and worm cultures.

F = Wind generator.

G = Solar-heated Quonsets for rearing warm-water fishes and crustaceans.

H = Outdoor fish ponds and row-crop fields.

I = Agriculture insectary for rearing beneficial insects.

J = Experimental vegetable/herb /flower/weed beds for investigating diverse cropping systems.

K = Poultry shelter with methane digesters, sludge ponds, and gas storage tanks.

L = Dome-greenhouses.

M = Solar-heated laboratory/ homesite.

and computers. But as agriculture reaches its limits of space, resources, and pollution, the course of agriculture will come to depend less on the "progress" of an industrial society and more on the resolution of three fundamental issues: First, there is the *relationship between food production and energy.* Agriculture has always depended on cheap energy: human labor, beasts of burden, or oil. As the world's fossil fuel reserves become scarce on a seller's market, industrial agriculture will be left with two alternatives—nuclear power or solar energy (direct solar, wind, and organic fuels). The use of nuclear energy in agriculture will accelerate the current monopoly of farmlands by corporations, industry, and their esoteric technologies; it will take farming out of the hands of farmers once and for all. Also the thought of radioactive wastes in the human food chain summons the belief that a solar energy technology for farming will return to agriculture its proper function—transferring the sun's power for human needs. Since solar energy is readily available and easily devised (compared to nuclear power), its application to farming practices will tend to keep agriculture in the hands of farmers and pace down the acceleration and monopoly of the high-energy farm factory. That is, a decentralized solar economy for agriculture would extend the role of the rural community from an independent social order and help keep it from degenerating into a total extension of the industrial one.

Next, there is the *relationship between science and human needs.* In recent years conventional science has come under increasing attack for the moral implications of its basic inquiries and the long-term significance of its applied tools. There has been relatively little criticism of the agricultural sciences along these lines since the total external costs of modern farming practices are just beginning to surface with a broad impact. In light of these "costs," what are needed are ecological tools that keep farming productive, and economic incentives that make them practical. To do this, new questions have to be asked in the research laboratories and, most important, in the experimental fields. What are the long-term effects of farm chemicals on human health, and what are the options? How can diversified farms be integrated with adequate markets? How can renewable energy sources be integrated with crop production? Such questions reflect a holistic and extended approach to agricultural science; that is, a science that controls the questions being asked. Most likely, new-farm research will be carried out at the local level, for local purposes. New models for a land-based agriculture are not apt to come from organized

FIGURE 8. Courses for future agricultures.

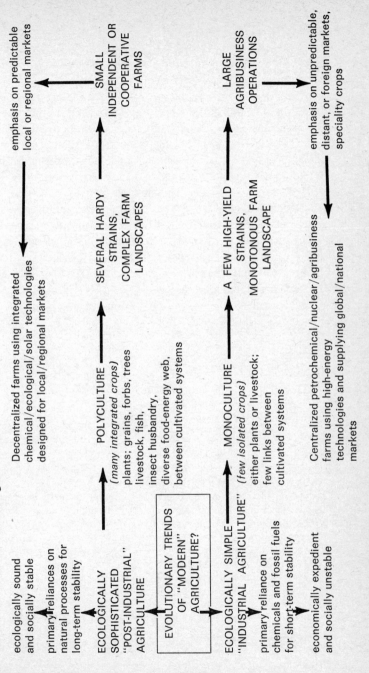

science, but from the ability of local groups to use their own kind of inquiry.

Finally there is the *relationship between people and land*. During the last century, the United States has experienced one of the largest internal migrations in human history, from the farm to the city. But during our flight from freedom to convenience there have been increasing signs that the changing relationship between people and land, plus the concentration of people into urban centers are at the heart of our most basic social problems. As a result people are again looking to the land. In the 50s and 60s a federal network of roads and reservoirs made it possible for people to retrace old trails to America's heartland with industry, recreation, second homes, and retirement communities. But beyond this, way beyond, is the need to make productive land available to all people who wish to farm. Since 1967 the value of farmland has increased by nearly 80 percent and there is no end in sight. As long as there is speculation and land monopolies, agriculture will continue its course to the industrial state, and all visions of a self-sustaining agriculture will lie fallow in our hopes and dreams.

## Notes

1. Karl Polanyi, *The Great Transformation: The Political and Economic Origins of Our Time* (Boston: Beacon Press, 1944).

2. P. G. Sears, *Deserts on the March* (Norman, Okla.: Univ. of Oklahoma Press, 1947); E. Hyams, *Soil and Civilization* (London: Thames & Hudson, 1952); V. G. Carter and T. Dale, *Topsoil and Civilization* (Norman, Okla.: Univ. of Oklahoma Press, 1974, rev. ed.)

3. U.S. Dept. of Agriculture, *Two-thirds of Our Land: A National Inventory*, SCS Program Aid 984 (Washington, D.C.: USDA, 1971).

4. The amount of U.S. farmland under cultivation has actually decreased over the last generation. Unfortunately the urbanization of prime farmlands is beyond control. Opening up marginal lands for farming requires high-energy technologies and increases the threat of pollution and exploitation of finite resources. It is a vicious problem with no obvious solution under present priorities of uncontrolled growth and development.

5. U.S. Tariff Commission, *Synthetic Organic Chemicals, U.S. Production and Sales* (Washington, D.C.: Government Printing Office, 1919 Annual Reports).

6. In his book, *Defoliation*, Thomas Whiteside notes a 1968 statement by Samuel Huntington, Southeast Asia advisor to the State Department: "In an absent-minded way the United States in Vietnam may have stumbled

upon the answer to 'wars of National Liberation' . . . forced draft, urbanization and modernization which rapidly brings the country in question out of the phase in which a rural revolutionary movement can . . . come to power."

7. Tariff Commission, *Synthetic Organic Chemicals.*

8. Environmental Protection Agency, *The Use of Pesticides in Suburban Homes and Gardens and Their Impact on the Aquatic Environment*, Pesticide Study Series #2 (Washington, D.C.: EPA, 1972).

9. U.S. Dept. of Agriculture, *Quantities of Pesticides Used by Farmers in 1966*, Economic Research Service, Agricultural Economic Report 179 (Washington, D.C.: Government Printing Office, 1970); U.S. Dept. of Agriculture, *Extent of Farm Pesticide Use on Crops in 1966*, Economic Research Service, Agricultural Economic Report 147 (Washington, D.C.: Government Printing Office, 1968).

10. USDA, *Quantities of Pesticides.*

11. J. Cohen and C. Pinkerton, "Widespread Translocation of Pesticides by Air Transport and Rain-out," in *Organic Pesticides in the Environment*, American Chemical Society Advances in Chemistry, Series 60 (1966): 163–176. L. Weaver et al., *Chlorinated Hydrocarbon Pesticides in Major U.S. River Basins*, Public Health Report 80 (1965): 481–493; R. J. Anderson, statement in *Interagency Coordination in Environmental Hazards (Pesticides)*, hearings, part 3 (Washington, D.C.: Government Printing Office, 1963).

12. R. L. Rudd, "Pesticides in the Environment," in *Environment: Resources, Pollution and Society*, ed. W. W. Murdoch (Stamford, Conn.: Sinauer Assoc., 1971).

13. D. Pramer, "The Soil Transforms Environment," *Environment* 13, no. 4 (1971): 42–46.

14. W. J. Hayes, "Pesticides in Relation to Public Health," *Annual Review of Entomol.* 5 (1960): 379–404.

15. U.S. Dept. of Health, Education and Welfare, *Report of the Secretary's Commission on Pesticides and Their Relationship to Environmental Health*, Mrak Report, (Washington, D.C.: Government Printing Office, 1969).

16. Morton Bisland, "Public Health Aspects of the New Insecticides," *Amer. Jour. Dig. Dis.*, November 1953, pp. 331–341.

17. HEW, Mrak Report.

18. V. Fiserova-Bergerova et al., "Levels of Chlorinated Hydrocarbon Pesticides in Human Tissue," *Ind. Med. Surg.* 36, no. 65 (1967).

19. The Mrak Report concluded that the majority of counties in the United States are not equipped to determine a pesticide as cause of death.

20. Committee on Labor and Public Welfare, "Pesticides and the Farm Worker," hearings before the Subcommittee on Migratory Labor, parts 6A-C (Washington, D.C.: Government Printing Office, 1970).

21. E. M. Boyd, *Protein Deficiency and Protein Toxicity* (Springfield, Ill.: Charles C. Thomas, 1972).

22. O. B. Cope, "Interactions between Pesticides and Wildlife," *Annual Review of Entomology* 16 (1971): 325–364; C. A. Edwards, *Environmental Pollution by Pesticides* (New York: Plenum Publishing, 1973); N. W. Moore, (ed.), "*Pesticides in the Enviornment and Their Effect on Wildlife.*" Journal Applied Ecology 3, (suppl.), 1966; Hunt, Eldridge, "Biological magnification of pesticides." Symposium on scientific aspects of pest control. National Academy of Sciences, National Research Council, 1966.

23. E. C. Martin and E. C. McGregor, "Changing Trends in Insect Pollination of Commercial Crops," *Annual Review Entomol.* 18 (1973): 207–226.

24. F. E. Todd and C. B. Reed, "Pollen Gathering of Honey Bee Reduced by Pesticide Sprays," *Journal Economic Entomology* 62, no. 4 (1969): 865–867.

25. H. G. Barre, "The Effects of an Acute Insecticide Stress on a Semi-enclosed Grassland Ecosystem," *Ecology* 49 (1968): 1019–1035.

26. C. S. Helling, P. C. Kearney, and M. Alexander, "Behavior of Pesticides in Soil," *Advances in Agronomy* 23 (1971): 147–240.

27. N. A. Krasil'nikov, *Soil Microorganisms and Higher Plants*, Academy of Sciences, USSR (Washington, D.C.: Office of Technical Services, U.S. Department of Commerce, 1958).

28. W. Kreutzer, "Selective Toxicity of Chemicals to Soil Microorganisms," *Annual Review Phytopathology* 1 (1963): 101–126.

29. G. P. Georghiou, "The Evolution of Resistance to Pesticides," *Ann. Rev. Ecol. Syst.* 3 (1972): 133–168.

30. C. B. Huffaker, M. van de Brie, and J. A. McMurty, "Ecology of Tetranychid Mites and Their Natural Enemies, A Review II: Tetranychid Populations and Their Possible Control by Predators," *Hilgardia* 40 (1970): 391–458.

31. U.S. Dept. of Agriculture, *Losses in Agriculture*, Agric. handbook 291 (Washington, D.C.: Government Printing Office, 1965).

32. The postwar farm policy in which controls were placed on acreage instead of production forced farmers to use pesticides on their remaining land to increase yields.

33. J. von Liebig, *Organic Chemistry in Its Application to Agriculture and Physiology* (London: Taylor & Walton, 1840).

34. This is an artificially refined version of the natural process in which acids from the metabolism of plant roots and soil microbes release nutrients held in reserve in the organic matter and minerals of the soil.

35. U.S. Dept. of Agriculture, *Fertilizer Situation*, Economic Research Service Annual Report (Washington, D.C.: Government Printing Office, 1971–1973).

36. Ibid. See also Tennessee Valley Authority, *Fertilizer Trends, 1971* (Muscle Shoals, Alabama: National Fertilizer Development Center, 1972).

37. H. Humphrey, *Fertilizer Crisis Presents Latest Threat to Farm Production*, Congressional Record 119, S16625 (17 September 1973): 134.

38. Lewis B. Nelson, "Advances in Fertilizers," *Advances in Agronomy* 17 (1965).

39. C. C. Delwiche, "The Nitrogen Cycle," in *The Biosphere, Scientific American* (San Francisco: W. H. Freeman & Co., 1970).

40. Some claim that carbon, rather than nitrogen or phosphorus, is the primary cause of eutrophication. Although there is evidence to the contrary, this notion has been encouraged by the soap industry, which has high stakes in the effects of phosphorus on water supplies.

41. M. J. Wright and K. L. Davison, "Nitrate Accumulation in Crops and Nitrate Poisoning in Animals," *Advances in Agronomy* 16 (1964): 197–247; H. Walters, "Nitrate in Soil, Plants and Animals," *Journal Soil Association* U.K. 16, no. 3 (1970): 1–22.

42. W. Lijinsky and S. S. Epstein, "Nitrosamines as Environmental Carcinogens," *Nature* 225 (1970): 21.

43. "Common Compounds Linked to Cancer," *Journal of the American Medical Association* 216, no. 7 (17 May 1971): 1106.

44. R. H. Harmeson, F. W. Sollo, and T. E. Larson, "The Nitrate Situation in Illinois," *Journal American Water Works Association* 63 (1971): 303–310; F. T. Bingham, S. Davis, and E. Shade, "Water Relations, Salt Balance, and Nitrate Leaching Loss of a 960-Acre Citrus Watershed," *Soil Science* 112 (1971): 410–418; D. H. Kohl, G. B. Shearer, and B. Commoner, "Fertilizer Nitrogen: Contribution to Nitrate in Surface Water in a Corn Belt Watershed," *Science* 174 (1971): 1331–1334.

45. F. G. Viets and R. H. Hageman, *Factors Affecting the Accumulation of Nitrate in Soil, Water and Plants*, U.S. Dept. of Agriculture, Agricultural Research Service, Agr. Handbook No. 413 (Washington, D.C.: Government Printing Office, 1971).

46. National Academy of Sciences, *Genetic Vulnerability of Major Crops* (Washington, D.C.: Government Printing Office, 1971); W. J. Mistric, "Effects of Nitrogen Fertilization on Cotton under Boll Weevil Attack in North Carolina, *Journal Econ. Ent.* 61, no. 1 (1968): 282–283.

47. H. Walters, "Aspects of the Use of Organic and Inorganic Fertilizers," *Journal Soil Assoc.*, April 1970.

48. S. Beck, "Resistance of Plants to Insects," *Annual Review of Entomology* 10 (1965): 207–232.

49. For example, the yield of corn has risen from about 40 bushels/acre in 1950 to nearly 100 bushels/acre in 1970.

50. For instance, many patterns of human migrations reflect the gradual reoccurrence of wheat rust epidemics. Today, the survival of a given wheat variety in the Pacific Northwest is about five years before a disease will force development of a new resistant variety.

51. As to the large world collections of germ plasm stored for future use at the National Seed Storage Laboratory, one agronomist writes: "If you are willing to entrust the fate of mankind to these collections, you are living in a fool's paradise. . . . [They are] enormously redundant. . . .

Some races are hardly represented at all and the wild and weedy gene pools are conspicuously missing. In no collection is there an adequate sampling of the spontaneous races that are the most likely sources of disease and pest resistance. On the whole, the collections we have are grossly inadequate for the burden they will have to bear." J. R. Harlan, "Genetics of Disaster," *Journal of Environmental Quality* 1, no. 3 (1972): 212–215.

52. NAS, *Genetic Vulnerability*.

53. D. Pimentel et al., "Food Production and the Energy Crisis," *Science* 182 (1973): 443–449.

54. Cervinka et al., *Energy Requirements for Agriculture in California*, California Dept. of Food and Agriculture, Sacramento, Calif.; K. Blaxter, "Power and Agricultural Revolution," *New Scientist* 61 (14 February 1974): 400–403. See also Slesser, M., "Energy Analysis in Policy Making," *New Scientist* 60 (1 November 1973): 328–30.

55. J. N. Black, "Energy Relations in Crop Production: A Preliminary Survey," *Annals of Applied Biology* 67, no. 2 (1971): 272–278.

56. In the United States it takes about 6500 Kcal to produce one pound of beef or about 38,000 Kcal per pound of protein. In contrast, one pound of corn fed to cattle requires from 500 Kcal to 650 Kcal to produce depending on the extent of energy one puts in.

57. In 1970, the proportion of energy inputs into U.S. corn production included the following: chemical fertilizers, 36 percent; pesticides, 1 percent; maintenance and operation of farm machinery, 42 percent; electricity, 11 percent; labor, seeds, irrigation, crop drying, and transportation, 10 percent. Pimentel et al., "Food Production."

58. National Academy of Sciences, *Report of the Committee Advisory to the U.S. Department of Agriculture*, PE 21338, PE 21339 (Springfield, Va.: National Technical Information Service, 1972).

59. R. Levins, "Fundamental and Applied Research in Agriculture," *Science* 181 (1973): 523–524.

60. D. Pimentel, *Animal Populations, Plant Host Resistance and Environmental Control*, in Tall Timbers Conference on Ecological Animal Control by Habitat Management, Proceedings, No. 1 (Tallahassee, Fla.: Tall Timbers Research Station, 1969).

61. J. P. Dempster and T. H. Coaker, "Diversification of Crop Ecosystems as a Means of Controlling Pests," British Ecological Society, Symposium (Oxford, 1972); D. Pimentel, "Species Diversity and Insect Population Outbreaks," *Ann. Ent. Soc. Amer.* 54 (1961): 76–86; R. B. Root, "Organization of a Plant-Arthropod Association in Simple and Diverse Habitats: The Fauna of Collards (*Brassica oleracea*)," *Ecol. Monogr.* 43 (1973): 95–124; van Emden and G. F. Williams, "Insect Stability and Diversity in Agro-ecosystems." *Annual Reviews of Entomology*, 1974, pp. 455–475.

62. P. H. McGauhery, *American Composting Concepts*, Environmental Protection Agency, Solid Waste Management Office Publication, SW-2r (Washington, D.C.: Government Printing Office, 1971); A. W. Breidenbach,

*Composting of Municipal Solid Wastes in the United States,* Environmental Protection Agency, Solid Waste Management Series Publication, SW-47r (Washington, D.C.: Government Printing Office, 1971).

63. C. Golueke, *Composting* (Emmaus, Pa.: Rodale Press, 1973).

64. O. H. Frankel and E. Bennett, eds., *Genetic Resources in Plants: Their Exploration and Conservation* (Philadelphia: F. A. Davis Co., 1970).

65. H. T. Odum, *Environment, Power and Society* (New York: Wiley Interscience, 1971).

66. R. L. Rudd, "Pesticides in the Environment."

67. K. Leius, "Attractiveness of Different Foods and Flowers to the Adults of Some Hymenopterous Parasites," *Canadian Entomologist* 92 (1960): 369–376.

68. K. Leius, "Influence of Wild Flowers on Parasitism of Tent Caterpillars and Codling Moth," *Canadian Entomologist* 99 (1967): 444–446.

69. G. S. Fraenkel, "The *Raison d'être* of Secondary Plant Substances," *Science* 129 (1959): 1466–1470.

70. J. O. Tahvanainen and R. Root, "The Influence of Vegetational Diversity of the Population Ecology of a Specialized Herbivore, *Phyllotreta cruciferae,*" *Coleoptera Oecologia* 10 (1972): 321–346.

71. V. Stern, *Interplanting Alfalfa in Cotton to Control Lygus Bug and Other Insect Pests,* in Tall Timbers Conference on Ecological Animal Control by Habitat Management, Proceedings, No. 1: (Tallahassee, Fla.: Tall Timbers Research Station, 1969), pp. 55–69.

72. H. F. van Emden, "The Role of Uncultivated Land in the Biology of Crop Pests and Beneficial Insects," *Scient. Hort.* 17 (1964).

73. Philbrick, H. and Gregg, R. B., *Companion Plants and How To Use Them* (Old Greenwich, Conn.: Devin-Adair Co., 1966).

74. There is some debate on the long-term effects of continuous rotations. From fifty years of six-year rotations (corn, oats, clover, and timothy), Albrecht noted that the problem of returning fertility to exhausted soils may be easier if one used one crop and grew it continuously: "That procedure would seem a logical one when the evidence shows that rotations were the quickest way of mining the soil by calling in several different crops in rapid sequence, each for its different and added exploitive effects." Ibid. W. A. Albrecht, *Soil Fertility and Animal Health* (Webster City, Iowa: Fred Hahue Printing Co., 1958).

75. Van Emden and G. F. Williams, "Insect Stability and Diversity in Agro-ecosystems." See note 61.

76. R. Merrill et al., eds., *Energy Primer: Solar, Wind, Water and Biofuels.* (Menlo Park, California: Portola Institute, 1974).

# 17

# AQUACULTURE: TOWARD AN ECOLOGICAL APPROACH

## William O. McLarney

It is generally accepted that man's primary nutritional problem is the shortage of high-protein foods. This is true not only for the "underdeveloped" countries, but for underdeveloped communities and homesteads in the overdeveloped world. It is true even for those who are comfortably entrenched within the system—ask providers about the role of meat in their food budgets.

A number of steps may be taken to solve the economic and physical problems of protein supply: one may advocate and pursue a more or less vegetarian diet;[1] or, attempts may be made to increase the production and lower the cost of common farm animals. Both of these approaches should be pursued, subject to certain limitations, but we shall also have to make more use of aquatic animals.

This latter approach has not been neglected, and for more than a few years we have been advised to "turn to the sea." The advice has been well heeded, notably by such people as the Japanese who depend for their very survival on continually increasing the effort and efficiency of their fisheries. It is clear, however, that we cannot support very many Japans; the natural productivity of the sea and the hunter-gatherer approach to its utilization have absolute limits,

William O. McLarney is a director and cofounder of the New Alchemy Institute, Woods Hole, Mass.

and there are indications that we may be approaching these limits much more rapidly than seemed possible a few years ago.[2]

So the aquatic hunter must be phased out the way his terrestrial counterpart was many years ago—by the development and refinement of animal husbandry. There has been much discussion of aquaculture versus agriculture and fisheries, in terms of economics and productivity. This discussion has been marked by a lack of meaningful data on all sides. Rather than become embroiled in partisan debate, I would simply like to state that aquaculture has been shown to be a practical means of protein food production, and that it is more akin to agriculture than to ocean fisheries in that the means of production (water, land, stock, and a minimum of equipment) are accessible to a large percentage of the world's population and may be relatively easily controlled by an individual, community, or small group.[3] There is also an ecological argument for aquaculture as an integral part of any food production scheme, which I trust will be made clear by the discussion of Oriental aquaculture that follows.

The best way to explain the present and possible role of aquaculture in the world is to describe the status quo in two very different systems—American and Chinese. In the United States, the very word "aquaculture" is new. There have long been hatcheries devoted to the rearing of game fish for sportsmen, and the usually futile attempts to augment commercial fisheries by stocking, but commercial aquaculture has assumed importance only in the last decade, and still cannot be regarded as a major protein source. In China, on the other hand, aquaculture dates back thousands of years. It now provides at least 1.5 million tons of high-protein food annually.

Chinese aquaculture is rooted in the realization that a pond is an ecosystem and should be treated as such. Chinese fish ponds contain a variety of species, so that as many as possible of the feeding and habitat niches are filled by food fish. Production is enhanced by the application of a variety of manures and by feeding vegetable wastes and the like. To attempt to describe the precise techniques used is impossible; Chinese aquaculture tends to treat each pond as a unique living system and to prescribe treatments accordingly.

The concept of polyculture (the rearing together of several

TABLE I

APPROXIMATE YIELD OF VARIOUS AQUACULTURE SYSTEMS

| Species Cultured | Location | Culture Methods | Annual Yield* (lb/acre) |
|---|---|---|---|
| Channel catfish | South-Central United States | Commercial pond culture with heavy use of high-protein feeds | 2,000–3,000 (best growers) |
| Channel catfish | Alabama | Experimental monoculture using commercial catfish feed | 1,260 (see next entry |
| Channel catfish and *Tilapia mossambica* | Alabama | Experimental polyculture using same kind and amount of feed as in monoculture experiment (see entry above) | 1,411, catfish; 200, *Tilapia*; 1,611, total |
| Chinese carps (several species) | China and Southeast Asia | Polyculture in heavily fertilized ponds, often integrated with terrestrial agriculture | Best ponds 6,300–7,200; average, 2,700–3,600 |
| Common carp | Japan | Intensive pond culture with heavy feeding | 4,500 |
| Common carp | Indonesia | Growth in cages in streams polluted with sewage; no feeding | 500,000–750,000 |
| Common carp | Japan | Culture in closed recirculating systems with heavy feeding | Up to 4 billion |
| Milkfish | Taiwan | Intensive commercial culture with manuring—no feeding | 1,700 (average) 2,700 (best growers) |
| Milkfish | Philippines | Intensive commercial culture with manuring or chemical fertilization—no feeding | 900 (best growers) |

TABLE 1 (cont'd)

APPROXIMATE YIELD OF VARIOUS AQUACULTURE SYSTEMS

| Species Cultured | Location | Culture Methods | Annual Yield* (lb/acre) |
|---|---|---|---|
| Tilapia | Congo | Pond culture with virtually no management | up to 17,000 (mostly very small fish) |
| Wolffia | Thailand | Growing in unfertilized ponds | 9,000 |
| Yellowtail | Japan | Intensive commercial culture in floating cages in the sea; heavy use of high-protein foods | up to 56,000 |

* It must be noted that per acre yields for culture in streams, cages, and recirculating systems are computed on the basis of the area actually devoted to growing enclosures; real per acre yield would be considerably lower.

species), that prevails in Chinese fish ponds may be extended to the surrounding land by integration of the fish pond ecosystem into the neighboring agricultural ecosystem. Livestock may be penned and gardens planted so that wastes and excess nutrients are washed into the pond to serve as fertilizers. Vegetable wastes may be used directly as fish foods and, in turn, enriched materials from the pond bottom may be applied to the surrounding land. Some ponds are periodically drained and planted to crops, often legumes, which are plowed back into the soil before reflooding.

The average annual production of Chinese fish ponds is believed to be in the neighborhood of 3,000 pounds per acre; many growers do better, particularly in Southeast Asia where the growing season is year-round. Under favorable conditions, yields of 8,000 pounds per acre may be achieved (see Table I).

Chinese fish ponds may function on a family subsistence basis, but most are community or commercially oriented. Since production is high and technological inputs minor, and since feeding involves the use mainly of "waste" materials, the cost to the consumer of most of the types of food fish produced is rather low.

Most aquaculture in the United States, on the other hand, is already locked into the agribusiness syndrome. Massive monocultures, mainly of channel catfish, rainbow trout, and oysters are the rule. Starting with this fundamental denial of the ecosystem concept, it is necessary to feed heavily if respectable production levels are to be maintained. This usually involves the conversion of relatively expensive protein (in the form of commercial "fish chows" containing such ingredients as fish meal, beef and other meats, and high-quality grains) into more expensive protein. In the rare cases in which ponds are fertilized at all, chemical fertilizers are the rule. (Oyster culture is an exception to the discussion thus far, as cultured oysters derive all their food from natural sources.) Seldom is any effort made to integrate aquaculture into the surrounding landscape. "Efficiency" is the byword, so that plants are kept out of the water and off the banks in order not to interfere with mechanized management.

While American aquaculture is in no way yet as dependent on chemicals as terrestrial agriculture, history does seem to be repeating itself. Algae blooms and weeds are treated with chlorinated hydrocarbon herbicides. Fish food pellets routinely contain antibiotic drugs for disease control. In an extreme case, predators of cultured oysters in Long Island Sound are controlled with massive doses (as high as a ton per acre) of Polystream, a "witches" brew of chlorinated benzenes so deadly its use was prohibited in most states before the controversy over chemical pesticides began.[4]

The attitude of too many American aquaculturists is illustrated by a catfish farmer who asked me to recommend a chemical herbicide to eliminate cattails and rushes in one of his ponds. On viewing the pond, I saw that the offending plants occupied two corners of the pond to the extent that one afternoon's manual labor could have disposed of them.

American aquaculture is almost exclusively a commercial proposition, aimed at the supermarket trade. The conventional measure of production is pounds per acre, but the mentality is dollars per acre. In terms of the former criterion, American fish farmers, for all their technological and chemical aids, are somewhat behind their Asian counterparts. The better catfish farmers manage to produce about 3,000 pounds per acre per year. And, of course, most of what is produced is relatively expensive and seldom reaches the tables of those who most need additional protein.

The American aquaculture situation is disturbing enough, but to make matters worse we are still effective in our role as the world's chief exporters of cultural influence, good or otherwise. Just as in agriculture, pressures are being brought to bear on fish culturists throughout the world to emulate our technological and unecological approaches. This is most apparent in Latin America, where aquaculture is even newer than in the United States; however, agribusiness is creeping into Asia, also. For example, there is this story of milkfish culture in the Philippines:

The milkfish is a very important brackish-water food fish, cultured chiefly in Taiwan, the Philippines, and Indonesia. Although the climate in the latter two countries is much more conducive to milkfish culture (fish culture), the Taiwanese have had greater success, due in part to the strong program of research in milkfish culture supported by the government. Since cultured milkfish are dependent on an algae-microbenthos complex known as "lab-lab," produced by careful fertilization of the growing ponds, much research has naturally been devoted to pond fertilization. A great variety of organic and inorganic fertilizers were tested by Taiwanese fisher-biologists, and in every experiment involving a chemical fertilizer, natural manures were found to be superior.[5]

Some of the biologists involved in this generally excellent research were eventually sent as FAO (Food and Agriculture Organization of United Nations) "experts" to the Philippines to attempt to upgrade milkfish culture in that country. After some study, they made a number of recommendations, most of them good. However, there was one peculiar recommendation, with no explanation attached: the use of rather heavy dosages of several chemical fertilizers—the very thing that the same researchers had found ineffective and had advised against in their own country. A little checking disclosed that one of the FAO experts was also in the employ of Standard Oil of the Philippines—the leading supplier of chemical fertilizers in the country.[6]

There is some evidence that the ingrained conservatism of Philippine fish farmers will resist the influence of the chemical pushers, but there are many such holding actions which will have to be fought in Asia. Meanwhile, in the West, there is a tide to be turned. Since aquaculture here is so new, and by no means economically stable, it is not at all unrealistic to hope that this will happen. For one thing, there have always been a few holdouts who have resisted the drive toward bigness and persisted in operating small, ecologically sound fish farms aimed at local markets. And there does seem to be the start of a trend toward one aspect of the Chinese model—polyculture. A number of catfish farmers in the South have found that by stocking *Tilapia* or buffalofish with their catfish, they not only increased the total production of their ponds at no added cost in feed, but actually increased the production of catfish. Recently, Roy Prewitt, president of the American Fish Farmers Federation, and one of the pioneers in fish farming in the United States, has come out strongly for polyculture and the utilization of organic wastes.[7]

An even more important countertrend may be beginning among the growing number of Americans who, as individuals or members of small communities, are inching toward self-sufficiency. Many of the people who are planting "survival gardens" or starting organic farms are inquiring about aquaculture. As director of the New Alchemy Institute's Backyard Fish Farm project,[8] I receive hundreds of letters asking, How? Were I to refer these correspondents to the American aquaculture literature, they would be misled. The state of the art here is such that a prospective small-scale aquaculturist delving into the literature is in the same position as a beginning gardener would be if all the material on agriculture in the library dealt only with truck farming.

In fact, it is not possible to say, this is how. The Chinese know how, but they have three thousand years experience in a very different ecosystem, and the best we can do is emulate their general approach.[9,10] In so doing, we should remember three general principles of aquaculture.

1. *A pond is an ecosystem.* Therefore, maximum production can be achieved only if all feeding and habitat niches are filled. Intrinsic to the Chinese genius is the ability to fill a large percentage of the available niches with animals that may be directly utilized by man. Plankton feeders, bottom feeders, algae feeders, predators, all have their place in a polyculture system. American commercial aquaculturists on the other hand do not pursue polyculture to its logical extreme because the necessity for sorting or for separate harvesting and processing procedures is likely to result in increased

expenses and not increased profits. This constraint need not apply
to the culturist who is mainly interested in *food* production.

2. *The lower an animal is on the food chain, the more abundantly
   and cheaply it can be produced, and the fewer poisons it gathers
   into its tissues.* Thus we should concentrate on herbivores, such as
   *Tilapia*, which feed on a variety of animal and plant foods, most
   of them unusable by man. Application of this principle will require the restructuring of some of our attitudes about fish.
   Americans, perhaps as a result of our sport fisherman heritage,
   have been conditioned to place a premium on "game" fish, all of
   which are high on the food chain, and to underrate equally edible
   fish which, because of their dietary preferences, are seldom attracted to the angler's lure.

3. *It is easier, cheaper, and faster to produce X pounds of small fish
   than the same weight of large fish.* Thus, we should harvest and
   eat smaller fish. This too will require some changes of attitude.
   Americans, unlike Asians, are taught to reject fish that are "too
   small" from a sporting point of view. Conventional commercial
   aquaculture comes under the same restraints since it is inefficient
   to process small fish for the supermarkets. However, I frequently
   eat four-inch bullheads, and consider them superior to larger
   specimens in that they cook up faster and more evenly, and can
   be eaten bones and all. Steinbeck and Ricketts on their famous
   Sea of Cortez voyage dined well on two-inch whitebait cooked
   whole.[10]

Bearing these principles in mind, some of the native American
fishes that are best suited to homestead and community culture are
carp,[11] the bluegill and other sunfishes,[12] the bullheads, the
buffalofish. A number of exotic tropical species, particularly herbivorous *Tilapia*, may be added to the list where greenhouse facilities can be provided.[13] It is not improbable that most or all of
these species, along with such invertebrates as crayfish and freshwater clams, could be combined to good effect. Nor need aquaculture be strictly a source of animal food. For example, in Thai-
land an aquatic plant known as khai-nam (*Wolffia*) is grown in
shallow ponds for human consumption; yields (protein dry weight)
exceed those of soybeans, peanuts, corn, and rice.[14]

A word of caution: The Chinese have had thousands of years to
develop their polyculture methods. We should not expect to match
their success overnight. The beginning aquaculturist should usually

start out with a near-monoculture and gradually build up to a complex pond community. In this way, he will get to know each species well and, if anything goes wrong, it will be easier to identify the weak link.

There are a couple of areas in which Americans might undertake to improve on existing methods if aquaculture is to become a valuable tool in the pursuit of alternative life-styles. One of these areas is the feeding of fish. The role of pond fertilization and vegetable wastes in feeding has already been mentioned, but faster growth, particularly of very young fish, can often be achieved if animal food is supplied in ample quantities. This is not ecologically or economically inconsistent if the food animals are of sorts which are not easily usable as human food.

One of the preferred foods of many fish is midge larvae, which are cultured as food for mullet fry in Israel, where the late Dr. Abraham Yashouv was able to harvest 250–375 $g/m^2$/week from a 1 $m^2$ tray. These brilliant red larvae were believed by Yashouv and others to contain a growth-promoting substance effective on all cold-blooded organisms.[15] Culture of midges is simplicity in itself, as they naturally colonize any pool of fertilized water.

Other dipteran larvae may also be fed to fish. For instance, some midwestern pond owners grow maggots by suspending rotting meat over their ponds in a wire frame. The maggots drop off and are consumed by the fish. More sanitary and aesthetic ways of growing fly larvae have been designed for laboratory purposes and could easily be adapted to fish culture.

Mosquito larvae are another excellent food for young fish. If properly handled, they do not create a pest problem. Suggestions for their culture, and for culture of many other potentially useful invertebrates, may be found in *Culture Methods for Invertebrate Animals*.[16]

Earthworms have become recognized as the classic fish bait for the excellent reason that they are eagerly accepted by most fish. Instructions for culturing earthworms are readily available from state conservation departments, university extension services, etc.

Tropical fish hobbyists routinely culture a variety of other food organisms such as brine shrimp, tubifex forms, and daphnia. Most of these animals are usable by cultured food fish, particularly in the early life stages.

Insects for fish food may be captured from the wild as well as cultured. A number of efficient insect capture devices were developed about the time that chemical pesticides became generally available, and so they never became really popular. Among these devices is the ultraviolet "bug light." A study by Dr. Roy Heidinger of Southern Illinois University suggests that a series of these lights on a one-acre bluegill pond could increase fish production by about 500 pounds per year at a cost of about one cent per pound.[17] The cost covers mainly the need for electrical power. This factor could be eliminated, along with any ecological objections, if the lights were run off, say, a wind generator.

Another way of increasing fish production which is, in my opinion, critical to the development of fish culture on a small scale is recirculation of water through a bacterial filter modeled after the sub-sand filters used by aquarists.

In addition to removing pollutants and buffering pH (calcium carbonate is used as the filter substrate), these filters are believed to neutralize chemical growth inhibitors produced by fish at high population densities.

The first commercial scale recirculating system was developed in 1951 by Dr. A. Saeki of Tokyo University and Mr. I. Motokawa of Maebashi City, Japan.[18] Such systems are now quite widely used in the culture of carp and eels in Japan, and have been adapted for carp culture in Germany,[19] and trout and catfish culture in the United States. Indeed, one American corporation now sells trout culture kits in any capacity from family size to giant commercial units based on Motokawa's design. However, there is nothing about recirculating systems which would prohibit any moderately handy person from building his own, as a quick perusal of Spotte's book[20] will show.

An effect similar to that achieved by closed recirculating systems is the simple expedient of growing fish in floating cages anchored in a larger body of water. This method, also pioneered in Japan, confines the fish and their management to a small area while permitting the natural circulation of great volumes of water through the cages. A possible adaptation of this method would involve stocking a natural pond with a prolific species such as the bluegill. The pond itself would then be left unmanaged, and would function as a natural hatchery. Periodically, small fish would be removed from the pond and stocked in the cages for brief, intensive culture.

I have barely begun to outline the possibilities for ecologically sensitive, people-oriented aquaculture. But if the possibilities are virtually unlimited, the perfected techniques are very few. In a sense, the would-be aquaculturist, as compared with the gardener or farmer, is at a disadvantage; he is starting from a position of relative ignorance. But in another sense, he is in a favorable position. Agribusiness has not had time to tighten its grip on aquaculture. Aquatic crops are not yet "hooked" on chemicals. The schools and government propagandists have not yet had time to promote blind acceptance of agribusiness methods. There is still plenty of room for aquaculture to move in *any* direction. If enough of us who are interested, not in dollars per acre, but in ecological, economic, and social sanity are willing to experiment, aquaculture in the West can become a powerful force in altering the relationship of land, food, and people as it did in China thousands of years ago.

## Notes

1. F. M. Lappé, *Diet for a Small Planet* (New York: Ballantine Books, 1971).

2. J. H. Ryther, "Photosynthesis and Fish Production in the Sea," *Science* 166 (1969): 72–76; W. E. Ricker, "Food from the Sea," in *Resources and Man: A Study and Recommendations by the Commission on Resources and Man*, Division of Earth Science, Natural Resources Council (San Francisco: W. H. Freeman & Co., 1969), pp. 88–108.

3. J. E. Bardach, J. H. Ryther, and W. O. McLarney, *Aquaculture: The Farming and Husbandry of Freshwater Organisms* (New York: John Wiley & Sons, 1972).

4. T. A. Gaucher, *Potential for Aquaculture*, General Dynamics Electric Boat Division Report U413–68–016 to Connecticut Research Commission, 1968, pp. 4–1–4–22.

5. S. Y. Lin, *Pond Fish Culture and the Economics of Inorganic Fertilizer Application*, Chinese-American Joint Committee on Rural Reconstruction, Fisheries Series No. 3, 1968.

6. J. R. Rabanal and Y. A. Tang, "Improved Techniques of Milkfish ('bangos') Culture," *Esso Agroservice Bulletin* 8 (1967): 17–22.

7. R. Prewitt, "Rambling Along," *American Fish Farmer* 3, no. 5 (1972): 19–20.

8. J. H. Todd and W. O. McLarney, "The Backyard Fish Farm," *Organic Gardening and Farming*, January 1972, pp. 99–109.

9. Ibid.; W. O. McLarney, "Pond Construction: First Step in Successful Aquaculture," *Organic Gardening and Farming*, April 1972, pp. 116–123.

10. J. Steinbeck, *Log from the Sea of Cortez* (New York: Viking Press, 1951).

11. W. O. McLarney, "Why Not Carp?," *Organic Gardening and Farming*, February 1972, pp. 76–81.

12. W. O. McLarney, "The Farm Pond Revisited," *Organic Gardening and Farming* 18, no. 11 (1971): 88–95.

13. Todd and McLarney, "Backyard Fish Farm." *The Journal of the New Alchemists* 2 (Woods Hole, Mass.: New Alchemy Institute).

14. K. Bhanthumnavin and M. G. McGary, "Wolffia Arrhiza as a Possible Source of Inexpensive Protein," *Naute*, no. 732 (1963), p. 495.

15. A. Yashouv in an unpublished interview with J. E. Bardach.

16. P. S. Galtsoff et al., *Culture Methods for Invertebrate Animals* (New York: Dover Pubs., 1937).

17. R. C. Heidinger, "Use of Ultraviolet Light to Increase the Availability of Aerial Insects to Caged Bluegill Sunfish," *Progressive Fish Culturist* 33, no. 4 (1971): 187–192.

18. K. Kuronuma, *New Systems and New Fishes for Culture in the Far East*, FAO World Symposium on Warm Water Pond Fish Culture, 1966, FR: VIII-IV/R-1.

19. C. Meske, "Breeding Carp for Reduced Number of Intermuscular Bones, and Growth of Carp in Aquaria," *Bamidgeh* 20, no. 4 (1968): 105–119.

20. S. H. Spotte, *Fish and Invertebrate Culture: Water Management in Closed Systems* (New York: Wiley Interscience, 1970), p. 145.

### References

*American Fish Farmer*. P. O. Box 1900, Little Rock, Arkansas 72203. Monthly periodical.

*Aquaculture*. New York: Elsevier Scientific Publishing. Bimonthly periodical.

*Amidgen*. Nir-David, Israel. Monthly bulletin of fish culture in Israel.

Hickling, M. *Fish Culture*. London: Faber & Faber, 1971.

Huet, M. *Textbook of Fish Culture: Breeding and Cultivation of Fish*. London: Fishing News Ltd., 1970.

Mehdola, D. "Aquaculture" in *Energy Primer: Solar, Wind, Water, and Biofuels*. Menlo Park, California: Portola Institute, 1975.

# 18

## INSECT POPULATION MANAGEMENT
## IN AGRO-ECOSYSTEMS

### Helga and
### William Olkowski

#### Introduction

The era of heavy reliance on chemical pesticides[1] for insect control may be drawing to a close, at least in the United States. There are two general reasons for this: (1) the growing concern over the potential of widespread damage to the environment from the dispersion of poisons through food chains leading to valuable predators and humans, and (2) the fact that for ecological reasons pesticides commonly fail to control pests. When pesticide use leads to more pesticide use, the manufacturer increases his sales and profits, the users suffer greater costs, while environmental contamination increases. The reluctant recognition of these problems associated with the pesticide treadmill is helping to force farmers toward the reconsideration of alternative methods of pest control.

Unfortunately, the fact that new tools are needed has not always gone together with the understanding that an entirely new approach, with radically different conceptual underpinnings is required. For example, one finds even in the "organic" gardening

Helga Olkowski is chairwoman of the Ecology and Natural Systems Program at Antioch College/West in San Francisco. William Olkowski is assistant research entomologist in the Division of Biological Control at the University of California, Berkeley.

and farming literature evidence of abundant searches for new, or sometimes old, products that will serve the same obsolete push-button eradication philosophy, but with hopefully fewer consequences in terms of public health. The "organic" grower who anticipates or observes insects present or insect damage is urged to fight back with an arsenal of simplistic remedies (garlic and pepper sprays), the release of praying mantids or ladybird beetles,[2] and the magic of "companion" plants that will mysteriously repel undesired wildlife if the right combination is planted. This is not to say that such remedies never work; in fact many of them have some of the desired effects of crop protection, though not always for the reasons given. However, the framework within which these solutions are offered and applied is basically fallacious.

The agribusiness "establishment" grower, equally enamored of pest *eradication* as a goal, is not helped to any new understanding of the problem by the rigid government-consumer demands for foods unblemished by insect damage or insect parts. The farmer is caught in an economic squeeze between a pesticide-altered agro-ecosystem on the one hand and cannery standards, agricultural codes, and marketplace entomophobia on the other. As a society we are all frightened of insects to some extent. They are culturally taboo as food and, with the exception of the uneasy tolerance of honeybees, are generally all lumped together as potential sources of human disease and competitors for our food and fiber. This general attitude makes rational observations and decisions about our relationship to insects and similar-appearing wildlife difficult for most people. The farmer is no exception.

But even where insects have been justly labeled injurious we must give up the naïve assumption that crop protection can be dealt with by the simple-minded formula of an All-Purpose Spray, regardless of whether the recommended panacea is garlic emulsion or DDT. In its place must come an awareness of the enormous complexity of the natural world, a willingness and ability to correctly identify the animals involved, as well as research into their different life cycles and population interactions, and a determination of what population sizes are acceptable. In addition, a system for analyzing and managing complex plant ecosystems is needed that can be applied by field personnel.

The science of ecology offers the basic principles for a functional model of crop protection. This alternate approach can be called ecosystem management, or integrated control. It rests upon two central techniques: monitoring pest populations and determining economic (or, in the case of ornamental plants, aesthetic) injury levels . . . that is, the level above which damage cannot be tolerated for economic reasons.

This chapter will explore some concepts needed for environmentally sound insect management and some examples of successful application of this approach.

### The Impact of Pesticides

The excessive, injudicious use of pesticides has had a number of disturbing consequences. It is worth examining these effects in greater detail for what they reveal about wildlife systems.

#### PUBLIC HEALTH EFFECTS

First there is the problem of residue. The label on a poison spray may lead you to believe that you are killing a target insect.[3] In fact, an entire ecosystem is being affected. Not only do problems arise when nontarget animals (such as humans and farmworkers) come in direct contact with the poison residues, but a number of these materials may accumulate in food chains, the amount becoming greater and potentially more lethal as it is transferred from organism to organism, eventually causing irreversible destruction on the highest trophic (eating) levels.

This bio-magnification, as it is called, is familiar to most of the lay public as the story of DDT and its effect on top carnivores such as the peregrine falcon. Since man is ultimately at the end of most of the food chains on this planet, both terrestrial and marine, the importance to man of minimizing the use of pesticides that accumulate in this manner is obvious.

There are other public health aspects of pesticide use which are important, such as the unknown synergistic effect of absorbing small amounts of many different poisons, each dose of which is not too toxic in itself or in amounts below a toxic dose, and accumulating these in the body over a period of many years. However, it is the effects of these materials on insect ecosystems that is least known or understood by the general public and it is important to mention them here.

### EFFECTS ON AGRO-ECOSYSTEMS: THE PESTICIDE TREADMILL

*Resurgence*:—Predators and parasites that control pests[4]—are frequently more susceptible to the effects of pesticides than are the target animals. Many factors may account for this (see Chapter 16). A few pesticides may actually be selectively more toxic to some beneficial insects; for example, Sevin® on certain hymenopterans such as bees and beneficial wasps.

The result is that after the use of a pesticide it may take the beneficial insects a long time to restore their population numbers. In the absence of insect controls the pest insects that escaped death can multiply unchecked. Soon there may be a worse pest problem than before the pesticide was used, and the farmer may have no other recourse but to spray again or see his crop totally destroyed. He is caught in a sequence of events of his own creation—the pesticide treadmill.

*Resistance*: Repeated spraying may lead to a treadmill effect through another route.[5] In every natural insect population there is an enormous genetic variation. Some members of that population will be far less susceptible to the fatal effects of any particular pesticide than others. Each time the pesticide is used, the most susceptible individuals will be "weeded out," leaving the resistant insects to multiply and replenish the population with their kind. A pesticide that seems to be miraculously efficient the first time it is used, if used repeatedly, will become less and less effective, until finally it may have little or no effect upon the pest population. In such cases the insects are said to have become *resistant*.

It is a tragic fact that the overuse of pesticides has caused many of man's most medically and economically important pests to become resistant on a world-wide basis. This is the story behind the Western equine encephalitis mosquito, *Culex tarsalis*, in California. The species cannot be controlled in many areas with any known pesticide. The phenomenon of insect resistance is also behind the bankruptcy of the cotton industry in northeastern Mexico, and the Matamoros district of Texas.[6]

*Secondary Pest Outbreaks*: In every agro-ecosystem, no matter how simple it may seem compared to a wild area, there are enormous numbers of insects and other organisms living together in a close and intricate association. Normally only one or two may become noticeable enough to achieve the label of "pest." The many other herbivores present are largely under control by their predators and parasites. Pesticides used against target pests frequently interrupt this well-balanced scene by killing off the beneficial insects that control the other potential, equally pestiferous animals present. These liberated species, or newly created pests, are then free to multiply until they achieve the distinction of being identified as highly injurious. Thereafter, these insects may be sprayed regularly.

The secondary pest outbreak may appear to the farmer to be totally unrelated to his attempts to eliminate his first pest insect. He may imagine he has been invaded by this new menace, while in fact, that very insect was residing unnoticed and in innocuously low numbers in the fields all the time. Numerous instances have been documented in which pesticides have actually "turned on" relatively harmless animals in this manner.[7]

## The Philosophy of Eradication

As mentioned previously, the farmer's desire totally to eliminate insects has been encouraged by the consumer's hysterical aversion to finding insects, or even evidence of previous insect presence, on fresh vegetables or fruits in the market. A basic confusion in distinguishing between disease-transmitting, stinging, and/or biting insects and those likely to inhabit the vegetable farm or orchard products may have led to the official government classification of insect or insect parts as "filth," a category that brings to mind unsavory items such as rodent droppings. Thus, pressure from government regulations and food processors to eliminate or reduce insect parts encourages and forces farmers to spray. The consumer today unknowingly chooses food with poisons over food with insects. The time has come for this trade-off to be brought to light and reconsidered.

While extermination may be the justified approach in dealing with termites burrowing into the foundations of your house, there are few other situations where this approach can be understood as anything other than a basic fear of insects often abetted by the chemical industry's interest in profits. Because an insect is present does not mean it should be eliminated. In fact, a careful study of the relationships between an animal and the factors that control its numbers will reveal the importance of maintaining some of the undesired populations so as to encourage the constant presence of its predators and parasites. This can result in a more stable situation over a period of time.

Since no insect is known to have been deliberately eradicated from the earth by man, regardless of vast attempts to do so, one can argue that the task is impossible. In addition, a basic understanding of insect problems indicates the futility of elimination when all that may be needed is temporary suppression.

The eradication approach usually involves vast poison campaigns (against such diseases as malaria; and such insects as the fire ant, gypsy moth, pink bollworm), utilizing costly technology over a short period of time. The alternative approach, called "integrated control," involves training specialists in a wide variety of population suppression techniques. This represents a basically different philosophy developed from systems-thought and ecosystem management concepts, all organized around a population monitoring system with emphasis on minimal impact to the environment. The integrated control specialist's central model is based on the concept of natural control.

### Natural Control: An Alternate Philosophy

Insects are very prolific (Table I lists the insect orders). The common housefly will not uncommonly lay 1,000 eggs. If 100 flies occupy one cubic inch, then in a single summer one pair of flies might produce enough progeny to cover the planet with a layer of insects 4,000 feet thick. This has never happened, because insects, like all animal populations are under "natural control." Their numbers rise and fall as one element of their environment becomes favorable or limiting.

With plant-eating animals, the amount of food plants available will have a great influence on population size. A monoculture of

TABLE I
THE INSECT ORDERS

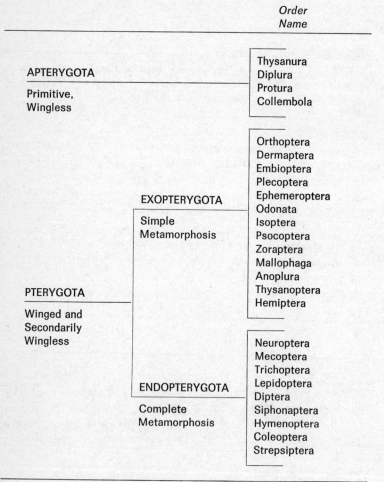

| | | Order Name |
|---|---|---|
| **APTERYGOTA**<br>Primitive,<br>Wingless | | Thysanura<br>Diplura<br>Protura<br>Collembola |
| **PTERYGOTA**<br>Winged and<br>Secondarily<br>Wingless | **EXOPTERYGOTA**<br>Simple<br>Metamorphosis | Orthoptera<br>Dermaptera<br>Embioptera<br>Plecoptera<br>Ephemeroptera<br>Odonata<br>Isoptera<br>Psocoptera<br>Zoraptera<br>Mallophaga<br>Anoplura<br>Thysanoptera<br>Hemiptera |
| | **ENDOPTERYGOTA**<br>Complete<br>Metamorphosis | Neuroptera<br>Mecoptera<br>Trichoptera<br>Lepidoptera<br>Diptera<br>Siphonaptera<br>Hymenoptera<br>Coleoptera<br>Strepsiptera |

SOURCE: Adapted from D. J. Borror and D. W. DeLong, *An*

| Common Name | No. World Species | No. Species North America |
|---|---|---|
| Silverfish | 350 | 20 |
| Japygids | 400 | 30 |
| Proturans | 100 | 30 |
| Springtails | 2,000 | 325 |
| Grasshoppers | 22,000 | 1,100 |
| Earwigs | 1,100 | 20 |
| Webspinners | 150 | 10 |
| Stoneflies | 1,500 | 350 |
| Mayflies | 1,500 | 550 |
| Dragonflies | 5,000 | 425 |
| Termites | 1,700 | 45 |
| Booklice | 1,100 | 150 |
| Zorapterans | 19 | 2 |
| Bird Lice | 2,600 | 320 |
| Sucking Lice | 250 | 65 |
| Thrips | 3,000 | 625 |
| Bugs, Aphids | 55,000 | 8,750 |
| Lacewings | 4,700 | 350 |
| Scorpionflies | 350 | 70 |
| Caddisflies | 4,500 | 950 |
| Butterflies | 200,000 | 10,500 |
| Flies | 85,000 | 16,700 |
| Fleas | 1,100 | 250 |
| Bees, Ants, Wasps | 105,000 | 14,600 |
| Beetles | 277,000 | 27,000 |
| Strepsipterans | 300 | 100 |

*Introduction to the Study of Insects* (Holt, Rinehart and Winston, 1971).

10,000 acres of cotton, for instance, provides a tremendous opportunity for the build-up of pest populations. One useful alternative may be diversifying the crops in an area, thereby limiting the amount of a food plant a potential pest population may specifically depend upon. However, many herbivores can survive equally well by feeding on a variety of plants, and factors other than availability of food may then become limiting.

Weather, natural catastrophies, competition within and between species, disease, predators, parasites, and availability of habitat both for the herbivore and its parasites and predators, all play their part in animal population control (see Figure 1). Each of these limiting factors offers an opportunity to devise a suitable management strategy, once the larger picture of the ecosystem interrelationships is visualized (see Figure 2). The costs of such manipulation must be estimated not only in terms of immediate time and energy requirements, but also in long-term impact upon the rest of the total system.

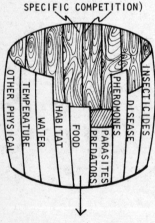

FIGURE 1. Hypothetical barrel used to convey concept of limiting factors, where water level in barrel indicates population size of particular insect which in turn is determined by height of shortest stave. Each barrel stave actually changes in height as the season or time period progresses. (H. O. Buckman, & N. C. Brady, 1969. *The Nature and Property of Soils*, Macmillan & Co., London.)

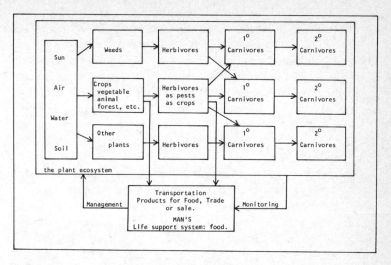

FIGURE 2. The agricultural ecosystem components and major relationships. The arrows indicate paths of material or energy flow.

### An Insect Ecosystem Management Example

The complexity of insect ecosystems is hard to imagine. An insight into the complexity of such a system may be gained from an examination of the one surrounding the linden aphid (*Eucallipterous tiliae*) on the linden trees (*Tilia* spp.) in Berkeley, California, a system which has been studied for several years.[8]

Linden trees are widely planted as a shade tree in this country, but they, like many other common plants, were brought over originally from Europe by people nostalgic for the familiar landscapes and food plants of their youth. In Europe the tree is valued not only as a bee plant, but for its flowers, from which a tea is made; for its wood; and for its ornamental qualities.

With these imported trees came the aphid that lives in close association with the linden and which obtains its food by sucking the juices of this kind of tree and no other. Unfortunately, the beneficial insects, predators and parasites, that feed upon *E. tiliae* were not imported with the aphid so that the linden aphid multiplied largely uncontrolled.

Actually, this by itself did not constitute a problem since linden trees can withstand horrendous populations of aphids. However, aphids, as do many of their relatives such as scales and mealybugs (order Hemiptera), produce a sticky substance called honeydew. This honeydew, a rich protein and sugar mixture, normally serves to attract parasites or predators that may be present in the vicinity. Frequently, an unaesthetic black "sooty" mold grows upon it. This combination of unsightly sooty mold and honeydew dropping upon parked cars, sidewalks, and driveways can cause considerable annoyance. People believe the trees are "sick" and want a pesticide to cure them.

Although the aphid is an introduced species there are many native predators that will feed upon it, such as various kinds of ladybird beetles (*Coleoptera: Coccinellidae*) and lacewings (*Neuroptera: Chrysopidae* and *Hemerobiidae*). There are also syrphid or flower flies (*Diptera: Syrphidae*) that in the larval stage will prey upon aphids. Like many other beneficial insects that are carnivorous at certain periods of their development, the adult syrphid fly requires flower pollen. Thus, a pollen source is also a component of this system.

Also involved is another introduced insect, the Argentine ant (*Iridomyrmex humilis*). This ant feeds on honeydew produced by the aphid and protects the aphids by either killing the aphid predators or otherwise interfering with their actions. Therefore, one can see that the ant is another very important component; it occurs in many insect ecosystems that include honeydew-producing herbivores.

Of course, this is a very incomplete picture of the total linden tree system, since there are numerous other plant-eating or juice-sucking insects with their predators and parasites that inhabit the many different areas of the tree from root to leaf. There are also numerous levels of carnivorous spiders about which little is known. To give a more exhaustive description one would need to include disease organisms such as the fungi, which may occasionally cause epidemics among the aphids.

However, even such a simplistic model should give the reader a fair impression of the complexity involved. In the system just described three basic techniques have been used very successfully. The first was the introduction from Europe of a parasitic insect that attacks the aphid. The second was the control of the Argentine ant by banding a few troublesome trees with a nontoxic sticky material that prevents the ants from interfering with the actions of the parasites and predators. The third strategy was an infrequent hosing down of certain trees with plain water, or a water and soap solution. The water sprays not only knock many aphids to the ground where they fall prey to other predators, but also reduce the amount of honeydew, dirt, and industrial pollutants on the leaves, branches, and trunks. This tree-washing may help the beneficial insects directly, since cleaner leaves are less likely to hamper their movements.

### Aesthetic and Economic Injury Levels

The purpose of the linden program was not to *eliminate* a pest but to *suppress* it below the level at which unsupportable economic or aesthetic injury to the plant occurred. Logically then, when one uses a technique to reduce a pest population, the savings in plant damage should be worth the costs of treatment. Thus, for every case, knowledge of unacceptable injury levels is essential. The fact is that knowledge of such levels is nonexistent or rare for most plant ecosystems. Trained pest managers who can make such evaluations are needed.

Determining economic injury levels is difficult but essential for the farmer who is currently relying on chemical pesticides and either wishes to reduce their use or seeks alternate methods of pest management. First he needs to know that the least amount of pesticide used in an ecosystem makes for easier and cheaper management. There are a number of approaches that a grower may use to study the situation in his particular case.

One possibility is to set aside an area where no pesticides will be applied and where no drift will reach. Then a regular inspection and counting in this area must be followed. When unacceptable damage to the product is threatened or occurs it should be documented along with the population size of "pests" and their natural enemies. In subsequent crops when such levels are inevitable treatments can be applied. Possibly treatment-free or "control" areas could be maintained and monitored by researchers with a reporting service to the grower, or with field scouts hired by the grower.

Another approach would combine a monitoring procedure with treatments when some known but low number of pests per plant, per row length, or some other spatial unit occurs. In subsequent years this known density of "pest" organism could be adjusted upward until light or moderate damage occurred. Again, crop damage levels could be associated with pest and beneficial insect population sizes, with treatments applied to minimize damage without knowing whether yields will be affected. When plant damage levels can be associated with reduction in yield, a major component will have been identified which would allow for a proper biological and economic determination of the injury level. Additional knowledge concerning costs of treatments will also help define the injury level. Thus a treatment would be used if the potential yield loss would be greater than the cost of treatments. The issue is resolved when estimates of loss can be predicted with a degree of certainty. However, even these estimates, however precise, do not measure external costs such as food chain contamination, death to various nontarget species, etc., that occur with the use of pesticides.

## The Ecological Basis for Pest Problems

The original geographical source of plant and animal agricultural components is a question that is often difficult to answer but provides an important ecological perspective. Table II provides a list of major crop plants and animals. The area of origin of these plants clearly indicates that agricultural crops are not native in most areas of the world where they are grown. As was already mentioned, imported plants and animals may bear their pests with them but frequently without the natural controls important in reducing the pest population sizes in the native areas. The strategy of classical biological control is to import these natural enemies of pests as described in the linden tree example above.

According to van den Bosch,[9] seventeen out of twenty-seven species considered to be important major pests in the United States are of exotic origin (Table III). Native insect pests are also subject to control by imported enemies,[10] but the chances for success are low.

The remainder of this chapter will be devoted to a survey of some of the environmentally sound strategies for insect population management that are available to the farmer. These include cultural controls, the use of resistant plants, insect diseases, classical biological control, and the choice of a "safe" insecticide. However, before a suitable strategy can be chosen, developing a monitoring program may be desirable.

### Monitoring Insect Populations

The objective of a monitoring program is to gather information about potential pest population sizes and any other important ecosystem component in order to decide when and how to manipulate some part of the ecosystem. Monitoring also includes evaluation of a management practice and assessment of cost effectiveness.

Monitoring can be as simple as noting the date and describing the amount and type of damage of a particular insect. More complex programs of monitoring and advising, which are available commercially, are known as scouting or supervised control. The best monitoring programs are those operated by personnel who have no vested interest in pesticide use but are hired by a grower or group of growers to provide efficient pest control, minimize costs, and reduce pesticide use.

Field persons for chemical companies obviously derive benefits from selling pesticides, and cannot afford to minimize pesticide use. They thus make the worst possible advisers. Unfortunately such field representatives constitute the major advising group currently available. The connection of this vested interest group to pesticide overuse is elaborated by van den Bosch.[11]

A commercially operated monitoring program should provide information about both population sizes of the pests and their natural enemies. This information should be stated as a certain number of insects per leaf, per stem, per row, per quantity of fruit, per plant, or per unit of sampling time, etc. Hypothetically, if the economic injury level is X number of pest insects on the crop, then some lower number is set as an action threshold. Only when insect counts reach the action threshold is the control process set in motion. Ideally, population suppression should coincide with the the economic threshold.

Information about carnivorous insects is critical. Increasing

## TABLE II
## THE ORIGIN OF MAJOR DOMESTIC PLANTS AND ANIMALS

| *Grains* | *Scientific Name* | *Probable Origin* |
|---|---|---|
| Barley | *Hordeum vulgare* | Abyssinia, Nepal, and Tibet |
| Buckwheat | *Fagopyrum esculatum* | Central Asia |
| Amaranths | *Amaranthus spp.* | Orient |
| Maize | *Zea mays* | Central Amer.? |
| Oats | *Avena sativa* | Unknown |
| Pearl Millet | *Pennisetum glaucum* | Africa |
| Rice | *Oryza sativa* | S.E. Asia |
| Rye | *Secale cereale* | Asia Minor |
| Wheat | *Triticum vulgare* | Asia Minor |
| FORAGES | | |
| Alfalfa | *Medicago sativa* | Asia Minor |
| Bluegrass | *Poa pratensis* | Asia |
| Cowpea | *Vigna sinensis* | Africa, India |
| VEGETABLES | | |
| Beet | *Beta vulgaris* | Mediterranean |
| Broad Bean | *Vicia faba* | Abyssinia, Afghanistan |
| Cabbage | *Brassica oleracea* | Mediterranean |
| Carrot | *Daucus carota* | Afghanistan |
| Chick-pea | *Cicer arietinum* | Asia |
| Cucumber | *Cucumis sativus* | Himalayas, India |
| Eggplant | *Solanum melongena* | India |
| Kidney Bean | *Phaseolus vulgaris* | Unknown |
| Lentil | *Lens esculenta* | Asia |
| Lima Bean | *Phaseolus lunatus* | S. America |
| Mung Bean | *Phaseolus aureus* | India |
| Mustards | *Brassica spp.* | Orient |
| Pea | *Pisum sativum* | Mediterranean, Afghanistan |
| Radishes | *Raphanus sativus* | Unknown |
| Red Peppers | *Capsicum spp.* | C. Amer., Asia? |
| Scarlet Runner Bean | *Phaseolus multiflorus* | C. America |
| Soybean | *Glycine soja* | Orient |
| Spinach | *Spinacia oleracea* | Persia |
| Squashes, Pumpkins | *Cucurbita spp.* | N. & C. Amer., China? |
| Tomato | *Lycopersicon esculentum* | Peru |
| Turnip | *Brassica campestris* | Eurasia |
| Urd Bean | *Phaseolus mungo* | India |

TABLE II
THE ORIGIN OF MAJOR DOMESTIC PLANTS AND ANIMALS

| Grains | Scientific Name | Probable Origin |
|---|---|---|
| *FRUITS* | | |
| Avocado | *Persea americana* | C. America |
| Banana | *Musa sapientum* | Malay Pen.? |
| Citrus | *Citrus spp.* | S.E. Asia |
| Coconut | *Cocos nucifera* | Asia |
| Date Palm | *Phoenix dactylifera* | India, Persia |
| Fig | *Ficus carica* | Arabia, Mesopotamia |
| Grape | *Vitis spp.* | Transcaucasia-Turkestan |
| Guava | *Psidium spp.* | C. & S. America |
| Japanese Persimmon | *Diospyros kaki* | China |
| Jujube | *Zizyphus mauritiana* | China |
| Mango | *Mangifera indica* | S.E. Asia |
| Melon | *Cucumis melo* | Asia or Africa |
| Papaya | *Carica papaya* | Amazonian Basin |
| Peach | *Prunus persica* | China? |
| Pineapple | *Ananas comosus* | S. America |
| *POME FRUITS* | | |
| Apple | *Malus pumila* | ?, Europe?, Asia? |
| Pear | *Pyrus communis* | Eurasia? |
| Quince | *Cydonia oblonga* | Pre-European |
| Pomegranate | *Punica granatum* | Persia |
| Plum | *Prunus spp.* | Europe |
| Strawberry | *Fragaria grandiflora* | New World |
| Watermelon | *Citrullus vulgaris* | Africa |
| *ANIMALS* | | |
| Common Cattle | *Bos spp.* | Mesopotamia |
| Goat | *Capra spp.* | Asia |
| Sheep | *Ovis aries* | Asia |
| Horse | *Equus caballus* | Caucasus |
| Ass | *Equus spp.* | Africa |
| Chicken | *Gallus domesticus* | S.E. Asia |

SOURCES: E. Anderson, *Plants, Man and Life* (Berkeley: University of California Press, 1967); C. O. Sauer, *The Domestication of Animals and Foodstuffs* (Cambridge, Mass.: MIT Press, 1969); and D. B. Guralnik, ed., *Webster's New World Dictionary* (Cleveland: World Publishing Co., 1970).

## TABLE III

MAJOR PEST ARTHROPODS IN THE UNITED STATES IN
APPROXIMATE ORDER OF THE AMOUNTS OF
PESTICIDES USED ON THEM

| | Common Name | Scientific Name | Origin | Resist-ance |
|---|---|---|---|---|
| 1. | Corn Earworm = Bollworm = Tomato Fruitworm | Heliothis zea | N | + |
| 2. | Tobacco Budworm | Heliothis virescens | N | + |
| 3. | Cabbage Looper | Trichoplusia ni | N | + |
| 4. | Beet Armyworm | Spodoptera exigua | N | + |
| 5. | Armyworm | Pseudaletia unipuncta | N | |
| 6. | Codling Moth | Laspeyresia pomonella | E | + |
| 7. | European Corn Borer | Ostrinia nubilalis | E | |
| 8. | Oriental Fruit Moth | Grapholitha molesta | E | |
| 9. | Potato Tubermoth | Phthorimaea operculella | E | + |
| 10. | Pink Bollworm | Pectin gossypiella | E | + |
| 11. | Gypsy Moth | Porthetria dispar | E | |
| 12. | Boll Weevil | Anthonomus grandis | E | + |
| 13. | Alfalfa Weevil | Hypera postica | E | + |
| 14. | Plum Curculio | Conotrachelus nenuphar | N | |
| 15. | Japanese Beetle | Popillia japonica | E | |
| 16. | Green Peach Aphid | Myzus persicae | E | + |
| 17. | Rosy Apple Aphid | Dysaphis plantaginea | E | + |
| 18. | Greenbug | Schizaphis graminum | E | + |
| 19. | Pea Aphid | Acyrthosiphon pisum | E | |
| 20. | Housefly | Musca domestica | E | + |
| 21. | Apple Maggot | Rhagoletis pomonella | N | |
| 22. | Cabbage Maggot | Hylema brassicae | E | + |
| 23. | Lygus Bug | Lygus hesperus | N | + |
| 24. | San Jose Scale | Quadraspidiotus perniciosus | E | + |
| 25. | Western Flower Thrips | Frankliniella occidentalis | N | |
| 26. | European Red Mite | Panonychus ulmi | E | + |
| 27. | Two-Spotted Spider Mite | Tetranychus urticae | N | + |

SOURCE: Prepared by R. van den Bosch for the Environmental Protection
Agency.
NOTE: N = Native, E = Exotic.

numbers of predators and parasites may indicate that spraying can be delayed, perhaps until it is unnecessary. Any monitoring program conducted without population assessments of beneficial insects is headed for excessive pesticide use.

For the small grower who seeks to operate his own monitoring program, the first step is identification of the "pest" and its natural enemies. The next step is to learn the biology of these organisms and then to learn how to measure their abundance in the field situation.

Identification of the "pest" is important because the scientific name of the animal is the basic link to all research reports about it. The scientific name is also the key to understanding its evolutionary position, hence the biology and subsequent management of the problem to which the pest is central.

Insects can be identified by collecting specimens[12] and giving them to either an entomologist who will tell you their names or a county agent who will send them to appropriate specialists, or submitting them to the department of agriculture in your state. Identifying the insects by yourself requires a dissecting microscope, the proper taxonomic keys,[13] an entomological glossary,[14] and good morphological diagrams of the insect group in question. Frequently, once the crop is known and the damage caused by the pest is severe, that insect can be easily identified.[15] Beneficial insects are more difficult because they are poorly known. Debach has keys for identifying the important families of entomophagous (insect-eating) insects.[16]

Learning the biology of an organism can be done directly by observation, although this is usually time-consuming, difficult, and requires special knowledge. Basic information on the seasonal distribution of an insect, the number of generations per season, egg-laying habits, feeding patterns, mating behavior, resting or hiding places, and adult longevity all help to provide a perspective necessary for intelligent management. Knowing the seasonal mortality rates for the different factors in the environment of the insect is also essential in developing management strategies. Sometimes such information can be obtained from researchers, county agents, literature searches, and careful observations in field situations.

Measuring the abundance of a particular target insect and its natural enemies requires knowing what the different stages of the various insects look like, where they can be discovered, and a sampling program. The pest sampling program can be a relatively simple affair, (e.g., samples from a representative area, from a given number of plants, or number of plant parts). An average of all the samples, taken repeatedly through a season is the usual method used to follow an insect population. Knowing rates of parasitism, predation, or numbers of entomophagous insects present may also provide indications of population changes useful in forecasting herbivorous insect population sizes.

Many monitoring programs include trapping systems because they save time, provide an early warning, and sample hard-to-get-at stages (e.g., adults). Such traps usually have an attractive element such as a light, bait, or pheromone (chemical attractant produced by an insect), and a cage or enclosure lined with a sticky material. Operating a trapping system requires learning to "read" trap catches. Some insects for which traps are used in monitoring programs include mosquitoes, biting flies, the codling moth, houseflies, the walnut husk fly, hornworms, bollworms, and others.

### Cultural, Physical, and Mechanical Strategies of Insect Control

Some of the oldest, cheapest, and most environmentally sound strategies for pest control fall under this heading. By modifying the dates and methods of planting, growing, cultivation, irrigating, and harvesting crops, the build-up of many pest populations may be averted or reduced. Numerous ways exist for trapping, excluding and/or mechanically killing pest insects with little or no other detrimental effects upon the larger environment. All of these methods require knowledge of the particular biology, life cycle, and habitat requirements of the target animal. Cultural controls, particularly, involve a long-term management approach. Many of these methods are fairly labor-intensive and that alone may explain their neglect in favor of the seductive push-button appeal of modern pesticides.

We do not mean to minimize the importance of these alternative methods by giving them so brief a mention. For an excellent, more detailed survey of cultural, physical, and mechanical strategies of insect control the reader is referred to the book on insect pest management compiled by the National Academy of Sciences.[17]

## Use of Resistant Plants

It is important to choose the exact variety of plant that is most resistant to the potentially most numerous pests in the area; each micro-environment (see Figure 1) may be regarded as unique. Therefore considerable observation, experimentation, and consultation with agricultural authorities may be needed to discover precisely the best varieties of plants that will be relatively pest resistant for any particular area. Understanding this, one can see how much of the "organic" gardening mythology may arise from attempting to generalize from a particular experience without being able to discover the crucial limiting factors peculiar to a local condition.

Poor agricultural management of the crop may also have a considerable effect upon the pest management situations. Plants under stress, because they are not receiving the right amount of water drainage, light, temperature, or mineral nutrients, may be more susceptible to insect damage or less capable of repairing it. Overuse of nitrogen, for instance, may stimulate the production of a succulence that encourages sucking insects.

Some species of plants are just not suited to some localities. Rather than support a particular kind of plant that will survive only with continuous use of chemical crutches, it might be better not to grow it at all.

The reader is referred to the book on insect pest management for a fuller discussion of the use of plants resistant to specific insects and the genetic manipulation of plants to achieve this effect.[18]

## Classical Biological Control

If a study of a particular crop ecosystem leads to the conclusion that a crucial component of the system is missing, as in the linden aphid example, then it may be worthwhile to embark on a long-range program to find the beneficial insects in their native land and import them. There are numerous successful examples of this being done world-wide. See Table IV for examples of U.S. introductions.

This kind of aid must necessarily be sponsored by government or private groups such as growers' associations, since once the

TABLE IV

EXAMPLES OF BIOLOGICAL CONTROL IN THE U.S.

| Pest | Crop Attacked | Type of Natural Enemy | Degree of Control |
|------|---------------|----------------------|-------------------|
| Alfalfa weevil | Alfalfa in California | Parasite (Bathyplectes) | S |
| Alfalfa weevil | Alfalfa in Mid-Atlantic States | Parasites | C |
| Avocado mealbug | Guava, avocado, fig, mulberry in Hawaii | Parasite (Pseudaphytis) | S |
| Black scale | Citrus in California | Parasite (Metaphycus) | S |
| Brown-tail moth | Deciduous forest and shade trees in Northeastern United States | Parasites | C |
| California red scale | Citrus in California | Parasites (Aphytis spp.) | S |
| Cereal leaf beetle | Grain in Michigan | Parasites | S |
| Chinese grasshopper | Sugarcane in Hawaii | Parasite (Scelio) | S |
| Citrophilus mealybug | Citrus in California | Parasites (Coccophagus) | C |
| Clover leaf weevil | Eastern United States | Parasites | C |
| Coconut scale | Coconut and other palms in Hawaii | Predator (Telsimia) | S |
| Comstock mealybug | Apple in Eastern United States | Parasites (Allotropa) | C |
| Cottony cushion scale | Citrus in California | Predator (Vedalia beetle) | C |
| European larch sawfly | Northeastern United States | Parasites | C |
| European pine sawfly | Eastern United States | Parasites | C |
| European pineshoot moth | Northeastern United States | Parasites | S |
| European spruce sawfly | Northern United States | Parasites | S |
| European wheat stem sawfly | Eastern United States | Parasites | C |
| Florida red scale | Citrus in Florida | Aphytis holoxanthus from Israel | S |
| Greenhouse whitefly | Vegetables and ornamentals in New York | Parasite (Encarsia) from Israel | S |
| Japanese beetle | Turf in Eastern United States | Disease & parasites | S |
| Larch casebearer | Northeastern United States | Parasites (Tiphia) | S |
| Linden aphid | Linden trees in California | Parasites (Trioxys) | S |
| New Guinea sugarcane weevil | Sugarcane in Hawaii | Parasite (Ceromasia) | S |
| Nigra scale | Ornamentals in California | Parasite (Metaphycus) | S |

| Pest | Location | Control agent | |
|---|---|---|---|
| Olive scale | Olive, deciduous fruit trees, ornamentals in California | Parasite (*Aphytis*) | S |
| Oriental beetle | Sugarcane in Hawaii | Parasites (*Campsomeris*) | S |
| Oriental moth | Shade trees in Massachusetts | Parasite (*Chaetexorista*) | S |
| Pea aphid | Alfalfa in North America | Parasite (*Aphidius*) | S |
| Pink sugarcane mealybug | Sugarcane in Hawaii | Parasite (*Anagyria*) | S |
| Purple scale | Citrus in Texas and Florida | Parasite (*Aphytis*) | C |
| Rhodes grass scale | Grass in Florida | Parasite | C |
| Rhodes grass scale | Grass in Texas | Parasite | S |
| Satin moth | New England, Pacific Northwest | Parasite (*Apanteles*) | S |
| Spotted alfalfa aphid | Alfalfa in Southwestern United States | Parasites & resistant varieties | S |
| Sugarcane aphid | Sugarcane in Hawaii | Parasite, various predators | S |
| Sugarcane leafhopper | Sugarcane in Hawaii | Predator (*Cyrtorhinus*) | C |
| Taro leafhopper | Taro in Hawaii | Predator (*Cyrtorhinus*) | S |
| Torpedo bug plant hopper | Coffee, mango, citrus, etc., in Hawaii | Parasite (*Aphanomerus*) | S |
| Walnut aphid | English walnut in California | Parasites (*Trioxys*) | C |
| Western grape leaf skeletonizer | Grapevine in United States | Parasites (*Apanteles*) | S |
| White peach scale | Mulberry, papaya, etc., in Puerto Rico | Predator (*Chilocorus*) | S |
| Yellow scale | Citrus in California | Parasite (*Comperiella*) | S |
| **WEED** | | | |
| Alligator weed | Southeastern United States | *Agasicles* beetles | S |
| Klamath weed | Pacific states | *Chrysolina* beetles | C |
| Lantana rangeweed | Hawaii | Several moths and beetles | S |
| Prickly pear | Santa Cruz Island, California | Cochineal scale and coreid bugs | S |
| Puncturevine | California and Hawaii | *Microlarinus* beetles | S |
| Tansy ragwort | Pacific states | Cinnabar moth | P-C |

SOURCES:: From Debach, *Biological Control of Insect Pests and Weeds*; and the Council on Environmental Quality, *Integrated Pest Management* (Washington, D.C., 1972).

NOTE: S = Substantial Control, C = Complete Control, P = Partial Control.

beneficial insect is introduced (preceded, of course, by a period of quarantine and testing) and becomes adjusted to its new habitat, control of the pest may be so complete that nothing further needs to be done. In fact, pest and parasite may seem to disappear from view, as was the case with the successful introduction of the walnut aphid parasite (*Trioxys pallidus*) into California.[19] This means that there are no high and/or continuing profits to be made, as may be expected by a pesticide company where sprays can be sold to the grower year after year. One cannot expect private business enthusiastically to promote this permanent method of pest management. In fact, "success may be the Nemesis of biological control . . . once the problem is solved, through biological control, it is then forgotten and the benefits no longer tallied."[20]

It must be understood, however, that bringing in beneficial insects is not a simple push-button affair equivalent to spraying poisons around. One is dealing with complex living systems and this must be done with care and intelligence. Only certain insects are good candidates for this method of management, usually imported pests with specific parasites or predators capable of surviving only on the pest insect or its close relatives, and usually responsible for significant population mortality of the pest in its native area. The biology of these beneficial insects is so closely tied to the life cycle of their host that any small aberration of the latter in the new area, due to some factor such as weather, for instance, may make the survival of the beneficial impossible. There is always the possibility that the introduction effort may fail to establish the missing component.

If the beneficial insect is established, then time and energy must be devoted to monitoring the populations so that experience will be gained that can be applied to integrating the beneficial insect and cultural methods used on the crop.

### Mass Release of General Natural Enemies

In contrast to classical biological control, which is particularly suited to tree crops and other situations afforded stability over a long period of time, under certain conditions large-scale periodic releases (inundative or inoculative) of beneficial insects is proving a valuable technique. For example, there is the use of parasitic wasps (*Muscidifurax raptor, Spalangia endius,* and *Tachinaephagus zealandicus*)[21] to control flies on poultry farms. For some additional examples see Table V. There are already in the United States a number of firms providing beneficial insects for sale, along with information on their proper use. The importance of timing and the proper integration of these releases with other

TABLE V

EXAMPLES OF PARASITES AND PREDATORS OF POTENTIAL VALUE IN PEST SUPPRESSION THROUGH INUNDATIVE RELEASES

| Biological Control Agent | Pest |
| --- | --- |
| Nematodes: | |
| DD-136 (Biotrol NCS) | codling moth; European corn borer |
| *Heterotylenchus autumnalis* | faceflies |
| *Reesimermis nielseni* | mosquitoes |
| Parasitic insects: | |
| *Apanteles* species | various caterpillars |
| *Bracon kirkpatrick* | pink bollworm |
| Cuban fly | sugarcane borer |
| *Lysiphlebus testaceipes* | aphids |
| *Macrocentrus ancylivorus* | Oriental fruit moth |
| *Macroterys flavis* | brown soft scale |
| *Micropletis* | bollworm complex |
| *Pediobius foviolatus* | Mexican bean beetle |
| several tachinid flies | bollworm complex |
| *Trichogramma* | various moths and butterflies |
| Phytophagous insects: | |
| *Agasicles* (beetle) | alligator weed |
| *Bactra veratana* (moth) | nutsedge |
| Predacious insects: | |
| *Coccinella* (ladybugs) | aphids |
| *Cryptolaemus montrouzieri* | mealybugs and soft scale |
| *Hippodamia* (ladybugs) | aphids; bollworm complex |
| Other: | |
| *Cyprinidon variagatas* (saltwater fish) | mosquitoes |
| dung beetles | hornflies |
| *Gambusia* (freshwater fish) | mosquitoes |
| *Marisa* (snail) | aquatic weeds |
| *Mollienesia latipinna* (saltwater fish) | mosquitoes |
| white amur | aquatic weeds |

SOURCE: Council on Environmental Quality, *Integrated Pest Management* (Washington, D.C., 1972).

farm management strategies cannot be overstressed if success with this technique is desired.

The green lacewing (*Chrysopa carnea*) is a valuable carnivore that feeds on a wide range of potentially injurious pests such as aphids, scales, mealybugs, mites, and caterpillars. Green lacewings can be released in the egg stage and the larvae will not migrate from the field. A food spray can be obtained that will encourage adult lacewings to remain in the area and lay eggs.[22] This is essentially a technique for increasing the predator population at the time the pests are most likely to be present in damaging numbers, without having to wait for the normal lag time to elapse while predator populations catch up to their prey, and during which time considerable damage to the crop may occur.[23]

This food spray (wheast) is a by-product of the cheese industry. Wheast and sugar are mixed together (1 pound wheast, 1 pound sucrose, 1 gallon water) and applied to plant leaf surfaces or feeding stations as an artificial honeydew. A check should be made first to be sure the mixture does not harm the plant.[24] If the solution is found to be toxic to the plants, flat vertical boards, supported by stakes banded by a sticky material to exclude ants, may be placed at intervals in the field and serve to hold washes of the solution. Wheast and sugar solutions have been shown to be commercially useful on cotton and alfalfa.[25]

### Flowers as Food for Natural Enemies

On much the same principle as the use of food sprays, one may wish to have specific blooming ornamentals available to provide pollen and/or nectar for various beneficial insects. There are some striking examples of this being a successful component of an insect management program. For instance, flowering shrubs have been used to feed the beneficial wasps *Apantales schizurae* and *Hyposoter fugitivus*, parasites on the red-humped caterpillar (*Schizura concinna*), in the insect management system devised for the California Highway Department by Dr. Dudley Pinnock (University of California, Berkeley, Department Entomological Sciences). Also, Russian experiments have been reported with a similar beneficial wasp feeding on mustard flowers in orchards.[26] Generally speaking, however, the precise relationships between many beneficial insects and the plants most likely to encourage their presence have not been worked out. Here is a fertile area for research for anyone willing to acquire the necessary insights into both botany and entomology.

## Using Disease Pathogens

The use of disease pathogens against pest insects has been proved effective in a number of cases, especially with caterpillars. These biological control agents offer great promise since they are generally specific, are usually harmless to vertebrate life, degrade quickly, and can be applied with conventional spraying equipment. Many of these agents, particularly viruses and fungi, are actually missing components of the insect ecosystem, and once established may function like an imported natural enemy. Table VI lists important pathogens currently under development.

Best known and easiest to obtain is *Bacillus thuringiensis*, used successfully in different strains and strengths against many types of caterpillars. It is sold under such names as Biotrol, Thuricide, Dipel, etc. Since it is a live material it is important to obtain a batch that has been stored under the proper conditions and not for too long. The wettable powder stores the longest. It is also desirable to test the various strains to find which is most effective against the target caterpillar.

The microbial insecticides are a natural part of the environment. This is important in two respects. As the insect population becomes composed of many resistant individuals there is a good chance that a new mutant strain of the disease organism will also appear, much as new flu virus strains constantly appear and sweep through human populations that have not previously been exposed to them.

In addition, the dangers to humans and other animals from these diseases are very slight. Not only are insects far removed from man in the evolutionary sense, but they are also much more ancient in their development. Man and other modern mammals evolved in an environment in which these other organisms were probably already present and our nonsusceptibility is built in, as it were.

TABLE VI
EXAMPLES OF PATHOGENS UNDER DEVELOPMENT BY GOVERNMENT AND/OR INDUSTRY FOR USE AGAINST AGRICULTURAL AND FOREST PESTS

| PATHOGENS UNDER COMMERCIAL PRODUCTION | IMPORTANT AGRICULTURAL PESTS CONTROLLED |
|---|---|
| *Bacillus popilliae* (Trade name—Doom) | Japanese beetle |
| *Bacillus thuringiensis* (Trade names—Biotrol® BTB Thuricide® Dipel, Parasporin®, Bakthane® L69, Agritrol®) | Lepidopterous pests (larvae of moths and butterflies) |
| Nuclear polyhedrosis virus (temporary label) (Trade names—Biotrol VHZ, Viron/H®) | Cotton bollworm (corn earworm) |

| PATHOGENS UNDER SERIOUS DEVELOPMENT | IMPORTANT AGRICULTURAL PESTS INVOLVED | |
|---|---|---|
| Viruses (Product names—Polyvirocide Biotrol VPO Biotrol VSE Biotrol VTN) | Cabbage looper Diamond back moth Beet armyworm Tobacco budworm Pink bollworm Cotton leaf perforator Alfalfa looper Fall armyworm Saltmarsh caterpillar | Douglas fir tussock moth Gypsy moth Codling moth Red-banded leaf roller European pine sawfly Pine sawfly Spruce budworm Soybean looper Citrus red mite |
| | Mosquitoes | |

Fungus—*Hirsutella thompsonii*  
            *Metarrhizium anisopliae*

            *Beauveria bassiana*  
            (Produce name—Biotrol FBB)

Protozoan—*Nosema locustae*

Citrus rust mite  
Pecan weevil, corn borer, leafhoppers, sugarbeet curcuilio, cutworm froghopper, rhinoceros beetle, wheat cockchafer  
Corn rootworm, white fringed beetle, Colorado potato beetle

Major range grasshopper species in Montana

PATHOGENS KNOWN, BUT NOT YET UNDER SERIOUS DEVELOPMENT

IMPORTANT AGRICULTURAL PESTS INVOLVED

Nuclear polyhedrosis viruses

Yellow striped armyworm  
Almond moth  
Indian meal moth  
Cotton leaf worm  
Alfalfa looper

IMPORTANT FOREST PESTS CONTROLLED

Nuclear polyhedrosis viruses

Great Basin tent caterpillar  
Western tent caterpillar  
Eastern tent caterpillar  
Hemlock looper (Western and Eastern)

SOURCE: National Research Council, *Principles of Plant and Animal Pest Control*, 1969.

## What Is a "Safe" Insecticide?

It is important to realize that all agricultural situations are "unnatural" from certain points of view. Not only the growing of vast areas of a single type of plant together in one area to the exclusion of others, the exotic origin of the plants and animals, and the regular cultivating, fertilizing, and watering program that maintains them, but also the domesticated plant itself differs greatly from its wild ancestors or living wild relatives. Among other characteristics that may have changed are its proportion of tough cellular material to the succulent, edible flesh that makes it prized as human food. A modification of originally strong oils and juices may have occurred. These changes, which produce the palatability we have come to expect of our modern vegetables and fruits, have simultaneously reduced the plants' defenses against insects.

In most cases the modern food plant and the agro-ecosystem of which it is a part are so totally artificial that in order to produce sufficient food for expanding human populations some major manipulations occasionally will be necessary to substitute for the "natural" defenses long lost. The sheer impossibility in terms of human labor of handling serious insect problems on a commercial farm, city park system, or highway landscape plantings, as one would tend a private home garden, means that sometimes it will be necessary to resort to some kind of spray or dust.

Although one's choice may be a botanical, or plant-derived, pesticide, regarded by most people as relatively innocuous, it is important to distinguish the effects of these substances upon the health of humans and other mammals from the results when they are used on insect ecosystems. Garlic, for instance, much touted by the "organic" farming literature, is a nonselective insecticide which can be deadly to both beneficial and pest insect alike. Used injudiciously it may cause the same side effects of pest resurgence and secondary pest outbreak as DDT or any synthetic broad-spectrum poison.

Nevertheless, in selecting treatments to control pest animals, toxicity to mammals should be a major consideration. The toxicity of any material is usually expressed as LD 50, or the *L*ethal *D*ose necessary to kill *50 percent* of the animals used (usually rats), when given by mouth (oral) or drops on the skin (dermal). It is expressed as milligrams per kilogram of body weight of the test animals.

The limitations of this information are obvious. Rats are not humans, and our reactions may be quite different from those of the test animals. Also, as observed earlier, long-term exposure to small amounts of pesticides may be quite different from immediate reactions in a test situation. Some compounds accumulate in various organs or body fat, while others are quickly metabolized and excreted. Also, the synergistic effects of these various poisons used in combinations are almost completely unknown.

However, as a very rough guide one can obtain considerable information from knowing the LD 50 measure of any pesticide. Pyrethrum, for instance, a pesticide derived from plants, has an LD 50 of 1,500, while parathion, a highly toxic artificial pesticide, has an LD 50 (oral) of 3.6 (see Figure 3). This means that less than half a gram could kill a 150-pound person. (More than one million pounds were used on crops in California in 1970.)[27] The higher the number, the less toxic to mammals.

An additional factor to consider when choosing a pesticide is how quickly it breaks down and whether it breaks down to non-toxic compounds. Botanicals and naturally-occurring insect diseases are very good in these respects since generally they are quickly broken down by bacterial action or ultraviolet light, or are destroyed by heat and humidity.

The least toxic spray is plain water. It has been used successfully in many ecosystems to knock aphids off leaves[28] and to suppress mite populations (from overhead sprinklers).[29]

The addition of less than a 1-percent solution of an old-fashioned, fat-based soap such as green soap (potassium stearate) may help to wash insects off plants, but should be tested first on a small scale to check for plant toxicity.

Oil sprays have long been used effectively against scale insects when plants are in a dormant condition. Inert dusts such as diatomaceous earth and silica gel are useful in certain situations and have a mechanical or desiccating effect upon insects.

After the above materials, which have a more or less mechanical action, one might consider the botanical poisons such as pyrethrum, garlic, rotenone, and nicotine. Not all insects are

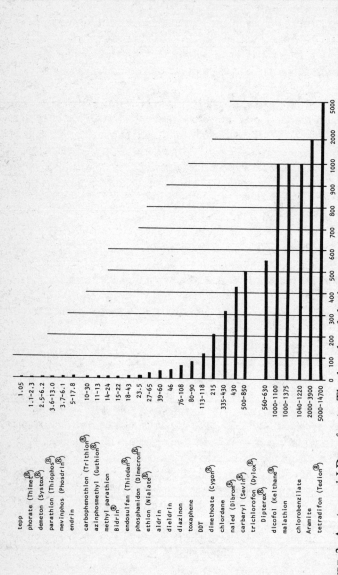

FIGURE 3. Acute *oral* LD$_{50}$ of rats. The lengths of the bars are expressed in milligrams of pesticide per kilogram of body weight of rat and indicate the relative toxicity to humans of certain common insecticides when swallowed. The shorter the bar, the more toxic the pesticide. (Redrawn from J. Blair Bailey and John E. Swift. "Pesticide Information and Safety Manual," Division of Agricultural Sciences, Univ. Calif. Agric. Extension Service, Berkeley, 1968.)

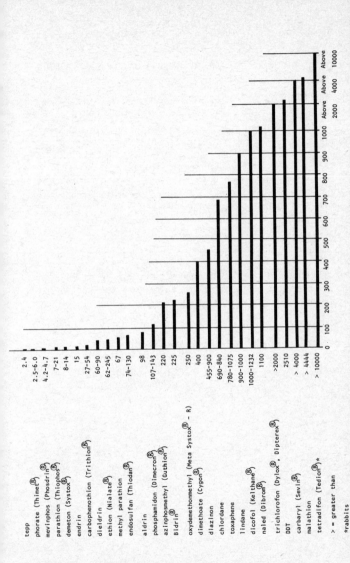

FIGURE 4. Acute *dermal* $LD_{50}$ of rats. The lengths of the bars are expressed in milligrams of pesticide per kilogram body weight of rat and indicate the relative toxicity to humans of certain common insecticides when exposure is by the skin. (Redrawn from J. Blair Bailey and John E. Swift. "Pesticide Information and Safety Manual," Division of Agricultural Sciences, Univ. Calif. Extension Service, Berkeley, 1968.)

susceptible to each of these. A good procedure is to consider the least toxic first and then, if that is known or proved not to be effective against the target pest, move up to more toxic materials. Nicotine, very toxic to humans, was used extensively prior to the development of modern synthetic pesticides. It breaks down rather rapidly. Rotenone, moderately toxic to most animals, is very toxic to fish and should not be used where it might get into aquatic systems. Of course, one should use even relatively nontoxic materials very carefully.

Environmentally sound management of wildlife in agro-ecosystems requires that one manipulate the animal populations only when it appears that the damage will be economically insupportable. One should choose the least toxic pesticides, use them in spot application only when and where they are needed, and resort to their use only after all other methods have failed.

## Notes

1. The vast majority of poisons used in agriculture are not specific as to what they kill. The word "pesticide" implies that only pests are killed, which is clearly not the case.

2. The latter is quite a racket in California where most ladybird beetles are scooped up while hibernating in huge masses in the Sierra Mountains and then released with body fat deposits that must be burned off before they can begin to eat; this virtually guarantees that they will fly away from the point of release.

3. A target insect is the pest for which a pesticide is intended. The word "pest" is subjective and often misleading, and we hesitate to use it unless it is properly defined.

4. Natural enemies include predators (carnivores or meat-eaters), parasites, and disease. "Parasite" is a term used to distinguish insects that attack and lay eggs on other insects (usually one parasite developing on one host) from predatory insects that feed on many prey. Sometimes the word "parasitoid" is used to make this distinction.

5. R. van den Bosch and B. Messinger, *Biological Control* (New York: Intext Educational Publ., 1973).

6. C. B. Huffaker, *Biological Control* (New York: Plenum Press, 1971).

7. Ibid.

8. W. Olkowski, "A Model Ecosystem Management Program for Street Tree Insects in Berkeley, California" (Ph.D. thesis, Entomology Department, Univ. of California, Berkeley, 1973).

9. R. van den Bosch, "Biological Control of Insects," *Annual Review of Ecology and Systematics* 2 (1971): 45–66.

10. P. Debach, *The Use of Imported Natural Enemies in Insect Pest Management*, in Tall Timbers Conference on Ecological Control of Animals by Habitat Management, 3 (Tallahassee, Fla.: Tall Timbers Research Station, 1971), pp. 211–234.

11. Van den Bosch, "Biological Control of Insects."

12. H. H. Ross, *How to Collect and Preserve Insects*, Natural History Survey Division, State of Illinois, Urbana, Ill., Circular 39; Borror and DeLong, *Introduction to the Study of Insects* (New York: Holt, Rinehart & Winston, 1971); J. W. Knudsen, *Biological Techniques* (New York: Harper & Row, 1966).

13. Borror and DeLong, *Introduction to the Study of Insects*; H. E. Jacques, *The Beetles* (Dubuque, Iowa: W. C. Brown Co., 1951); R. H. Arnett, *An Introduction to the Study of Beetles* (Washington, D.C.: Catholic Univ. of America Press, 1963); P. R. Ehrlich and A. H. Ehrlich, *The Butterflies* (Dubuque, Iowa: W. C. Brown Co., 1961); J. R. Helfer, *The Grasshoppers, Cockroaches and Their Allies* (Dubuque, Iowa: W. C. Brown Co., 1953); B. J. Kaston, *The Spiders* (Dubuque, Iowa: W. C. Brown Co., 1972); H. F. Chu, *The Immature Insects* (Dubuque, Iowa: W. C. Brown Co., 1949).

14. J. R. de la Torre-Bueno, *A Glossary of Entomology* (Brooklyn, N.Y.: Brooklyn Entomological Society, 1937).

15. L. O. Essig, *The Insects of North America* (New York: Macmillan Co., 1926); C. L. Metcalf and W. P. Flint, *Destructive and Useful Insects* (New York: McGraw-Hill, 1962); W. Ebeling, *Subtropical Fruit Pests* (Berkeley: University of California Agricultural Extension Service, 1959); D. J. Borror and R. E. White, *A Field Guide to the Insects of America North of Mexico* (Boston: Houghton Mifflin, 1970).

16. P. Debach and I. Schlinger, eds., *Biological Control of Insect Pests and Weeds* (New York: Halsted Press, 1973).

17. National Research Council, Committee on Plant and Animal Pests, Subcommittee on Insect Pests, Agricultural Board, *Insect-Pest Management and Control*, vol. 3, *Principles of Plant and Animal Pest Control* (Washington, D.C.: National Academy of Sciences, 1969).

18. R. H. Painter, *Insect Resistance in Crop Plants* (New York: Macmillan Co., 1951).

19. Van den Bosch and Messinger, "Biological Control of Insects."

20. Huffaker, *Biological Control.*

21. E. F. Legner and H. W. Brydon, "Suppression of Dung-Inhabiting Fly Populations by Pupal Parasites," *Ann. Ent. Society of America* 59 (1966): 638–651.

22. K. S. Hagen and R. L. Tassan, "The Influence of Food Wheast® and Related *saccharomysetes fragilis* Yeast Products on the Fecundity of *Chrysopa carnea*," *Canadian Entomologist* 102, no. 7 (1970): 806–811.

23. K. S. Hagen, E. F. Sewall, and R. L. Tassan, *The Use of Food Sprays to Increase Effectiveness of Entomophagous Insects*, in Tall Timbers

Conference on Ecological Animal Control by Habitat Management, 2 (Tallahassee, Fla.: Tall Timbers Research Station, 1970).

24. Unfortunately, recent changes in the formulation (the addition of salts) necessitates further evaluations. However, the salt can be decanted, with the wheast becoming useful again.

25. Hagen et al., "Use of Food Sprays."

26. National Research Council, *Principles of Plant and Animal Pest Control.*

27. Gordon L. Berg, "Farm Chemical Handbook," *Farm Chemicals,* 1971.

28. Olkowski, "Model Ecosystem Management Program."

29. D. Flaherty et al., "Spider Mite Populations in Southern San Joaquin Vineyards," *Calif. Agri.* 4 (1972): 10–12.

## References

Clausen, Curtis. *Entomophagous Insects.* New York: McGraw-Hill, 1940.

Clausen, Curtis. *Biological Control of Insect Pests in the Continental United States.* Washington, D.C.: Government Printing Office, USDA Bulletin No. 1139. 1956.

Davidson, R., and Peairs, L. *Insect Pests of Farm, Garden and Orchard.* New York: John Wiley & Sons, 1966.

De Bach, Paul. *Biological Control by Natural Enemies.* Cambridge: Cambridge University Press, 1974.

Dietrick, Everett. *Private Enterprise Pest Management Based on Biological Controls.* Tall Timbers Conference on Ecological Control of Animals by Habitat Management, Proceedings, No. 3, pp. 7–20. Tallahassee, Florida: Tall Timbers Research Station, 1972.

Olkowski, H. *Common Sense Pest Control.* Richmond, California: Consumers Cooperative of Berkeley, 1971.

Rabb, R. L. and Gutherie, F. E. eds. *Concepts of Pest Management.* Proceedings of Conference, North Carolina State Univ. Raleigh, N.C.: North Carolina University Press, 1970.

Westcott, C. *Handbook on Biological Control of Plant Pests.* Brooklyn, N.Y.: Brooklyn Botanical Garden, 1960.

Westcott, C. *The Gardener's Bug Book.* Garden City, N.Y.: Doubleday & Co., 1964.

# 19

# SMALL-SCALE UTILIZATION
# OF SOLAR ENERGY

*John F. Elter*

> If man's ingenuity through the years had been
> directed to the utilization of solar energy instead
> of to the development of devices to consume fossil
> fuels, it is quite conceivable that we might today
> have a solar economy just as effective and just as
> efficient as our fossil-fuel economy. Ultimately
> man will probably be driven to turn to the sun.
> —L. P. GAUCHER
> *Energy Sources of the Future for the U.S., 1965*

Awareness of the energy problem has resulted in a phenomenal
rebirth of interest in the age-old dream of making direct use of
the sun's energy. It is no surprise that various proponents of solar
energy have expressed vastly different philosophies regarding the
manner in which this abundance of energy ought to be used. To
some, solar energy ought to be used in order to guarantee the
continuation of present economic, social, and political patterns,
and to meet the energy demands of the magnitude encountered in

---

John F. Elter is a former scientist for the Xerox Corporation.

power plants using natural forms of stored energy. Others, however, regard the energy problem primarily as an environmental issue, and their concern relates to the present and proposed technological methods of using the sun's energy to supply *needed* energy. Of equal importance is their fear, supported by history, of the effects of further continued "growth" once these demands have been temporarily satisfied. Generally speaking, methods of using the sun's energy to cope with this aspect of the energy problem are characterized by their small-scale, decentralized nature.

The energy from the sun, although abundant, is dilute. This fact strongly affects the manner in which the sun's energy is used. Large-scale schemes for using solar energy require large land areas. Recent proposals ranging from 5,000-acre solar "farms"[1] to huge satellite power stations beaming microwave energy to terrestrial antennas[2] are thus plagued by the inherent nature of the sun's energy at the earth's outer atmosphere, the solar constant (the solar constant $= 1.94$ calories/cm$^2$/minute; value varies slightly with seasonal changes in the distance to the sun). Such schemes could very well affect the local weather and ecology.

Small-scale methods of using the sun's energy, on the other hand, utilize the sun's energy as it naturally arrives. They do not attempt to collect vast amounts of sunlight and generate power at a single location, only to redistribute it with complex transmission lines back ino its originally dilute form. Rather, they are designed to intercept only that portion of the sun's energy that is needed in order to accomplish individual small tasks at hand. As such, small-scale solar energy devices are particularly well suited for supplying the energy demands of single dwellings, small farms, and decentralized rural communities.

To gain perspective regarding the possibilities of using solar energy, let us consider the task of supplying the hot water needs for a family of five people. Assume that daily each person uses twenty gallons of water at 120°F. If this water is heated from 65°F, the total amount of energy required would be 45,833 BTUs per day.[3] Now the average annual U.S. solar radiation intensity is 1,400 BTU per ft$^2$ per day. Consequently, a collector 8 feet by 16 feet would have nearly four times the energy falling on it as would be needed to supply the energy for the hot water alone. This hot water would normally require the burning of 244 gallons of fuel oil (at 70 percent efficiency) each year. With fuel oil costing about $.40 per gallon, the use of solar energy would represent a savings of $100 per year. The savings would pay for the cost of a $400 solar water heater in slightly less than four years.

I intend to discuss the underlying principles and techniques whereby the sun's energy can be put to use in relieving some of the normal demands placed upon conventional sources of stored energy, especially for rural areas or groups seeking self-sufficiency. My purpose is not to speculate what *could* be accomplished in the future, but rather to discuss what *can* be accomplished today. I shall, therefore, emphasize the low temperature applications. Solar energy conversion is particularly adaptable to such tasks as supplying domestic hot water needs, easing the burden placed upon fossil fuels in supplying space heat for homes, livestock shelters, etc., providing hot air for the drying of crops, and even providing fresh water and food. Furthermore, the needed technology for the most part already exists. High temperature applications, on the other hand, are just not competitive with the wall plug at the present time. This area is in need of more basic research, and some interesting possibilities will be pointed out.

## Solar Energy Collection

The vital link between the fusion energy of the sun and its earthly application is the solar collector. It is the heart of any system utilizing the sun's energy, whether it is generating power or providing home heat. The principles by which solar collectors are designed to intercept the sun's energy are consequences not only of the quantity of sunlight available, but also its quality.

Most of the sunlight reaching the earth's surface is in the form of short wavelength radiation that is capable of photosynthesis, of interaction with solar cells, or of transmitting heat to solar collectors.

### THE FLAT-PLATE COLLECTOR

The simplest, and by far the most important for farm use, is the so-called flat-plate collector. As shown in Figure 1, it usually consists of a blackened plate that conducts heat well, covered with one or more layers of a transparent material, usually glass. This

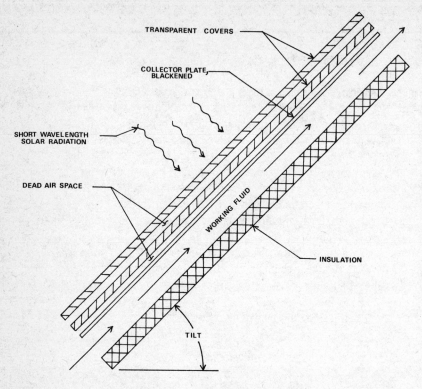

FIGURE 1. Working parts of a typical flat-plate collector.

simple construction makes use of the greenhouse effect in which the energy absorbed by the black plate is trapped because the energy it reradiates is of longer wavelengths[4] (infrared) that cannot penetrate the glass plate. Thus, the glass covers serve not only as an insulation against heat loss by convection, but also as a selective filter to the heat energy reradiated by the collector plate.

The basic purpose of the collector is to heat the working fluid circulating over or through the collector to temperatures suitable for intended needs. Sometimes the working fluid, usually air or water, is passed over the backside of the collector plate to which fins may be attached to augment the heat transfer. The fluid may also flow through tubes built into or on the plate itself. As the result of controlled experiments, the operating characteristics of

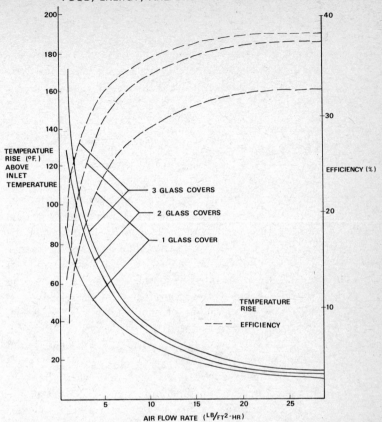

FIGURE 2. Typical operating curves for flat-plate collectors. The left vertical axis corresponds to the rise in the air temperature above its inlet temperature. The right vertical axis represents the collector efficiency, or the fraction of the total energy falling on the outermost cover which is transferred to the fluid of the collector. The curves hold only for the case when (1) the air entering the collector is 50°F above the outside temperature, and (2) when the solar energy flux incident on the collector is 250 BTU/ft²-hr (a value typical of near-noon conditions on a reasonably clear day). Flow rates on the horizontal axis are in terms of square feet of absorber surface.

this type of conventional collector are fairly well understood. Some typical operating curves for collectors are presented in Figure 2 in order to illustrate the air temperatures and efficiencies that can be achieved with the sample flat-plate collector like that shown in Figure 1.

For example, referring to Figure 2, suppose that the outside temperature is 20°F and air enters the collector at 70°F, at a rate of 5 lb/ft²-hr. If the collector is tilted toward the sun so that the incident flux is 250 BTU/ft²-hr, the curves apply and they show that the air can leave the collector at 115°F (45° + 70°) if one cover is used, or 137°F (67° + 70°) if three covers are used. A rough calculation can serve to put these numbers in perspective. A typical home with 1,000 ft² of floor space has a space heat demand of about 28,000 BTU/hr. If the home had 350 ft² of collector surface, 1,750 pounds per hour of air would pass through the collector. With one glass cover, the air temperature rise would be 45°F, so that the total amount of heat collected would be 18,890 BTUs.[5] If two covers were used, the air temperature rise would be 61°F but the heat collected in this case would still not meet the demand. If, on the other hand, the flow rate of air through the collector were doubled, lower air temperatures would result but the collector would operate at a higher efficiency.[6] That is, the twice-covered collector with double the air flow rate would result in an air temperature rise of 35°F, and the total heat collected would meet the demand.

The curves in Figure 2 clearly indicate that when the fluid entering the collector is at a temperature much higher than the outside air temperature, the number of transparent covers should be selected with care. Since the addition of another cover can increase the cost per square foot of collector surface by as much as 15 percent, the extra cover should be added only if it results in an increase in energy output by at least an equivalent amount. On the other hand, when the temperature of air entering the collector is the same as the outside temperature and is heated up to 50°F above the surrounding temperature, then it can be shown that the number of covers, provided there is at least one, is not important.

Plastic materials for use as covers should also be selected carefully since in general they have poor weathering properties. Some plastics, depending upon their thickness and type, can be nearly transparent to the long wavelengths emitted by the heated black-plate, thus decreasing their ability to trap reradiated solar heat. In addition, most plastics have the tendency to lose their transparency to sunlight and become foggy as a result of ultraviolet waves. Some plastics have ultraviolet absorbing stabilizers, and an extended operating life.

Plastic materials would be appropriate for fruit-drying cabinets[7] with operating temperatures from 110°F to 150°F, and the expected life of the dryer would not be great. For high temperature or extended applications, they should not be used at all.

## IMPROVEMENTS TO THE FLAT-PLATE COLLECTOR

The most important characteristic of the solar collector is that its efficiency decreases as the fluid flow rate decreases, even though the exit temperature increases. Whether you want a little bit of hot fluid or lots of warm fluid, it is important to design solar collectors in order to minimize the causes of heat loss: radiation (by heat waves); conduction (directly through a medium); convection (indirectly through the mixing of a medium).

*Reducing Radiation Losses*: Radiation is the source of most of the heat losses in conventional collectors. The radiant heat loss depends not only upon the collector plate temperature, but also the degree to which the surface is capable of reradiating the heat of longer wavelengths. A perfect absorber, or black body, is also a perfect radiator of heat. What is desired is a surface that is a good absorber of the short wavelength solar radiation but a poor emitter of the longer wavelengths. Such surfaces are called "selective radiation surfaces," and a great deal of research has been done to achieve their dual property of appearing black to sunlight yet shiny to the longer wavelengths.

There are various designs for selective radiation surfaces, but not all are suitable for localized, low-cost applications. This is particularly true of surfaces that are prepared by electrochemical or vacuum deposition techniques.[8] Although such methods have produced surfaces that absorb 90 percent of the sunlight and yet reradiate only 10 percent of that of normal black painted surfaces,[9] the cost and know-how involved prevent their practical use.

One particular method of obtaining a certain degree of selectivity seems adaptable to small-scale applications and yet has received little attention from researchers. The method depends upon the fact that heated shiny metal surfaces give off very little radiant energy. It has been found that if shiny metal surfaces are coated with thin layers of carbon black, they will absorb the sunlight and yet still be poor emitters of the long wavelengths.[10] Such selective surfaces can be prepared by coating them with the carbon black given off during the burning of styrene candles, kerosene, turpentine, or other suitable material. The selectivity depends upon the conditions under which the carbon black is deposited. High flames result in large particles and poor selective properties. Low flames, however, result in small particles and a uniform velvety layer. It is impossible to determine if the deposited layer is too thick (in which case the selective property can be lost even with small particles), since in all cases the surfaces look equally black to the eye. Their selective properties can be determined, however, by simply comparing its temperature rise with that of a normal black surface under identical conditions. It would seem that this simple method could be perfected and thus enable selective solar collectors to be fabricated at low cost and on the local level.

Selective surfaces have advantages over the normal black painted absorber. From various experiments it has been found that at least two glass covers are usually required with a normally blackened collector base. If instead a base with a selective surface is used, even smaller heat losses occur with just one cover. The selective surface thus saves the cost of one cover and its transmission losses, not to mention the additional labor of installation. The advantages that selective surfaces offer are even more significant with higher collector temperatures. As pointed out by Hottel, a normally blackened surface with three glass covers for producing low pressure steam would have no useful output, or zero efficiency, at solar intensities of 135 $BTU/ft^2$-hr.[11] The same system with a selective absorber would, on the other hand, have an efficiency of about 25 percent.

Other types of flat-plate collectors can be made that reduce the radiant heat losses and yet depend primarily upon the geometry and, to a lesser extent, the selective nature of the surfaces. These "honeycomb" absorbers make use of transparent or reflective vertical spacers placed between the base and the glass cover that reduce the transfer of radiant heat between these surfaces. Of course, the walls would not have to be transparent or reflecting if the collector could be mounted so as to follow the sun. In most cases, however, this is impractical for small-scale applications.

Recent descriptions of honeycomb absorbers have appeared in the literature. In one case the absorber cells were made from glass tubes,[12] although the expense involved seems to preclude their use at present. In another case, the honeycomb was made from commercially available aluminized Mylar sheets that were cut into strips, notched, and assembled in an egg-crate fashion to form a rectangular honeycomb.[13] The assembly was then dipped into clear polyurethane to obtain a coating approximately one mil (.001 inches) thick. This coating, being transparent to sunlight and opaque to the long wavelengths, resulted in the cell walls having the desired properties. The cold air entered the collector through a space provided between the glass cover and the honeycomb structure. Efficiencies ranged from 67 percent to 78 percent for flow rates ranging from 6.7 lb/hr to 20 lb/hr per square foot of collector surface. Exit temperatures ranged from 100°F to 250°F.

The design of this air heater was optimized in order to minimize the cost of crop drying.[14] For example, in order to dry 250 pounds of sorghum per hour, the optimum design resulted in the heater having 303 ft² and exit temperatures of 228°F. In order not to damage the crop with excessive temperatures near noon, the air flow rates and the processing rate had to be varied throughout the day.

The use of a porous absorber in the honeycomb collector described above illustrates an important aspect of solar collector design. Higher collector efficiencies always result whenever the heat transfer between the fluid and the absorber can be improved. Buelow[15] has tested air heaters using corrugated sheets and heaters having various degree of surface roughness on the underside to promote turbulence and thereby improve the heat transfer. As expected, the roughest surfaces gave the highest efficiencies by exposing the fluid to more heated surface area per square foot of collector surface.

Such techniques for reducing radiation losses or for improving the heat transfer, however, are not without their limitations. Gains achieved by the use of honeycomb devices can easily be lost if the weight of the collector is excessive. Likewise whenever the heat transfer goes up, so does the power required to pump the fluid. Consequently, there is no "optimum" design in a technical sense, and a trade-off between improved efficiency, pumping power, and cost has to be determined.

*Reducing Convection and Conduction Losses*: It is well known that *still* air forms an effective insulating barrier because of its low heat conductivity. Thus, in the confined air spaces between the absorber plate and the various covers any movement of air will lead to an increase in the rate of heat transfer by convection and lower collector efficiencies.

The degree of convection depends upon the size of the air gaps, the temperature difference between the surfaces of each air layer, and the angle of tilt. In the flat-plate collector, convection can be suppressed by reducing the size of the air spaces. However, whereas convection losses decrease with decreasing air gaps, the opposite is true with regard to conduction losses. Practically speaking, this means that air gaps of about one-half inch or less should be used. Since heat conduction in the air spaces will always be present, a very low heat transfer across the air gap cannot be obtained. Of course, evacuating the air from the spaces between the various covers and the absorbing plate would eliminate both convection and conduction, but maintaining this condition is difficult. Alternatively, various solar scientists have experimented with the idea of again introducing vertical cell walls between the absorber plate and its innermost cover. However, this method of suppressing natural convection only works when the collector is *horizontal*. Recent data indicate that for *inclined* collectors the air flow in spaces is inherently unstable and convection will always be present.[16]

*Thermal Traps—Solids as Collectors*: One other method of achieving higher collector temperatures makes use of the so-called thermal-trap effect. It has been observed that some lakes possess the unusual property of having higher temperatures with increasing depth. Because of natural salt deposits, lower portions of the lakes remain more dense than portions near the surface, so that natural convection is prevented. In still lakes temperatures as high as 150°F have been reported at depths of about four feet. The "solar pond" has been studied by the Israel National Research Council which is investigating its potential for generating power at a cost that is competitive with commercial fuels.[17] We shall delay a discussion of the solar pond for a moment and examine its implications in flat-plate collector design.

It has been found that when a thick, transparent solid is irradiated by the sun, the interior blackened side, insulated by the low conductivity material ahead of it, becomes hotter than the side directly exposed to the sun.[18] In addition, the temperatures achieved are greater than those obtained with a normally blackened opaque plate. The temperature rise at the interior side of the transparent solid (or at the bottom of a solar pond) over that of the opaque plate is called the thermal-trap effect. Its magnitude depends upon the transmission characteristics of the material and its thickness. But the theoretical and experimental work that has been performed to date leaves some questions as to the practicality of the thermal-trap effect for small-scale applications. Thus, although temperatures as high as 340°F can be obtained with a 2½-inch slab of methyl methacrylate (plexiglass), it takes about four hours of direct exposure to solar intensities of near-noon conditions.[19] Still, this technique appears to warrant further investigation, since it affords a means of achieving high temperatures without the use of expensive solar concentrating devices.

### THE SOLAR STILL AND SOLAR HEATED PONDS

The basic characteristic of the flat-plate collector is that it heats a stream of fluid that then exchanges that heat at a remote point. There also exist special applications in which the collector and the working fluid are linked together, as with the solar still and the solar heating of fish ponds. I will consider these two applications simultaneously, because the factors involved in their successful operation, although similar, are sometimes at odds.

*Solar Still*: A typical solar still consists of a shallow basin of brine that is covered with a layer of glass (Figure 3). When plastic covers are used they are normally treated to make them "wettable," thus preventing condensation. Most still basins are lined with a black waterproof material such as butyl rubber to aid in the absorption of sunlight. The slope of the cover, normally ten to twenty degrees from ground, is adjusted to enable the vapors condensing on the underside of the cover to drain into troughs. An interesting feature of the still is the fact that evaporation can

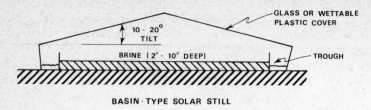

BASIN-TYPE SOLAR STILL

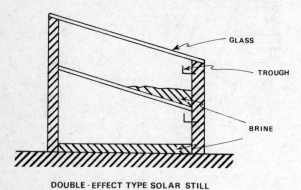

DOUBLE-EFFECT TYPE SOLAR STILL

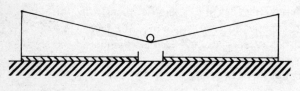

WEIGHTED V-TYPE SOLAR STILL

FIGURE 3. Various types of solar stills.

only take place when the cover temperature is lower than that of the brine. Otherwise, the air space between the cover and the brine tends to become saturated and the rate of evaporation decreases. Thus, a still will have a temporary increase in output as the wind speed increases. Normally, evaporation proceeds at a usable rate when the cover is 30°F to 40°F lower than the brine. However, evaporation increases rapidly with temperature so it is desirable to have as high a brine temperature as possible. For example, the output of a still having a brine temperature of 130°F and a cover temperature of 100°F is three times as great as one having a brine temperature of 90°F and the cover at 60°F. Since the output depends greatly on the brine temperature, every precaution should be taken to eliminate heat losses to the ground. Insulating the bottom of the basin can increase the still output by as much as 35 percent. Still outputs are best correlated with the intensity of solar radiation, and to a lesser extent by the ambient air temperature, wind velocity, and still design. A well-designed still has an output of roughly six to ten gallons of water per day for each 100 square feet of evaporating surface when the solar radiation intensity is 2,000 BTU/ft$^2$ per day (a value typical for southwestern America).

*Fish Ponds*: The potential of the backyard fish pond offers a tremendous incentive for the small-scale utilization of solar energy. The fish pond and the still are essentially governed by the same thermodynamic principles. However, the purpose of the dome-covered fish pond is to *limit* all heat losses, including evaporation, so as to provide water temperatures high enough to promote the growth of algae for fish food. Normally, this means water temperatures of 80°F for periods of at least 150 days, although year-round 80°F pond temperatures would certainly be more desirable.

In building a solar-heated fish pond, various factors should be taken into account. A typical dome-covered aquaculture pond is normally about sixteen feet in diameter and three to four feet deep. Consequently, about 200 ft$^2$ of the pond surface is capable of losing heat by convection, evaporation, and radiation to the cover, and then to the outside, whereas 350 ft$^2$ can lose heat by conduction to the ground. The sides and bottom of the pond should therefore be insulated with low-conductivity material such as sawdust in order to keep these losses at a minimum. Having taken care of the sides and bottom, the next step is to reduce the heat losses from the surface of the pond. On a clear night with an outside air temperature of 40°F, the combined heat losses from an 80°F pond surface are about 100 BTU/hr/ft$^2$ of pond surface. The reason for this rather high loss is that on a clear night the effective sky temperature for radiation can be 50°F.[20] (This explains why oranges can freeze even though the evening air temperature is above freezing.) During the eleven hours of darkness, 1,100 BTUs would be lost for each square foot of pond surface. This value is nearly double the amount of solar energy falling on a horizontal surface for many regions of the country during the latter part of October. During the twenty-four-hour period, the temperature of a pond sixteen feet in diameter and three feet deep would drop more than 3°F. Thus, unless preventative measures are taken, the pond could drop to temperatures unsuitable for algae growth in a period of a few days.

Attempts to use the sun's energy to heat backyard fish ponds are fairly recent.[21] As a result, very little experimental data exist on their performance over extended periods of time. However, some techniques for reducing the heat losses from the pond can be inferred from the operation of the solar still. First, the dome or other cover structure of the pond should be provided with a double "skin" in order to reduce nighttime heat losses. If plastics are used, they should be impervious to infrared heat waves radiated from the pond inside. The use of a portable insulating pond cover with a reflective coating (facing upward) would reduce heat losses further during periods of low sunlight. In addition, heat losses due to evaporation could be reduced by supplying heated air to keep the inner cover of the dome at a higher temperature. This additional heat could also be supplied by solar energy collected and stored during the day. However, more work needs to be done in order to understand better the physics of the solar heated fish pond. This is particularly true with respect to the interplay between the pond and ground temperature variations during the seasons and the influence of this on the time of optimum yield.

*Solar Ponds*: It is appropriate at this point to return to a discussion of the solar pond. Recall that a solar pond remains in a quiescent state even though its temperature at the bottom can be considerably higher than at the top. Salt concentrations that increase with depth maintain the proper density gradients in order to achieve this effect. In practice these conditions have to be maintained artificially.

Temperatures near boiling have been achieved in experimental ponds about three and one-half feet deep.[22] The main problems involve maintaining the proper salt concentrations at various depths, removing energy from the bottom of the pond, and keeping the pond clean. Each of these has been examined in some detail. The proper salt concentrations can be maintained by adding salt (usually $MgCl_2$) or highly concentrated solutions to the bottom and removing salt that has diffused to the surface with streams of fresh water. The most obvious method of removing energy from the bottom of the pond is to use an array of pipes as a heat exchanging device. Besides its high cost, this method has certain limitations. The effectiveness of the heat transfer depends upon the presence of convection outside the pipe, but this same convection could destroy the pond's stability. Experiments, however, have shown that whole layers of the bottom part of the pond can be removed and replenished without disturbing the salt concentrations in the layers above it.

The solar pond has certain technical difficulties that still have to be overcome. Even so, it has such tremendous potential for supplying the energy required for small-scale tasks that further research and individual experimentation is more than appropriate.

### High-Temperature Collectors and Power Generation

Thus far I have examined the flat-plate collector and the principles underlying various schemes for improving its performance. Still, the maximum temperatures achievable with flat-plate collectors are limited. In order to attain even higher temperatures, and a higher quality of energy, it is necessary to make the area that intercepts the sun's rays large in comparison with the area from which this energy can be lost. For flat-plate collectors the ratio of the collecting area is equal to the absorbing areas. For concentrating collectors, by definition, the collecting area is greater than the absorbing area.

Concentrating collectors can provide enough heat for the generation of mechanical power and electricity. In fact, since the last half of the nineteenth century there have been numerous attempts to build solar power plants using parabolic reflectors to focus the sun's energy onto boiler surfaces (Figure 4). However, there are several important considerations involved with their use. For one thing, unlike flat-plate collectors, concentrating collectors are not able to use that portion of the total sunlight which is diffuse, the so-called sky radiation. In many instances sky radiation can ac-

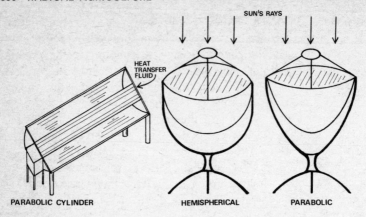

FIGURE 4. Types of focusing solar collectors.

count for more than 50 percent of the available sunlight. Although inexpensive tracking mechanisms can be built, the collector/tracking device must be capable of withstanding wind forces. This requirement can add to the weight and cost of the collector to the point where some trade-off in optical concentration can be easily justified.

One trade-off tactic would be the use of cylindrical parabolic reflectors (Figure 4) mounted on an east-west axis. This reduces the efficiency of solar collection, but it removes the necessity of having to track the sun across the sky. Only the seasonal variation in the sun's angle has to be accommodated, and this can be done manually.

Attaining high temperature energy is only one problem. Two problems remain: the storage of that energy and its conversion into mechanical or electrical power. Direct storage of high temperature energy is impractical. The energy can be more easily stored in the mechanical form of pumped water or compressed gas, or as electrochemical energy in batteries. Fortunately, there exist numerous agricultural tasks that can be timed with periods of sunshine so that storage is unnecessary (see Chapter 20).

The conversion of thermal energy into power requires the use of some sort of heat engine. Most small engines have efficiencies around a few percent, an order of magnitude smaller than their theoretical maximum. Tabor has suggested the use of high-molecular-weight fluids, in order to reduce the speed of single-stage turbines and thereby increase their efficiencies.[23] He found that chlorobenzene, could result in efficiencies of 15 percent to 20 percent for turbines of the size of a few horsepower. These systems are still in the experimental stage and generally are not available to the small farmer.

A major objective of small-scale solar power generation should be over-all simplicity, even at the expense of reduced efficiency. Here again the flat-plate collector can be put to work. Hottel has optimized the design of the flat-plate collector for producing low-temperature steam.[24] He concluded that for climates similar to that of El Paso, the optimum design would be a collector with three to four glass plates and working fluid temperatures of 240°F. With an efficiency of 4 percent (10 percent if selective surfaces are used) the system would deliver 4.25 kwh/day for each 100 ft² of collector surface. The cost of this power could be less than 4¢/kwh, if the collector cost could be kept to $2.50/ft.². Similarly, d'Amelio has shown that a straight reaction turbine run directly with hot water furnished at atmospheric pressure by a flat-plate collector is only slighly less efficient than more complex systems involving the production of steam or the evaporation of special fluids.[25] The power system is simple. It consists of a flat-plate collector, a simple turbine runner, a condenser, and a small feed pump. Similar systems can be built that use a free-piston Stirling engine. Such simple systems, although not necessarily having the highest efficiencies, are within the technical reach of most people. They are relatively easy to build and maintain and yet are readily adaptable to such tasks as pumping water and providing power for stationary farm equipment.

### Solar Energy Transmission and Storage

So far I have examined some of the techniques involved in the collection of the sun's energy. Of equal importance is the manner in which this energy, once collected, is transferred to and stored at a remote point. The way in which the collection, transmission, and storage systems are integrated depends upon the type of working fluid used. The choice of the working fluid is not always obvious, however, and for this reason I shall discuss the transmission and storage aspects separately. Then I shall indicate the factors involved in the integration of a complete solar energy system.

## TRANSMISSION

There are two basic methods of transporting heated fluid from the collector to its point of use: natural convection and forced convection. Natural convection is the natural movement of gas or fluids as the result of buoyancy forces (e.g., hot air or water rising). Forced convection, as the term implies, refers to the flow of a working fluid by fans or pumps.

## NATURAL CONVECTION

Natural convection is attractive for small-scale applications because it eliminates the need for auxiliary power. Efficient natural convection systems depend upon not only the fact that "warm air rises," but also other factors involved in the rate at which the flow proceeds.

Natural convection systems are simple in construction but com-

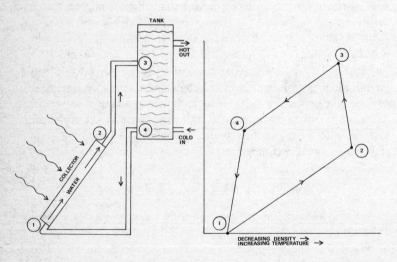

FIGURE 5. A simple solar water heater with water circulating by natural convection (the thermo-syphon effect). The relationship between water temperature and density at different points in the circuit is also shown.

plicated in design. An example can be found in the "thermo-syphon" effect in solar water heaters. A system for supplying hot water is shown in Figure 5. It consists of a flat-plate collector, an insulated storage tank, and connecting pipes. Also shown is a schematic diagram of the water density (and temperature) at various points in the system during daylight hours. Hot water leaving the collector is replenished by the cooler water entering from the bottom of the tank. Similarly, the water can reverse its flow direction at night as the water in the storage tank cools down. This heat loss can be prevented with the use of one-way check valves.

The solar water heater offers an example of the coupling between the buoyancy forces, flow rates, and temperatures that is characteristic of any system using natural convection. For example, if the bottom of the tank is raised with respect to the top of the collector, higher flow rates result, but water temperatures leaving the collector go down. The opposite occurs if the tank is lowered. There is also a question of trade-off between flow rate and temperature in the design of air ducts for transferring energy in space heating systems.

## FORCED CONVECTION

Solar energy devices using forced rather than natural convection are generally easier to design. Although they require the use of fans or pumps, they also enable flow rates to be maintained irrespective of the temperatures or any special geometrical relationships between the collector and the storage. Now the crux of the design problem is to match the pressure losses in the transmission network with those developed by the pumping equipment. Forced convection in tubes and ducts has received considerable attention because of its widespread application. The handbooks put out by the American Society of Heating and Refrigeration Engineers contain a wealth of information concerning frictional losses for various duct geometries, pumping power requirements, and the thermal properties of various building materials. Accordingly, we shall not elaborate on forced convection systems except to note that even in these systems the design is by no means straightforward. In general, there is always an optimum design with respect to flow rates, temperature, pumping power, and cost.

## STORAGE

The need for solar energy storage is a result of the fact that the sun's energy is intermittent, its peak intensity often coming at times when it is least needed. Therefore, providing for one or more points of storage in a system's "flowsheet" can often increase its performance and lower its cost. The principal methods of storing low-temperature energy involve the heating, melting, or vaporizing of a suitable material. The energy stored in an inert material at a high temperature is referred to as its "sensible heat." Similarly, the heat aborbed when a material (e.g., salt) melts is referred to as its "latent heat of fusion." In either case energy becomes available when the process is reversed. Some practical heat storage materials for these purposes are listed in Table I.

## SENSIBLE HEAT STORAGE

The effectiveness of sensible heat storage depends upon the heat storing capacity of the material. Water is an outstanding material for this purpose. The amount of energy stored in 200 cubic feet of water when raised 80°F is equal to that needed by

TABLE I

HEAT STORAGE MATERIALS FOR SOLAR COLLECTORS

| Sensible Heat Storage | | Heat Capacity $BTU/°F\text{-}ft^3$ |
|---|---|---|
| Water | | 62 |
| Scrap Iron | | 59 |
| Concrete (30% void space) | | 37 |
| Stone (40% void space) | | 21 |
| Latent Heat Storage | Melting Point °F | Heat of Fusion $BTU/ft^3$ |
| Sodium Carbonate (decahydrate) | 90–97 | 10,400 |
| Glauber's Salt | 90 | 9,700 |
| Hypophosphoric Acid | 131 | 8,700 |
| Calcium Chloride (hexahydrate) | 84–102 | 7,900 |
| Paraffin Waxes | 100–130 | 3,600 |

a modest dwelling on a typical winter day. The major disadvantage of water is usually the cost of the storage tank. Nevertheless, the technologies of heat transfer, container materials, and controls place water in a position of immediate advantage with respect to small-scale applications.

When air is used as the working fluid, water can still be used as a storage material, but it becomes less attractive because of the additional cost of a heat-exchanging device. In this case crushed stone or gravel is unique in offering an effective and inexpensive heat transfer surface. However, according to Table I a stone bin would require nearly three times the space needed by water.

An additional measure of the usefulness of stone or gravel is the pressure force required to pass air through its porous structure. This pressure is strongly dependent upon the size of stone, the density with which it is packed in the bin, and the rate of air flow. As an example, consider a stone bin with a cross-sectional area of 100 ft². At a volume flow rate of 840 ft³/min, the pressure drop across one foot of one-inch stone is nearly four and a half times that across a foot of stone about three inches across. It is clear that the bin geometry and stone size can have a significant effect on the performance of natural convection systems. The effect of flow rate on the resistance to the flow is even more pronounced so forced convection systems are affected as well. There exist pressure drop correlations that can serve as a guide in selecting the proper bin and stone size.[26] Generally, however, this particular form of energy storage has not received the attention it deserves.

## LATENT HEAT STORAGE

Many materials undergo physical changes at temperatures normally encountered in small-scale solar energy systems. These changes are accompanied by the absorption (during melting) and liberation (during freezing) of a considerable amount of heat, the so-called latent heat of fusion. These materials can be used to store solar energy in the form of this latent heat; they have the advantage of storing a much larger amount of energy than an inert material occupying the same volume. For example, 9,700 BTUs can be stored in one cubic foot of Glauber's salt at 90°F (Table I). This feature has been widely appreciated, and latent heat storage has been examined in some detail by solar scientists. Most of the inexpensive latent heat materials, and in particular the salt hydrates, do not really undergo a true melting but rather a separation into various phases of different density and composition. Organic materials do not have this problem but their heat of fusion is normally not high enough to offset their increased cost.

The lack of a compact and inexpensive storage system can only serve to limit the range of solar energy technology and delay its general acceptance as a primary source of energy. More emphasis has to be placed upon finding not only new methods of storage but also new ways of using those that already exist. Thermodynamically speaking, water is the best sensible heat storage material. Concrete, on the other hand, has a larger useful temperature range. In space heating applications, it can be integrated into the home structure.

### Solar Energy Systems

There are three steps involved in the design of a solar energy system. First, the actual energy needs have to be determined. This normally includes the range of allowable temperatures. Space heating requirements, for example, are usually expressed in terms of the required number of British thermal units per degree day. The number of degree days that characterizes the local climate is obtained by taking the temperature difference in degrees Fahrenheit between the daily average outdoor temperature and the heated interior (65° F) and adding these differences for each day throughout the heating season. Only those days for which daily average outside temperatures are less than, say, 65° F are included. In this way, the heating requirements of homes, expressed as BTU/DD, involve and reflect the size of the house, number of windows, type of insulation, etc.[27] For example, a modest home with 1,000 ft² of floor space, rated at a reasonable 15,000 BTU/DD would demand 75,000 BTUs for each day when the average temperature is 60° F (heat needed $= 15,000 \times [65\text{-}60]$). But such a value should not be accepted as minimum. The use of new materials in the building trades often lags behind their acceptance in other applications. The next step involves assessing the amount of sunshine available at the time of need. This information, including the percentage of cloud cover and other climatic data, can be obtained from local weather stations or existing compilations.[28] The final and most creative step involves the actual selection and integration of collecting, transmitting, and storage devices into an over-all system which, when combined with any other existing energy sources, results in the best performance at the least cost.

Integrating an appropriate combination of collecting and storage units into a solar system is a complex task. Problems become visible only as the design proceeds and they usually appear in the form of a whole series of trade-offs. Thus, the "hothouse" effect of the flat-plate collector can be enhanced with the use of more transparent covers, but after a certain point the cost becomes prohibitive. Determining at which point this occurs is part of the design process. Sometimes a particular collector tilt may be known to give higher temperatures, but it may do so at the expense of aesthetics. The major part of the total system design, however, has to do with the fact that the collection, transmission, and storage subsystems each have their own peculiar characteristics, although they do not function independently of one another. It is their mutual interaction that determines the performance of the system as a whole. It is through an examination of this interaction (i.e., a system analysis) that the greatest improvements in performance are to be found.

A great many systems analyses have been performed for a wide variety of solar energy applications,[29] but because of the unpredictable nature of climates few have been performed in great detail. An exception is the definitive work of Tybout and Löf.[30] Using actual hour-by-hour weather data, they examined the performance of a combined solar house and hot water heating system for eight different climates. The system consisted of a conventional rooftop flat-plate collector, a water storage system, and forced convection transmission. A conventional energy source was included in order to provide heat and hot water during cold periods. Water was chosen as the working fluid in order to eliminate the need for separate heat transfer equipment from the solar system to the living quarters. The system was examined for a full one-year period. On the basis of a twenty-year cost depreciation the optimum design was a collector with two glass covers (except for climates similar to Miami and Phoenix, which had one) and a tilt of 15 degrees more than the local latitude, facing south. The cheapest water storage capacity varied from ten to fifteen pounds of water per square foot of collector area. This same storage capacity was found for each of the eight different climates. The optimum collector area in this "least-cost" sense varied according to the type of weather. A 25,000 BTU/DD home in Boston, for example, was found to have an optimum collector size of about 350 ft². The appropriate storage capacity would therefore amount to 5,250 pounds of water (about 630 gallons) or its thermal equivalent. For the same home located in Omaha, the least-cost collector area was 520 ft² with a proportional increase in storage capacity. In either case the space occupied by the storage tank would be less than that taken up by the old-fashioned coal bin. The important point, however, is that in every case, the energy storage capacity amounted to roughly one to two days winter heat delivery, and never more than three. In addition, the best "mix" between solar and conventional energy always occurred when the solar energy system provided a good deal less than the total heat supply. The least-cost solar energy system in Omaha, for example, contributed 47 percent of the total heat needed.

The results of the Tybout and Löf study are important for two reasons. First, the work confirmed and strengthened similar results that had been obtained with monthly weather averages, rather than hour-by-hour data. This, of course, means that the performance of a particular solar energy system can be predicted reasonably well with data that is readily available. Second, the work constituted an important economic study. Assuming a collector cost of $2.00/ft², the study indicated that for all the locations considered (except Seattle-Tacoma), solar heat is quite a bit cheaper than electricity, and for climates and fuel costs like those of Santa Monica, California, solar energy is competitive with gas and oil.

There are several points with regard to the Tybout-Löf study that deserve comment. First of all, optimization was based only on cost. It did not reflect the fact that people might be willing to absorb a moderate cost increase in order to conserve fuel and reduce pollution. As an example, the Omaha least-cost system resulted in an overall cost of $2.45 per million BTUs of heat delivered. If the solar contribution to the total heat supply were increased from 47 percent to 69 percent, the cost would increase

22 percent to $3.00 per million BTUs. This energy cost is still cheaper than electricity. It is clear that consideration of cost only can be misleading; one can often reduce the cost of the system by making use of materials that are otherwise regarded as "junk." In addition, as pointed out in the study, their cost estimates have not included any special architectural considerations necessary to achieve the optimal tilt. At northern latitudes of 40 degrees, for example, the optimum tilt of 55 degrees would be appreciably greater than the slopes of the roofs of most homes. In certain cases it is sometimes more attractive to make the collector tilt 90 degrees so that it can be made a part of the home structure.

There are many other solar devices that could be discussed at length. I have attempted to touch upon the principles, and to a lesser extent the specific details, involved in the more practical solar energy systems. But if there exists one general feature that is characteristic of small-scale operations, it is their *simplicity*, and this makes the utilization of solar energy on a small scale practical. Since the testing and construction of solar devices requires neither expensive equipment nor exotic materials, they are within the technical and economic reach of the people whose needs are best suited to the capabilities of small-scale devices. For these same reasons the rural dweller should welcome the opportunity to examine the natural processes occurring around him and use these observations in creating alternative energy systems that can fulfill personal needs.

The present energy problem is a complex entanglement of social, economic, political, and environmental isues. In essence, however, the problem is one of *attitude*. It would be ironic if the energy from the sun were utilized merely to whet our growing appetite for more energy. The small-scale utilization of solar energy, however, has implicit in it a reordering of life-style priorities that involve a decentralization of human activities and a revitalization of the countryside.

### Notes

1. A. B. Meinel and M. P. Meinel, "Physics Looks at Solar Energy," *Physics Today*, February 1972, pp. 42–50.

2. P. E. Glaser, "Satellite Power Station." *Solar Energy* 12 (1969): 353–361.

3. A BTU, or British thermal unit, is the energy required to raise one pound of water one degree Fahrenheit. Since a gallon weighs 8.33 pounds, we need (8.33 pounds) $\times$ (20 gal.) $\times$ (5 people) $\times$ (120° F — 60° F) = 45,833 BTUs each day.

4. A black body emits 75 percent of its energy at wavelengths greater than the value given, in microns (L), by the expression L = (4,756/T + 460), where T is the temperature of the body, °F. Thus, for example, a blackened plate at 250° F radiates 75 percent of its energy at wavelengths greater than about 7 microns.

5. The number of BTUs required to raise the temperature of one pound of a material one degree Fahrenheit is referred to as its specific heat. For air this number is 0.24 BTU/lb-F°. Thus, the heat collected with one glass cover at 5 lb/hr ft² would be Q = (Mass) $\times$ (specific heat) $\times$ (Temperature Rise) or Q = (1750) $\times$ (0.24) $\times$ (45) = 18,890 BTU.

6. Efficiency $= \dfrac{\text{Total energy given up to working fluid}}{\text{Total solar energy falling on cover}} \times 100$

In the above example:

$$\text{Efficiency} = \frac{18{,}890 \text{ BTU/hr} \times 100}{250 \dfrac{\text{BTU}}{\text{ft}^2 \text{ hr}} \times 350 \text{ ft}^2} = 21.6\%$$

This value corresponds to that given in Fig. 2 for 5 lb/hr-ft² with one glass cover.

7. C. D. Davis and R. I. Lipper, "Solar Energy Utilization for Crop Drying," in *New Sources of Energy*, vol. 5, *Solar Energy II*, Proceedings U.N. Conference, Rome, 1961, pp. 273–282; T. A. Lawand, *A Solar Cabinet Dryer for Agricultural Producer* (Brace Research Institute, MacDonald College of McGill University, Ste. Anne de Bellevue 800, Quebec, 1966); F. H. Buelow, "Drying Crops with Solar Heated Air," in *New Sources of Energy*, vol. 5, *Solar Energy II*, Proceedings U.N. Conference, Rome, 1961, pp. 267–270.

8. L. F. Drummeter, Jr., and G. Hass, "Solar Absorptance and Thermal Emittance of Evaporated Coatings," in *Physics of Thin Films*, 2 vols. (New York: Academic Press, 1964), 2: 305–361.

9. B. P. Kozyrev and O. E. Vershinin, "Determination of Spectral Coefficients of Diffuse Reflection of Infrared Radiation from Blackened Surfaces," *Optics and Spectroscopy* 6 (1959): 345.

10. Ibid.

11. H. C. Hottel et al., "The Role of Scatter in Determining the Radiative Properties of Surfaces," *Solar Energy* 11, no. 1 (1967): 2–13.

12. G. Francia, "A New Collector of Solar Radiant Energy: Theory and Experimental Verification," in *New Sources of Energy*, vol. 6, *Solar Energy III*, Proceedings U.N. Conference, Rome, 1964.

13. H. Buchbert et al., "Performance Characteristics of Rectangular Honeycomb Solar-Thermal Converters," *Solar Energy* 13 (1971): 193–221.

14. O. A. Lalude and H. Buchbert, "Design and Application of Honey-comb Porous-Bed Solar-Air Heaters," *Solar Energy* 13 (1971): 223–242.

15. F. H. Buelow, "Corrugated Solar Air Heaters for Crop Drying," *Sun at Work*, 4th quarter, 8 (1962).

16. W. W. S. Charters and L. F. Peterson, "Free Convection Suppression Using Honeycomb Cellular Materials," *Solar Energy* 13 (1972): 353–361.

17. H. Tabor, "Solar Ponds: Large Area Solar Collectors for Power Generation," *Solar Energy* 7 (1963): 189–194.

18. E. Lumsdaine, "Transient Solution and Criteria for Achieving Maximum Fluid Temperature in Solar Energy Applications," *Solar Energy* 13 (1970): 3–19.

19. M. H. Cobble et al., "Verification of the Theory of the Thermal Trap," *Journal of the Franklin Institute* 282, no. 2 (1966).

20. The "effective sky temperature" for radiation is the temperature that the sky *appears* to have insofar as *radiant* energy exchange between it and a surface on the earth is concerned.

21. W. O. McLarney, "The Farm Pond Revisited," *Organic Gardening and Farming* 18, no. 11 (1971): 88–95.

22. Tabor, "Solar Ponds."

23. H. Tabor, "Solar Collectors, Selective Surfaces, and Heat Engines," *Proceedings National Academy Sciences* 47 (1961): 1271–1278.

24. H. C. Hottel, "Power Generation with Solar Energy," in *Solar Energy Research*, Daniels and Duffie, eds. (Madison, Wisc.: Univ. of Wisconsin Press, 1955).

25. C. d'Amelio, "The Hot-Water Thermal Cycle in the Utilization of Solar Energy," *Solar Energy* 7, no. 3 (1963): 138–143.

26. S. Ergun, *Chem. Eng. Progress* 48 (1952): 227.

27. Sometimes daily averages are not available. Then the daily outside temperature, averaged over the month, can be used.

Example: Albuquerque, N.M. (from note 24)

| Month | Av. Temp (°F) | DD/Month |
|-------|---------------|----------|
| Jan | 37.3 | 31 × (65 − 37.3) = 859 |
| Feb | 43.3 | 28 × (65 − 43.3) = 607 |
| Mar | 40.1 | 31 × (65 − 50.1) = 462 |
| April | 59.6 | 30 × (65 − 59.6) = 162 |
| Nov | 47.8 | 30 × (65 − 47.8) = 516 |
| Dec | 39.4 | 31 × (65 − 47.8) = 794 |
| | | Total # DD = 3,400 |

NOTE: May, June, July, August, September, and October had average outside temperatures >65°F and so they are not included.

28. Hottel, "Power Generation with Solar Energy."

29. B. Y. H. Liu and R. C. Jordan, "The Long-Term Average Per-

formance of Flat-Plate Solar-Energy Collectors," *Solar Energy* 7, no. 2 (1963): 53–74.

30. R. A. Tybout and G. O. G. Löf, "Solar House Heating," *Natural Resources* 10, no. 2 (1970): 268–326.

## References

Beckman and Duffie. *Solar Energy Thermal Processes.* New York: Wiley and Interscience.

Daniels, Farrington. *Direct Use of the Sun's Energy.* New Haven: Yale Univ. Press, 1964.

Landa, H. C. et al. *The Solar Energy Handbook* 3d ed. Milwaukee: Ficoa/ Seecoa.

Merrill, R. et al., eds. *Energy Primer: Solar, Wind, Water, and Biofuels.* Menlo Park, California: Portola Institute.

Proceedings of the World Symposium on Applied Solar Energy, Phoenix, Ariz. Menlo Park, Calif.: Stanford Research Institute, 1955.

*Solar Energy* (previously *Journal of Solar Energy Science and Engineering*). Tempe, Arizona: Solar Energy Society.

Total Environmental Action. *Solar Energy Housing Design in Four Climates.* Harrisville, N.H.: TEA.

Transactions of the Conference on the Use of Solar Energy. Tucson: Univ. of Arizona Press, 1958.

Zarem, A. M., and Eway, D. D. *Introduction to the Utilization of Solar Energy.* New York: McGraw-Hill, 1963.

# 20

## LOCAL ENERGY PRODUCTION FOR
## RURAL HOMESTEADS AND
## COMMUNITIES

### Donald Marier
### and Ronald Weintraub

Modern man lives isolated in his artificial environment, not because the artificial is evil as such, but because of his lack of comprehension of the forces which make it work—of the principles which relate his gadgets to the forces of nature, to the universal order. It is not central heating which makes his existence "unnatural," but his refusal to take an interest in the principles behind it. By being entirely dependent on science, yet closing his mind to it, he leads the life of an urban barbarian.

—ARTHUR KOESTLER
*The Act of Creation*

### Introduction

An ecologically centered farm or rural community can develop an integrated, small-scale power generation system just as it can develop an integrated, diverse agriculture. However, the setting up

Donald Marier is coeditor of *Alternative Sources of Energy* magazine.

of such a scheme is not an easy task. A new energy ethic is required that says that the energy used in a geographical region should be derived from that region and any energy which isn't should be obtained by exchange for real goods. The system should also be reasonable in price, easily built and maintained, and relatively nonpolluting.

The energy available to a small farm, intentional community, or homestead will most likely be in the form of solar energy, wind energy, water energy, or organic matter (fuels from animal and plant waste). Any one of these sources alone will probably be insufficient to supply the complete energy needs of a farm or community. But a *combination* of these sources, each sharing mutually the energy loads and waste products, will be the model for a power system in most climates and situations. The supply of each of these natural energy sources is intermittent so that some form of energy storage will be necessary. Furthermore, each energy source comes in a different form that makes it suitable for different uses with a minimum number of conversions. Solar energy comes as heat; wind and water energy come in mechanical form; fuels from organic matter such as methane gas and alcohol are generally more portable and versatile. Thus, an integrated energy system will consist of a diverse matrix of the various energy sources, their storage, interconnections, and final energy use.

### Alternative Energy Sources, Load-Sharing Potential, and Storage

#### SOLAR ENERGY

The amount of energy reaching the earth as solar radiation is extremely great, with more global energy influx from the sun every fifteen minutes than man currently uses in a year. The reason this large potential has not been put to widespread use is the difficulty involved in utilizing such a diffuse and intermittent energy source. Two general kinds of solar collecting devices hold real promise for small-scale applications.

#### SOLAR COLLECTORS

*Flat-Plate Collectors*: The flat-plate collector usually consists of a blackened back sheet of either metal, plastic, sections of glass, or cloth enclosed in a well-insulated container covered with one or more sheets of glass or transparent sun-resisting plastic (see Chapter 19 for more details).

*Focusing Collectors*: The focusing collector reflects sunlight onto a small area, resulting in greater heat and higher temperatures. The parabola is best for this purpose, but due to the difficulty of producing an accurate parabolic surface, hemispheres and cylindrical collectors are used when temperatures up to 350°F are desired (see Figure 4, Chapter 19).

The parabolic types can be made from a spun metal or a molded plastic backing to which a highly reflective coating or coated tape (such as aluminized Mylar tape) is applied. Small mirrors, when pasted on a parabolic or hemispherical shell and set to focus rays on a small area, can result in a very good collector. The parabolic metal reflecting mirrors of searchlights scrounged from surplus stores can also be used as reflectors for solar ovens. Similarly, if a cylindrical form is cut lengthwise or if sheet metal is attached to evenly spaced parabolic-shaped wooden frames the surface may be aluminized, or else long strips of thin mirror can be glued to the

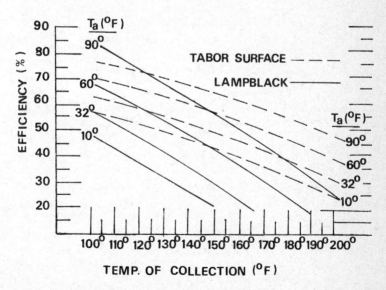

FIGURE 1. Efficiency of collector and one glass plate in relation to various ambient temperatures. (Redrawn from Hottel, *Solar Energy for Heating*.)

surface.[1] Uses of this type of collector include solar coolers, solar furnaces, solar ovens, and solar steam generation for heat engines.

## AVAILABLE ENERGY

Table I shows some typical values of daily averages in BTU/ft$^2$ of solar energy[2] received on a horizontal surface by months.[3]

If information is available concerning the efficiency of a collector under various conditions, the energy likely to be obtained can be calculated. Figures 1 and 2 illustrate these type of data graphically[4] for a one-glass and two-glass plate collector for various temperatures of the collector and surrounding air ($T_a$ or "ambient"). For $T_a$ of 70°F and a temperature of collection of 150°F, from Figure 1, the efficiency of a one-glass plate collector is approximately 50 percent. Suppose we needed information for Boston, Mass., in June; the average June solar insolation for Boston (from Table I) is 1,823 BTU/ft$^2$/day $\times$ (0.50 efficient) = 912 BTU/ft$^2$/day. A 1,000 ft$^2$ flat-plate collector would supply 91,200 BTUs of heat, enough to raise the temperature of about 137 gallons of water 80°F (see water heating below). If insolation values are not available from the Weather Bureau for a given loca-

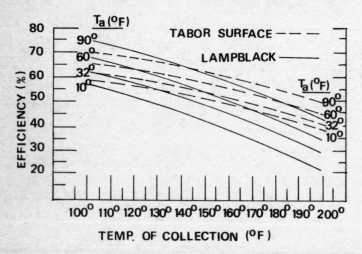

FIGURE 2. Efficiency of collector and two glass plates in relation to various ambient temperatures. (Redrawn from Hottel, *Solar Energy for Heating*.)

TABLE I

DAILY AVERAGES (BTU/FT$^2$/DAY) OF SOLAR ENERGY RECEIVED ON A HORIZONTAL SURFACE BY MONTHS

| City | Jan | Feb | Mar | Apr | May | Jun | Jul | Aug | Sep | Oct | Nov | Dec | Average |
|---|---|---|---|---|---|---|---|---|---|---|---|---|---|
| Santa Maria, Cal. | 1070 | 1380 | 1882 | 2251 | 2506 | 2399 | 2428 | 2369 | 1945 | 1594 | 1114 | 867 | 1817 |
| Grand Lake, Colo. | 790 | 1144 | 1624 | 2030 | 2177 | 2362 | 2236 | 1989 | 1720 | 1328 | 863 | 613 | 1373 |
| Miami, Fla. | 1100 | 1284 | 1535 | 1756 | 1852 | 1771 | 1749 | 1716 | 1528 | 1351 | 1232 | 1085 | 1497 |
| Griffin, Ga. | 1063 | 1107 | 1181 | 2103 | 2288 | 2273 | 2170 | 2066 | 1546 | 1358 | 1044 | 738 | 1578 |
| Twin Falls, Idaho | 613 | 827 | 1269 | 1705 | 2184 | 2303 | 2280 | 1985 | 1646 | 1255 | 738 | 450 | 1438 |
| New Orleans, La. | 756 | 915 | 1207 | 1487 | 1500 | 1697 | 1491 | 1439 | 1402 | 1258 | 952 | 804 | 1250 |
| Caribou, Maine | 531 | 745 | 1192 | 1697 | 1771 | 1983 | 2015 | 1690 | 1343 | 937 | 450 | 391 | 1227 |
| Boston, Mass. | 454 | 745 | 1085 | 1328 | 1661 | 1823 | 1690 | 1582 | 1184 | 937 | 502 | 406 | 1110 |
| E. Lansing, Mich. (low) | 384 | 649 | 945 | 1284 | 1395 | 1638 | 1653 | 1432 | 1048 | 819 | 380 | 343 | 998 |
| St. Cloud, Minn. | 627 | 878 | 1461 | 1734 | 2070 | 2066 | 2118 | 1667 | 1343 | 1048 | 646 | 561 | 1352 |
| Glasgow, Mont. | 576 | 900 | 1146 | 1852 | 2362 | 2494 | 2435 | 2011 | 1395 | 900 | 642 | 450 | 1455 |
| Lincoln, Neb. | 686 | 930 | 1247 | 1576 | 1852 | 2052 | 2122 | 1775 | 1506 | 1114 | 771 | 613 | 1354 |
| Las Vegas, Nev. | 963 | 1292 | 1956 | 2111 | 2362 | 2771 | 2539 | 2332 | 2044 | 1483 | 1166 | 845 | 1822 |
| Seabrook, N.J. | 686 | 908 | 1321 | 1668 | 1897 | 2007 | 1838 | 1771 | 1336 | 1052 | 771 | 535 | 1316 |
| Albuquerque, N.M. | 1133 | 1354 | 1834 | 2236 | 2494 | 2749 | 2502 | 2299 | 2018 | 1712 | 1284 | 1085 | 1802 |
| New York, N.Y. | 450 | 705 | 956 | 1339 | 1572 | 1646 | 1620 | 1351 | 1166 | 897 | 546 | 395 | 1054 |
| Hatteras, N.C. | 941 | 1063 | 1550 | 2103 | 2229 | 2266 | 2229 | 2125 | 1587 | 1269 | 1015 | 756 | 1594 |
| Cleveland, Ohio | 373 | 675 | 1030 | 1550 | 2140 | 2214 | 2192 | 1934 | 1705 | 1041 | 487 | 406 | 1312 |
| Stillwater, Okla. | 923 | 1004 | 1520 | 1801 | 1838 | 2196 | 1889 | 1937 | 1565 | 1255 | 900 | 775 | 1467 |
| Toronto, Ont. | 351 | 605 | 1084 | 1317 | 1668 | 1926 | 1756 | 1627 | 1144 | 797 | 399 | 347 | 1078 |
| Medford, Ore. | 391 | 768 | 1232 | 2107 | 2790 | 2590 | 2804 | 2494 | 1553 | 1063 | 550 | 362 | 1575 |
| State College, Pa. | 506 | 642 | 1015 | 1428 | 1572 | 1845 | 1889 | 1683 | 1321 | 915 | 605 | 424 | 1154 |
| Newport, R.I. | 583 | 845 | 1196 | 1535 | 1786 | 1963 | 1860 | 1683 | 1358 | 1092 | 668 | 524 | 1238 |
| Charleston, S.C. | 923 | 1232 | 1664 | 2059 | 2288 | 2166 | 1989 | 1945 | 1509 | 1203 | 1139 | 786 | 1575 |
| Nashville, Tenn. | 524 | 753 | 1089 | 1557 | 1838 | 1934 | 1867 | 1668 | 1439 | 1125 | 779 | 465 | 1253 |
| El Paso, Tex. (high) | 1328 | 1546 | 2125 | 2524 | 2716 | 2731 | 2531 | 2435 | 2403 | 1786 | 1428 | 1207 | 2037 |
| Seattle, Wash. | 229 | 328 | 1033 | 1823 | 1867 | 2286 | 2170 | 1753 | 1235 | 627 | 325 | 229 | 1160 |
| Washington, D.C. | 568 | 738 | 1225 | 1513 | 1716 | 1867 | 1808 | 1631 | 1373 | 1035 | 745 | 539 | 1234 |
| Madison, Wis. | 539 | 797 | 1166 | 1498 | 1727 | 1904 | 1993 | 1609 | 1321 | 959 | 557 | 443 | 1218 |

SOURCE: Heating & Ventilating Reference Data, 1954.

tion, values can be estimated from the difference "Δ" (degrees F.) between the dry-bulb temperatures in the sun and in the shade, by the following relation:[5]

Total solar intensity $= 1.6\Delta + 0.123\Delta^2 \pm 15$ BTU/sq.ft./hr.   **(1)**

## UTILIZATION OF SOLAR ENERGY

*Space Heating*: The largest potential of solar energy in terms of local energy needs is in space heating, which amounts to 20–25 percent of total energy requirements. Schemes for solar heating of buildings usually include flat-plate collectors tilted at an optimum angle to favor the low winter sun, with a fluid transporting the heat of the collector to the building and to some type of storage system. The heat in the storage unit is circulated through the house at night and on cloudy days. The collector area required to provide sufficient energy to heat a one-family dwelling varies from 450 ft² in the Southwest to 2,000 ft² in the North.[6]

The efficiency of such a heating system in winter in a cool temperate zone is around 40 percent. Some solar heating may be afforded by: (1) large south-facing windows; (2) walls containing low-melting salts or water placed behind large, glassed-in, southerly exposed areas; or (3) water pipe coils embedded in concrete floors that serve as solar collectors.[7]

*Miscellaneous*: As discussed in Chapter 19, solar energy can be used for drying farm products,[8] for distilling water, and for heating water and air. Applications such as cooling, water pumping, photoelectric cells, thermoelectric generators, or mechanical power via heat engines all currently possess obstacles to their locally derived use and are not discussed here.

## WIND POWER

### PHYSICAL PLANT

*Windmills*: The fan-type windmill with many wide blades is used for pumping water, and is designed to produce maximum power at low speeds. It is suited for water pumping because of its low-starting torque, which allows it to be connected directly to the pump. The speed of the blade tips is about the same as that of the wind so the "tip-speed ratio" (ratio of tip speed to wind velocity) is about one or two. Efficiency of this type of windmill is about 30 percent.[9]

TABLE II

ANNUAL HOUSE HEAT AND WATER HEAT LOADS FOR A
WELL-INSULATED HOUSE OF APPROXIMATELY 1,500 FT²

| Locality | Latitude | Annual House Heat (Therms) | Annual Water Heat (Therms) |
|---|---|---|---|
| Albuquerque, N.M. | 35.0° | 1238.80 | 306.36 |
| Atlanta, Ga. | 33.8° | 976.17 | 308.15 |
| Blue Hill, Mass. | 42.0° | 1693.91 | 319.10 |
| Central Mass. | 42.5° | 1685.75 | 320.17 |
| Charleston, S.C. | 32.8° | 690.55 | 302.28 |
| Columbia, Mo. | 38.9° | 1193.89 | 313.23 |
| Columbus, Ohio | 39.9° | 1403.75 | 315.09 |
| El Paso, Texas | 31.8° | 890.69 | 300.51 |
| Fresno, Calif. | 36.6° | 1064.69 | 309.22 |
| Ft. Worth, Texas | 32.8° | 688.82 | 302.28 |
| Grand Junction, Colo. | 39.1° | 1267.89 | 313.66 |
| Ithaca, N.Y. | 42.5° | 1512.47 | 320.17 |
| Lander, Wyo. | 42.9° | 1746.74 | 320.96 |
| Los Angeles, Calif. | 34.0° | 922.35 | 304.50 |
| Madison, Wis. | 43.1° | 1600.21 | 321.32 |
| Medford, Ore. | 42.2° | 1591.45 | 319.46 |
| Nashville, Tenn. | 36.1° | 976.17 | 308.15 |
| N.Y. City and Long Island | 41.8° | 1342.03 | 318.74 |
| Phoenix, Ariz. | 33.5° | 677.08 | 303.43 |
| Rapid City, S. Dakota | 44.0° | 1544.71 | 322.75 |
| Rhode Island | 41.7° | 1621.60 | 318.38 |

SOURCE: From Speyer, "Optimum Storage of Heat with a Solar House."

*Wind Generators*: For producing electricity, these have two or
three relatively narrow blades. Generally the blades are designed
to be most efficient at the average wind speed for the area where
the wind generator will be located. The aerodynamic shape that is
optimum is dictated by the range of wind speeds present and the
amount of power desired.[10] Tip-speed ratios as high as eight are
usually designed. Electrical generators produce their power at
higher speeds than the speed of the windmill blades. For example,
the blades may turn at 200 revolutions per minute and the generator may provide its maximum power at 1,000 rpm, calling for a
5:1 gear ratio between the generator and the propeller.

Hans Meyer has described a wind generator using an automobile alternator, a motorcycle chain for gearing (automobile differentials were also tried), and blades made of honeycomb cardboard (Hexcel) coated with fiberglass.[11] The Jacobs windmill, which was popular during the 1930s and 1940s, utilized a generator that was designed to give its maximum output at low speeds so that no gearing was necessary. This is a much more reliable design than most but more expensive initially. As in most designs, the Jacobs windmill had a mechanism to "feather" the blades or change their angle of attack in strong winds.[12]

*Panemones*: The moving surface of a panemone moves parallel to the direction of the wind instead of vertical. One of the most common types is called the Savonious. The Savonious rotor is limited to a theoretical efficiency of about 30 percent, but this can be offset by its ease of construction and low cost.[13]

## AVAILABLE WIND ENERGY

The *theoretical* power available in the wind ($P_1$) is:

$$P_1 = (5.3 \times 10^{-6}) AV^3 \qquad (2)$$

Where $P_1$ is in kilowatts, A is the area swept by the blades ($\pi r^2$) in square feet, and V is the wind velocity in miles per hour.

For an ideal windmill with 59.3 percent efficiency, the power ($P_2$) that could be extracted is, from Eq. 2:

$$P_2 = (3.1 \times 10^{-6}) AV^3 \qquad (3)$$

Assuming that about 80 percent of this can be extracted, a more reasonable power formula is:

$$P_3 = (2.5 \times 10^{-6}) AV^3 \qquad (4)$$

The amount of energy available from the wind in a year is not as easy to calculate. A good estimate is the *average* wind speed for an area.[14] These data can be obtained from the National Weather Service and are usually based on data from a local airport. Average wind speeds for geographical regions can be obtained from the Climatic Atlas.[15]

## WATER POWER

Water power sites are a valuable asset. Water power is definitely limited by geography, however, and in many instances building a dam can have far-reaching effects on the local ecology. The best sites are those where a natural falls or a dam already exists. In

TABLE III

WINDPOWER IN KILOWATTS, FROM EQ. 4, FOR VARIOUS WIND
SPEEDS AND DIFFERENT-DIAMETER WIND BLADES
(8,760 HOURS IN A YEAR)

| Wind Speed (mph) | Windmill Diameter (Feet) | | | | | |
|---|---|---|---|---|---|---|
| | 10 | 15 | 20 | 25 | 50 | 100 |
| 5 | 0.0247 | 0.0556 | 0.0987 | 0.154 | 0.617 | 2.47 |
| 10 | 0.197 | 0.455 | 0.790 | 1.23 | 4.94 | 19.8 |
| 15 | 0.666 | 1.50 | 2.67 | 4.16 | 16.8 | 66.6 |
| 20 | 1.58 | 3.55 | 6.32 | 9.87 | 39.5 | 158.0 |
| 25 | 3.09 | 6.94 | 12.3 | 19.3 | 78.1 | 308.5 |
| 30 | 5.33 | 12.0 | 21.3 | 33.3 | 133.4 | 533.1 |

most locations some type of government agency must approve a
water power site. In locations where heavy rain runoff occurs,
building a dam can be very expensive.[16]

The power (P) developed at a water power site is:

$$P = 1.89 \times .001 \, F \, H \tag{5}$$

F is the rate of flow of the water in ft$^3$/min, and H is the "head"
or vertical distance the water drops.

The flow rate can be measured by building a temporary Weir
dam.[17] For larger streams, the flow can be calculated by estimating
the cross sectional area of the stream and measuring the velocity
of the water with floats.

### TYPES OF WATER WHEELS

The undershot wheel (Figure 3b) is seldom used; the 80–90
percent efficient overshot wheel (Figure 3a) is a better choice[18]
since gravity capabilities of the wheel increase with its width and
diameter. The blades or buckets are shaped so that the water is not
splashed or spilled until emptied out at the bottom or tailrace.
Overshot wheels are simple and useful devices for powers of up to
about one horsepower.[19]

Pelton wheels are used at locations with a relatively high head
of twenty-five feet and up and with a small flow. The blades are
cup-shaped to minimize turbulence and hence loss of energy. Although not as efficient as a Pelton wheel, a centrifugal pump can
be used as a water turbine with some success.[20]

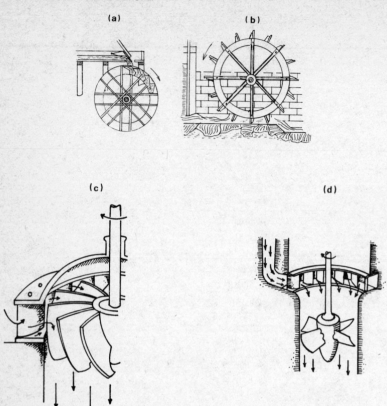

FIGURE 3. Various waterwheels and turbines for transferring the power of moving water into mechanical energy: (a) overshot wheel, (b) undershot wheel, (c) Francis turbine, (d) Kaplan turbine. (Redrawn from Carabateas, "More on Alcohol and Wood Gas," *Alternative Sources of Energy*, March 1973.)

Other types of turbines are the Francis turbine (Figure 3c), which is used for medium flow and heads of ten to 100 feet, and the Kaplan turbine (Figure 3d) for large flow and very low heads of one to ten feet.

Impressive advantages can spring from harnessing the water power in a small stream with water wheels. Descriptions and designs for selecting sites and for building various kinds of water wheels are given elsewhere.[21]

## METHANE DIGESTER

If fresh organic material is left to digest in an airtight container, anaerobic (without oxygen) bacteria break it down into simple, more stable substances. The ultimate products are a solid sludge with good fertilizing qualities, a sludge liquor, and gases including methane and carbon dioxide. Sludge gas is combustible and, depending upon the amount of methane,[22] has a fuel value of about 674 BTU/ft³. The idea of anaerobically digesting organic wastes has been around since at least 1900, but recently a renewed interest in this method of sewage and waste disposal and power production has been making itself evident.

The two basic types of digesters are batch digesters and continuous loading units, both of which have many possible constructions and advantages.[23] The energy available from different raw materials is variable. As a loose rule of thumb, the amount of digestible material is approximately one-sixth of the actual wet weight of the waste matter, and the production of gas per pound of digestible material (volatile solids) is about 5 ft.³

## ALCOHOL AND WOOD GAS

### AVAILABLE ENERGY FROM ALCOHOL

Alcohol is obtained by fermentation and distillation of various organic substances. The theoretical alcohol yields of several materials are given in Table IV.

TABLE IV
ALCOHOL YIELDS FROM VARIOUS MATERIALS

| Material | Gal/Ton | Gal/Acre |
|---|---|---|
| Wood | 70 | 70 |
| Corn | 84 | 89 |
| Potatoes | 23 | 178 |
| Sugar Cane | 15 | 268 |
| Sugar Beets | 22 | 287 |

SOURCE: From Ayres and Scarlott, *Energy Sources: The Wealth of the World* (New York: McGraw-Hill, 1952).

Ethyl alcohol, also known as grain alcohol or ethanol, has a heat value of 84,000 BTUs per gallon as compared to 135,000 BTUs for gasoline.

## ALCOHOL PRODUCTION

Many crops can be fermented to make alcohol, carbon dioxide, and water by the addition of yeast. Wood presents an interesting possibility since it is widely available in the form of scraps, sawdust, and paper products. Typical forest yields are 0.8 to 2 tons of wood per acre per year.[24] However, the cellulose in wood must first be hydrolyzed to glucose before it can be fermented. This is done industrially using strong acid, high temperatures, and high pressures.[25] This would be costly to do on a small scale, as there is as yet no process to convert wood scraps, sawdust, and paper products to glucose using a simple local technology.

The fermented solution contains about 12 percent alcohol, which must be distilled out of solution. The ethyl alcohol boils at 72°F and is then condensed and collected in a container (Figure 4).

Ethyl alcohol is flammable, especially mixed with air, and alcohol obtained from a still made of material other than copper, glass, or stainless steel would be fatal to drink. The alcohol is poisonous

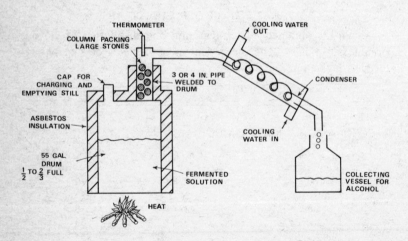

FIGURE 4. Diagram of a distillery used to extract alcohol from organic wastes.

even if solder is used. Also, the government requires that all stills be licensed, even if they are used for fuel or industrial alcohol.

## AVAILABLE ENERGY FROM WOOD GAS

The gases that go up the chimney in a wood stove or fireplace are referred to as wood gas and contain carbon monoxide, hydrogen, and a great deal of heat. A regular wood stove can extract only about 2,000 of the 7,000 BTUs available in a pound of wood. A wood gas burner, or "chimney in reverse," can extract 6,000 of those 7,000 BTUs.[26]

## ENERGY STORAGE

Inherent in any energy system with intermittent power sources is the need for various storage devices so that all of the energy received can be used whenever it is needed. The storage devices considered are batteries, hydrogen gas, pumped water, compressed air, heat storage systems, and flywheels.

## BATTERIES

A storage battery stores electrical energy as chemical energy and then converts it to electrical energy on demand. The emergency lighting plant batteries are generally used with a wind generator or heat engine-generator system. Automobile batteries are not suitable because they have small lead plates with close spacing between them, which allows a fast charge and discharge but a short life. A lighting plant battery, however, has large plates with wide spacing and a long life of from fifteen to thirty years or more, depending on care. Other heavy-duty batteries such as golf-cart, telephone, fork-lift truck, or train batteries can be used also.

The lighting plant battery can be charged at a maximum rate of only 10 or 20 amperes but will have a capacity of from 130 to 450 ampere hours. So, a 130-ampere-hour battery could maintain a 10-ampere discharge for 13 hours. The product of the battery voltage and the ampere-hour rating gives the watt-hour or energy storage capability of the battery. For example, a 12-volt, 130-ampere-hour battery could store 1.56 kilowatt-hours of energy.

Generally, a 6-, 12-, or 32-volt system is used for small lighting plants of one kilowatt or less. Two 12 volt, 180-ampere-hour batteries connected in parallel would give a 12-volt, 360-ampere-hour

(4.32 kilowatt-hour) capability. For larger systems, a 120-volt battery system is preferred. This voltage is compatible with most lighting, DC appliances, and DC motor applications. Many appliances can be converted to DC or have DC–AC capability. Where AC is needed, such as for certain communications equipment, a small electronic inverter can be used. In a typical setup, twenty 6-volt, 130-ampere-hour batteries connected in series would give a 120-volt, 130-ampere-hour (15.6 kilowatt-hour) system.[27]

## HYDROGEN GAS

Hydrogen is being viewed as a large-scale energy source both as a replacement for natural gas and for use in fuel cells to produce electricity.[28] However, hydrogen may have other applications as a storage medium and as a gaseous fuel for small-scale uses.

With present technology, the simplest way to generate hydrogen is to electrolyze water using an electrolytic cell. One kilowatt-hour of electricity electrolyzes water at a potential of 1.7 volts to produce 8.9 ft³ of hydrogen. The efficiency of the process is about 70 percent.[29]

The heat value of the hydrogen is only 274 BTU/ft³. However, hydrogen has 2.4 times the heat value of methane by weight.[30] Conceivably, hydrogen and methane could be mixed for use in direct burning or in internal combustion engines. Alternatively, hydrogen could be used in fuel cells.[31]

A completely self-contained energy system could utilize wind energy to generate electricity to power an electrolytic cell. The hydrogen and oxygen that was generated could then be stored in tanks and used in fuel cells to generate electricity. Such a system has been tested at Oklahoma State University.[32]

## PUMPED WATER

Water can be pumped to a reservoir for storage and later used to drive a water turbine just as for any water power site. This is generally possible only if a natural reservoir on a high hill is available. For reasonable amounts of power, a storage tank on a tower is generally unwieldy. Consider the following example: Water is stored in a 500-gallon (67 ft³) tank on top of a 20-foot tower. To generate 1 horsepower, the flow from the tank would have to be, from Equation 5:

$$F = 1 \text{ hp.}/(1.89 \times 10^{-3} \times 20 \text{ ft.}) = 26 \text{ ft}^3/\text{min.}$$

In other words, the tank would be emptied in 2.5 minutes! In addition, the water would weigh 4,200 pounds.

### COMPRESSED AIR

A rugged storage system capable of handling large amounts of energy relatively cheaply over many charge-discharge cycles would be an extremely valuable asset to any homestead, farm, or community power operation. Water storage fits this description well but the need for a naturally occurring storage site makes it impractical in most locations. The compression of air, its storage in tanks, and its subsequent use to drive compressed air tools or air turbine-driven generators is a possibility with the many advantages of pumped water storage but without the requirement of a naturally favorable site.

Air compressors can have one or more stages. A single-stage compressor produces pressures of 80 to 100 pounds per square inch (psi). This is the working pressure of many pneumatic motors and tools. Two-stage compressors can compress air from 80 to 500 psi; three-stage compressors can produce up to 1,200 psi; and four-stage units can produce 2,500 to 3,000 psi.[33] Mechanical efficiencies of piston compressors vary from 76 to 97 percent.

In conventional air power systems, the compressor is run semicontinuously, so the storage tank is limited in capacity. For a compressor driven directly by a windmill or indirectly by a wind generator, the storage requirements are considerably greater. The quantity of air that has to be stored in the compressed state to power a commercially available 2-hp. pneumatic motor is approximately 40–50 ft$^3$/min. at a pressure of 70–90 psi; or, for every 2 hp-hrs delivered from this motor, 60 minutes $\times$ 45 ft$^3$/min. or 2,700 ft$^3$ of air is needed at close to 80 psi.[34] A standard-size high-pressure tank of approximately 4/3 ft$^3$ can hold only about 300 ft$^3$ of gas at close to 3,300 psi. This will last approximately 6½ minutes and the tank is very expensive. Obviously a larger volume, lower pressure, lower cost arrangement is needed.

Assume that it is necessary to store enough compressed air to supply 2 hp-hrs of power from the pneumatic motor mentioned above. With 2,700 ft$^3$ of air delivered at a pressure of 80 psi, and the air stored in a tank at a pressure of 160 psi, then 5,400 ft³ of air would be needed; this is because as half of the air is used, the pressure of the remaining air will be half or 80 psi—the minimum pressure needed. The size of the tank needed would be:

$$\frac{14.7 \text{ psi (atmospheric pressure)}}{160 \text{ psi (tank pressure)}} \times 5,400 \text{ ft}^3 = 495 \text{ ft}^3$$

(or a cube tank about 8 feet on a side).

Instead of compressing air for storage, methods have been worked out to use the weight of a large body of water situated above a cavern or tank to maintain the desired pressure, no matter how much air has been removed.[35] In principle, a heavy floating cover similar to that used on many methane generators could furnish a constant pressure, but only with an almost impractical amount of weight (Figure 5).

Thus, for a compressed air storage system of this type to be practical, the tensile strength of various types of low-cost tanks must be investigated. Possibilities include the use of buried metal

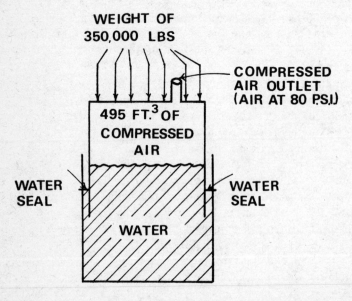

FIGURE 5. Compressed air storage system using weights rather than a compressor.

fuel tanks or pipeline, ferro-cement tanks, or rubberized tanks, gasoline storage tanks from old gas stations, etc.

## HEAT STORAGE

Heat may be stored and used in any of the following ways:

1. water heating to supply hot water for washing, space heating, pond warming, or temperature regulation of methane digesters
2. space heating or cooking in which the heat is retained in insulated solid blocks, rocks, water, or salts (see Table V)
3. soil warming for horticultural purposes
4. distillation of water
5. refrigeration
6. the generation of steam for heating purposes or for production of power in a steam engine or turbine

In a direct solar energy system the storage of heat is necessary to allow for nighttime and cloudy-day use. It would be best to store all the heat needed for any probable number of cloudy days, but many times this is impossible or too expensive. Because of the difficulty in storing *all* heat, most solar energy systems have some sort of conventional source of auxiliary heat.

Heat for solar space heating and water heating can be stored as sensible heat, that is, "specific heat" (concrete, water, stone, etc.) or latent heat, i.e., "heat of fusion" (salts, waxes). These are discussed in Chapter 19.

Experience and analyses of various collector-storage situations have been moving toward an optimum value of storage of two to three gallons of water or its crushed rock equivalent per square foot of collector.[36] If the heat storage medium is water, the most efficient maximum storage temperature is about 190°F. However, this assumes that a selective surface is used as a collector.[37] Regarding storage tanks, a spherical tank has the least amount of heat loss but usually costs about twice as much as a cylindrical tank. Steel tanks, lap-welded and painted with a rust inhibitor like calcium plumbate, are preferable to wooden tanks.[38] All storage tanks should be well insulated.

## FLYWHEELS

A rotating flywheel can be "charged" by an electric motor to store energy in a mechanical form. Recent interest in flywheels for automobiles has led to research toward developing flywheels that can store 40 watt-hours of energy per pound as opposed to 10 watt-hours per pound for present-day steel flywheels.[39]

## LOAD SHARING AND INTERCONNECTION OF ENERGY SOURCES

The energy sources proposed so far include wind energy, solar radiation, water power, and fuels derived from organic matter. The reason for combining the four is that no one source alone is either steady or sufficient enough to supply a complete energy need. Each of the sources has individual characteristics that make it specially suited for certain loads. Wind energy and water power come in the mechanical form; solar energy comes as heat; and methane gas, alcohol, and wood as a more portable and versatile chemical source. One of the aims in setting up a multiple energy source system is to use the available energy so as to minimize the number of energy conversions that take place.

Consider how the five basic forms of *available energy* interact with one another (Figure 6): solar, chemical, mechanical, electrical, and heat. Direct conversions of energy results when there is only one conversion (shown by a line) between any two basic energy sources (e.g., solar → electrical via photovoltaic cell; chemical → electrical via fuel cell, etc.). The more conversions that

TABLE V

AMOUNT OF HEAT (BTU/FT$^3$) THAT CAN BE STORED IN VARIOUS MATERIALS

| Material | *Temperature Difference Between Incoming and Outgoing Heat Transfer Fluid* | | |
|---|---|---|---|
| | *10 Degrees* | *20 Degrees* | *30 Degrees* |
| Water | 625 | 1,250 | 1,875 |
| Rocks | 360 | 720 | 1,080 |
| Heat of Fusion | 10,100 | 10,700 | 11,300 |

SOURCE: M. Telkes, *Solar House Heating*, in Proceedings of the World Symposium on Applied Solar Energy, Phoenix, Ariz. (Menlo Park, Calif.: Stanford Research Institute, 1955).

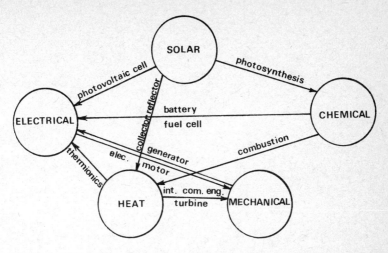

FIGURE 6. The conversion relationship between the five basic sources of energy.

occur between the five basic forms of energy, the less efficient and in the long run the more costly a process will turn out to be. However, in a few cases multiple conversion processes can be more efficient than single conversions, as we will see.

A most important characteristic of the four *conversion sources* (sun, wind, water, and organic material) is that the supply is limitless, and "costs attached to the production of usable energy from the free sources of wind, sun, and water are fixed for any particular power plant. The greater their output and the higher the proportion of it which is effectively used, the lower the cost per unit of energy."[40] This implies that the four sources must be fitted to the various loads so that each source-load pair results in the least number of energy conversions plus the matching of "time requirements," that is, the time patterns at which the conversion sources are most available (Table VI).

We can now see that both the sources of energy and the loads that use the energy have definite time requirements. The closer the matching of these supply and demand periods, the less storage will be required and the more efficient and inexpensive the system will be. When storage is required, a process such as pumping water, which is a random load (can be done any time there is an excess

TABLE VI

LOADS, ENERGY SOURCES, AND TIMING REQUIREMENTS OF
WITH AN ALTERNATIVE ENERGY POWER SYSTEM

| RANDOM (Day or Night) | DAY (No Precise Time) | DAY (Precise Time) | NIGHT (Precise Time) |
|---|---|---|---|
| Domestic Water | | | |
| Water Distillation | | | |
| Irrigation | | | |
| Water Heating, Steam-Raising | | | |
| | Threshing | | |
| | Grinding | | |
| | Food Mixing | | |
| | Fodder Chopping | | |
| | Small Indust. or Agri. Pwr. | | |
| | | Cooking | |
| | | Fans, Refrig., Air Cond. | |
| | | | Lighting |
| | | | Radio & Small Domes. Pwr. |

SOURCE: Golding and Thacker, *The Utilization of Wind, Solar*

## AN AGRICULTURAL COMMUNITY

| | ENERGY | | | |
|---|---|---|---|---|
| Wind | Solar | Methane | Water | NOTES |
| WD | | M | W | M—supplementary<br>W—hydro. ram |
| | S | | | |
| WD | | M | | M—supplementary |
| WD | S | M | | WD & M—supplementary |
| WD | | M | W | M—supplementary |
| WD | | | W | |
| WD | | | W | |
| WD | | | W | |
| WD | | M | W | In combination |
| WD | S | M | | In combination |
| WD | S | | | In combination |
| WD | | M | W | With bat. stor. |
| WD | | M | W | With bat. stor. |

*Radiation and other Local Energy Resources.*

of energy) and can store potential energy, is definitely favored over the use of a storage device, such as a battery, whose sole purpose is storing energy.

In any location where there is a reasonable wind climate (a few sites of 10 mph average wind speed) and where water power is not available, wind power will be used for practically all the non-heat loads. The optimum use of this source is therefore very important. One method of using the energy from the wind would be to generate electricity directly from the propeller shaft and either (1) store it all in batteries, (2) store it as hydrogen and oxygen electrolyzed from water, (3) store it as heat in water, or (4) run a compressor from the propeller shaft and store all the energy as compressed air. All mechanical and electrical energy needs would then be handled solely from this stored supply, with methane, alcohol, or wood as an auxiliary fuel used to run a compressor or generator connected to the same storage tank. This method is simple in that all the energy input and output flows through one terminal. The cost of the very large amount of storage needed, however, will usually offset the advantages of simplicity. Golding estimates that the energy supplied via a battery by this method would cost about six times as much as that supplied to the farm by random power.[41]

Another method would be to use the available wind energy as it is generated, utilizing some sort of multiple switching device to transfer the energy to different loads as they are needed or as a given load is taken care of. Some sort of minimal storage will always be needed for certain specific time requirements such as lighting.

A small-scale energy system must be designed so that no available incoming energy is wasted, all of it being made "firm" by either utilization or storage. For example, the immediate use of incoming wind energy to pump water or grind food is a sort of firming of that energy, since it is used directly as it is available. Having a number of large energy sinks, therefore, will help utilize all the energy that is generated. One such energy sink not usually thought of is a cold storage plant like that designed by the Brown, Bovery Company for the 6 kw standard wind power plants built by Allgaier of France. It is capable of storing 44,000 thermal units for long periods of time, requiring approximately 30 kw-hrs to completely freeze the eutectic brine in the cold storage plant.

Within ten days only 35 percent of the stored cooling power is lost to the outside.[42] A regular freezer can also serve the same purpose, though not as well.

As can be seen, the switching device needed has to be relatively sophisticated in that it must sense a surplus of available energy, route it to the neediest energy sink, and when that load is taken care of, reroute it to the next load. All the while, of course, the batteries, water electrolyser, or compressed air storage must be kept constantly charged or nearly so.

A system utilized in one Allgaier wind power plant consists of a three-phase alternator (380/220 volts) of 6–8 kilowatt capacity and a 220-volt DC exciter of about 2-kilowatt capacity that can be used to charge a battery serving lighting and small power needs. This way, when the wind is blowing, the battery is always kept charged (using a voltage regulator to prevent overcharging), while AC power is simultaneously supplied to the other loads. Through gearing with a unique speed regulator, it is claimed that the speed is kept so constant that directly connected incandescent lamps do not flicker.[43] Some available energy will definitely be lost by gearing the wind propeller to a constant speed and not taking advantage of very powerful strong winds. But if a total DC setup is used, an inverter of approximately 85 percent efficiency must be used for any AC load. This does not mean that a complete DC system is impractical; many appliances can be run on direct current and others may be converted.[44] In the first trial systems, manual switching of the loads and sources will probably be the case. An energy flow diagram (Figure 7) shows the possible inter-connections between energy sources, storage, and use.

Excess solar energy will usually be transferred to heat storage units due to the present inefficiency of solar-operated mechanical devices and the high cost of solar cells. In a very sunny climate, however, the possibilities of hot-air Stirling engines[45] and thermo-electric and photovoltaic devices are reasonable at least in terms of further experimentation.

Organic resources can be used in many ways. The number of loads this versatile source can handle depends on the availability of organic wastes in the particular location; but we can see how valuable the "reliable" sources of methane, alcohol, and wood can be in rounding out an energy system.

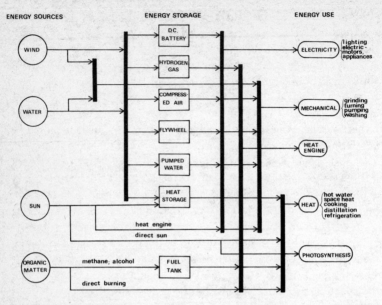

FIGURE 7. Some practical possibilities of load sharing for an integrated alternative energy system.

## Farm Homestead and Small Community Needs

### ESTIMATING ENERGY NEEDS

The energy needs and loads of a small farm or community vary greatly with personal preference, social organization, climate, and geography. Table VII, however, can be a useful aid. The table is based on the work of Golding and Thacker, who estimated the energy needs of a hypothetical community of forty to fifty families living in a semiarid area.[46] Timing requirements are noted so that energy demands do not occur simultaneously. Examples of calculating energy requirements are given below to illustrate the problems involved. For convenience, all energy requirements are expressed in kilowatt-hours (kw-hrs).

**Example 1.** *Pumping Domestic Water*: Assume a typical family uses 150 gallons per day. Assume the water pump can move 500 gal./hr. (1.1 ft³/min). Then the pump will run 3/10 hours per day or 110 hours per year. Assume the pump must lift the water against an effective "head" of 200 feet. Then, from Equation 5, the power required is: $P = .32$ kw, or (assuming 70 percent efficiency) 0.46 kw. The total energy per year is then: $0.46 \times 110$ hrs. $= 50$ kilowatt-hours.

**Example 2.** *Heating Domestic Water*: Assume a typical family uses 50 gallons per day and that the water is heated from 50°F to 130°F for a difference of 80°F. The amount of heat energy required to do this is 50 gallons $\times$ 8.34 lbs/gal. $\times$ 80°F $= 33,360$ BTU/day or 9.8 kw-hr/day. The annual energy requirement is then approximately 3,500 kw-hrs.

**Example 3.** *Pumping Irrigation Water*: Assume irrigation water is pumped from 200 feet below the surface and, using the same argument as Example 1: At 55 gal./hr. the pump will have to run a total of 650 hours consuming 0.46 kw of power to equal 300 kw-hrs/acre-ft. (1 acre-foot is about 325,000 gallons.)

**Example 4.** *Power for Cultivation*: The average U.S. farmer uses a tractor for 400 hours per year and the average tractor has 40 hp.[47] Assume that on a small farm a tractor rated at 25 hp is used 75 hours per year. Assuming an engine efficiency of 25 percent, 4 hp or 2.98 kw input power is required to reach 1 hp of developed power. The annual energy requirement for the tractor then is 5,587 kw-hrs.

**Example 5.** *Power for Transportation*: As in Example 4, assume 2.98 kw input for each 1 hp of developed power. If a small 25-hp vehicle were used for general transportation 200 hours per year, the annual energy expended would be 14,900 kw-hrs. Assuming an average speed of 40 mph, 200 hours of use per year would be equivalent to 8,000 miles of travel. The example points out the large amounts of energy used for individual transportation vehicles. Assuming that future mass transit for long-distance travel would cut the vehicle use to 100 hours per year, the annual energy use would then be 7,450 kw-hrs.

A similar table can be drawn up for any specific situation as an aid in estimating whether energy needs can be supplied by available local sources. In Table VIII, a list of typical amounts of available energy is given. These figures are meant to give an "order of magnitude" idea of the typical amount of energy available, and may or may not be typical for a given area.

TABLE VII

ESTIMATES OF LOCAL ENERGY REQUIREMENTS FOR A SMALL COMMUNITY
(1 FAMILY = 4 PEOPLE)

| Energy Use | Annual Energy (kw-hr)* | Timing Require-ments | Probable Energy Source | Basis of Estimation | Source |
|---|---|---|---|---|---|
| Cooking | 400/family | Day, Precise | Or, S | 1–1.5 kw stove used 1 hr/day | (a) |
| Domestic Water | 50/family | Random | W, Wa, Or | 150 gal/day. See Example 1 | (b) |
| Refrigeration | 200/family | Random | W, Wa, S, Or | 1/2 kw unit run 1 hr/day | (c) |
| Water Heating | 3,500/family | Random | S, Or, W | 50 gal/day. See Example 2 | (b) |
| Lighting & Domestic Power | 850/family | Night & Day, Precise | W, Wa, Or | Five 100-watt lights used 4 hrs/day; 1/3 kwh/day misc. items—power tools, stereo, small appliances, etc. | (c) |
| Water Distillation | 130/family | Day, Random | S, Or | 5 gal/day. Solar collector of 5m² giving 3/4 kw over 24 hours. | (a) |
| Solar House Heat | 4,100–15,000 | Day & Night, Precise | S, Or | 7.3–10 kw average heat requirements. 60–750 kw-hr heat storage based on a survey of existing solar houses. | (d) |
| Irrigation | 300/acre-ft | Random | W, Wa, S, Or | See Example 3 | (a) |

| | | | | |
|---|---|---|---|---|
| Cultivation Seeding | 5,600/25 hp/ 75 hrs | Variable | Or, W, S | 25-hp tractor used 75 hrs/yr. See Example 4 | (b) |
| Mech. Grinding, Food Mixing, & Fodder Chopping | 50/family | Random | W, Wa, Or | Average of 1–2 tons/family/hr including animals | (a) |
| Threshing | 50/family | Random | Or, W | 4 tons/family/year | (a) |
| Misc. Small Power | 100/family | Day & Night, Precise | W, Wa, Or | 5 kw total power | (a) |
| Transportation | 7,500/vehicle of 25 hp./ 100 hrs of use | Variable | Or, W | 25-hp vehicle used 100 hrs/yr. See Example 5. | (a) |

SOURCES: Table concept taken from Golding and Thacker, *"Utilization of Wind."* Specific credits follow:

(a) Golding and Thacker, *Utilization of Wind.*
(b) Edwin Anderson, *Domestic Water Supply and Sewage Disposal Guide* (New York: Theodore Audel & Co., 1967).
(c) Henry Clews, *Electric Power from the Wind* (Solar Wind Co., RFD 2, East Holden, Me. 04429, 1973).
(d) Daniels, *Direct Use of the Sun's Energy.*

NOTE: Or = Organic fuels; W = Wind; S = Solar; Wa = Water.
* 1 kw-hr = 3,413 BTUs.

## DESIGNING AN ENERGY SYSTEM BASED ON
## LIFE-STYLE AND RESOURCES

### PHYSICAL SITUATION

The first step in tackling an energy system is to become familiar with the physical structure of the farm, homestead, or community to be supplied. Two different physical situations and locales will be considered as examples. Many of the energy estimates will be based on the section, Estimating Energy Needs.

**Settlement #1:** A semifarming community of twenty people situated on seventy acres in the semiarid area around Albuquerque, New Mexico.

Of the land, forty acres are used for grazing, fifteen acres are irrigated for agriculture, and fifteen acres are for housing and living. Domestic animals include twenty goats, forty chickens, and two horses. Excess dairy products are sold or traded for other needs and for available items such as propane gas. The buildings include one small barn, one chicken house, one greenhouse (self-heated and partially underground), and housing with 4,500 ft² of floor area. Refrigeration, auxiliary space, and water heaters, washers, tools, vehicles, farm equipment, and most appliances are shared on a cooperative basis to cut down on energy and material consumption. There are two transportation vehicles—one truck and one car—and a tractor (25 hp) to take care of the fifteen acres of tillable land.

**Settlement #2:** A single family (four people) on a midwestern homestead of fifteen acres near Madison, Wisconsin.

Only one or two acres are cultivated but pasture is available on five acres, with the rest being woods and a homesite. The animals include one dairy cow, ten chickens, and one hog. There is one small barn, one chicken coop, and a house with approximately 1,500 ft² of floor area. All appliances and lights are kept to a minimum and utilize high-efficiency devices (e.g., fluorescent lights). There is one small truck and either a rototiller or a small tractor (6–10 hp) for cultivation.

### AVAILABLE ENERGY

The next step would be to assess the "raw" available energy before conversion by collectors, propellers, etc.

TABLE VIII

REPRESENTATIVE AMOUNTS OF LOCAL ENERGY AVAILABLE FOR A RURAL SETTLEMENT FROM BASIC CONVERSION SOURCES

| Source | Energy Available | Basis of Estimate |
|---|---|---|
| Solar | 65 kw-hr/ft²/yr | From Table I, the average daily solar energy received at Madison, Wisc., is 1,218 BTU/ft²/day. Assuming a 50% efficient solar collector, the useful energy would be 609 BTU/ft²/day. Converted to kw-hr/ft²: 609 BTU ft²/day × 2.93 × 10⁻⁴ kw-hr/BTU × 365 days/yr = 65 kw-hr/ft²/yr. |
| Wind | 4,000 kw-hr/yr | Assume a 15-ft-diameter windmill where the average wind is 10 mph. From Table III, the power generated = 0.445 kilowatts, or 0.445 kw × 8,760 hrs/yr = 3,898 kw-hr/yr. |
| Water | 9,800 kw-hr/yr | Assume a water power site has a flow of 100 ft³/min and a head of 8 feet. Then, from Equation 5, the power developed would be: 1.5 hp, or 1.5 hp × 0.745 kw/hr × 8,760 hr/yr = 9,789 kw-hr/yr. |
| Alcohol | 1,500 kw-hr/acre | From Table IV, alcohol yield from potatoes = 178 gal/acre. Assume that a distillery is able to obtain 60 gal/acre, or: 60 gal/acre × 84,000 BTU/gal × 2.93 × 10⁻⁴ kw-hr/BTU = 1,476 kw-hr/acre. |
| Wood | 3,000 kw-hr/acre/yr | Assuming a wood yield of 1 ton/acre/yr and 7,000 BTU per pound of wood, 14 million BTU/acre/yr would be available.* With a 75% efficient wood stove with an automatic damper, the amount of energy in kw-hrs would be: 0.75 × 14 × 10⁶ BTU/acre/yr × 2.93 × 10⁻⁴ kw-hr/BTU = 3,076 kw-hr/acre/yr. |

* From Glesinger, *Coming Age of Wood.*

*Solar Data*: Solar radiation for different parts of the country can be obtained from the National Weather Service, the American Society of Heating and Air Conditioning Engineers' Annual Guide, or from the Climatic Atlas. The information is usually given in BTU/ft²/day or in gram calories per square centimeter (Langleys) per day. For correctly tilted solar collectors the values may run as much as 15 percent higher in the winter months (see Table I for solar data of various U.S. areas).

*Wind Data*: Average wind speed in many parts of the country can be obtained from the National Weather Service, the Climatic Atlas, and from local airports. For Settlement #1 (Albuquerque, Southwest), assume 10 mph and for Settlement #2 (Madison, Midwest), assume 12 mph.

*Methane*: Available digestible wastes (volatile solids) per year per animal.[48]

### SOUTHWEST:

| | |
|---|---:|
| 20 goats @ 0.3 lbs/day/goat | 2,190 lbs |
| 20 humans @ 0.25 lbs/day/human | 1,820 lbs |
| 2 horses @ 5.5 lbs/day horse | 4,016 lbs |
| Plant wastes | 10,200 lbs |
| | 18,226 lbs/yr |

Assuming that all manures and excrement could be easily collected and that one pound of digestible matter yields 5 ft³ of gas, then 90,130 ft³ of gas would be produced. At a heat value of 700 BTU/ft³ this would total 63.1 million BTUs or 18,500 kw-hrs. If about 20 percent of this energy is used to keep the digester at 95°F[49] then about 14,800 kw-hrs would be available.

### MIDWEST:

| | |
|---|---:|
| 1 cow @ 8 lbs/day | 2,920 lbs |
| 1 hog @ 1.3 lbs/day | 475 lbs |
| 4 humans @ 0.25 lbs/day | 365 lbs |
| Plant wastes | 2,000 lbs |
| | 5,760 lbs/yr |

Using the same arguments as above, 5,760 pounds of digestible wastes would produce 18,500 ft³ of gas to equal 4,400 kw-hrs gross or 3,520 kw-hrs net.

### ENERGY NEEDS

*Irrigation* (Table VII): Water needs for crops such as sugar beets, potatoes, alfalfa, corn, wheat, oats, barley, peas, and beans vary from 1 to 3.4 acre-ft/acre/yr.[50] Water losses in hot, dry climates with efficient irrigation may be about 30 percent. If we asume a water need of 2 acre-ft/acre/yr and a 30 percent loss, 2.9 acre-ft will actually be required. Natural rainfall will contribute only about 0.7 acre-ft/yr in this area, leaving 2.2 acre-ft/acre to be supplied by irrigation.

*Southwest*: 2.2 acre-ft/acre/yr $\times$ 15 acres $\times$ 300 kw-hr/acre-ft = 9,900 kw-hrs.

*Midwest*: No irrigation is needed.

*Grinding, Fodder Chopping, and Food Mixing*: Assume the following grain requirements (most domestic animals do not eat grain exclusively; these figures are only for estimate purposes):

| | |
|---|---|
| Humans | 300 lbs/person/yr |
| Horses (medium, work) | 3,000 lbs/horse/yr |
| Goats | 548 lbs/goat/yr |
| Cows | 4,400 lbs/cow/yr |

A grain composed of corn and oats requires approximately 18 kw-hrs/ton for grinding and 3 kw-hr/ton for mixing or a total of 21 kw-hrs/ton. The energy needed in each location example is as follows:

SOUTHWEST:

| | |
|---|---|
| 2 horses | 6,000 lbs/yr |
| 20 goats | 10,095 lbs/yr |
| 20 humans | 6,000 lbs/yr |
| | 22,095 lbs/yr |
| | = 231 kw-hrs |

MIDWEST:

| | |
|---|---|
| 1 cow | 4,400 lbs/yr |
| 4 humans | 1,200 lbs/yr |
| 10 goats | 5,480 lbs/yr |
| | 11,080 lbs/yr |
| | = 116 kw-hrs |

*Domestic Water Pumping* (Table VII): For a 200-foot well, 50 kw-hrs would be needed for four people, or about 250 kw-hrs for the Southwest settlement, and 50 kw-hrs for the Midwest settlement.

*Lighting and Small Power* (Table VII):

*Southwest*: Assume 850 kw-hrs/family of four people or about 213 kw-hrs/person. Due to considerable savings of energy/person possible in a community situation, the estimated energy needed by the twenty-person Southwest settlement is assumed to be only three times the value for one family, or 2,500 kw-hrs.

*Midwest*: 850 kw-hrs.

*Miscellaneous Small Power* (Table VII):

*Southwest*: Again, assume the saving of energy in a community situation equals three times the energy required per family, or 30 kw-hrs.

*Midwest*: 100 kw-hrs.

*Refrigeration* (Table VII):

*Southwest*: Three times the power or 1½ kw/hr/day for 20 people equals 600 kw-hrs.

*Midwest*: About 200 kw-hrs.

*Space Heating* (Table II):

*Southwest*: The annual heat load of a building in Albuquerque is 124 million BTU/yr, for a dwelling with 1,500 ft² floor area (Table II); space of 4,500 ft² would require 369 million BTUs (108,000 kw-hrs).

*Midwest*: 1,600 therms, or 47,000 kw-hrs.

*Water Heating* (Table II):

*Southwest*: The annual water heating load is 918 therms (26,900 kw-hrs) for a 4,500 ft² dwelling, enough to heat 140,000 gallons of water by 80°F.

*Midwest*: The annual water heating load is 321 therms (9,400 kw-hrs), enough to heat 48,000 gallons of water by 80°F.

*Cooking* (Table VII): A number of sources consider a stove to be about 1 kw power unit.

*Southwest*: Assume that a meal for twenty people will require only about three times the energy used for four people; the annual energy needed will be about 1,200 kw-hrs.

*Midwest*: 400 kw-hrs.

*Cultivation*:

*Southwest*: A 25-hp tractor used 75 hours per year, along

with the use of the two horses equals 5,600 kw-hrs (Table VII).

*Midwest*: Assume a small 10-hp tractor or rototiller, 25 percent efficient, with a running time of 20 hrs/yr, the energy required is: 2.98 kw/rated hp $\times$ 10 hp $\times$ 20 hours = 596 kw-hrs.

*Transportation*:

*Southwest*: Assuming two vehicles as described in Estimating Energy Needs—15,000 kw-hrs (Table VII).

*Midwest*: Assuming one vehicle—7,500 kw-hrs.

### PHYSICAL PLANT

Now that both the energy needs and availability of each settlement have been calculated, the physical plant required to convert the incoming energy to a usable form can be designed. The available energy, of course, is supplied in a number of forms: wind and water come as mechanical, solar as heat, and methane, alcohol, and wood as a portable chemical energy. The energy requirements then, must be separated into these various categories and the size of each individual physical plant calculated.

*Mechanical Power Needs* (including electricity):

SOUTHWEST:

|  | kw-hrs/year |
|---|---|
| Refrigeration | 600 |
| Irrigation | 9,900 |
| Grinding, Mixing | 231 |
| Domestic Water Pumping | 250 |
| Lighting and Small Power | 2,550 |
| Misc. Small Power | 300 |
| Total Mechanical Load | 13,831 kw-hrs |

The refrigeration load could be classed under heat energy, since an absorption-type refrigerator would use heat and not mechanical energy for its power. The large amount of solar energy available in the Southwest might make this the wiser choice.

The mechanical load in the Southwest example is handled by wind energy, since water power will probably not be available in this semiarid zone.

The biggest load is the 9,900 kw-hrs needed for irrigation,

which points out the problems encountered in farming arid areas. Irrigation, though, probably uses less energy than trucking in produce grown elsewhere. The best arrangement would probably be to have one 20-foot and one 15-foot wind plant close to the area to be irrigated, providing a total of 10,906 kw-hrs/yr (from Table III) if there are suitable 10-mph sites in the vicinity. Another wind generator with a 15-foot propeller situated near the living area could supply the needs of the other five loads.

MIDWEST:

|  | kw-hr/yr |
|---|---|
| Refrigeration | 400 |
| Grinding, Mixing | 116 |
| Domestic Water Pumping | 50 |
| Lighting and Small Power | 850 |
| Misc. Small Power | 100 |
| Total Mechanical Load | 1,516 kw-hrs |

Assuming an average wind speed of 12 mph, the power in kilowatts (from Equation 4) and the energy available per year is:

| Propeller Size (Ft) | Power Produced (kw) | kw-hrs/year |
|---|---|---|
| 10 | 0.34 | 2,978 |
| 15 | 0.77 | 6,745 |
| 20 | 1.36 | 11,913 |
| 25 | 2.14 | 18,740 |

The modest mechanical-energy needs of the Midwest homestead can be handled very well by a wind generator with a 10-foot-diameter prop. As the output of the power plant is close to two times the expected mechanical loads, it would be wise to let this generator supply power to other forms of energy needs (such as heat) even though there will be larger losses due to the mechanical-heat conversion.

If water power is available the output of the water power plant would have to equal 0.17 kw to supply the total mechanical load. In this way, from Equation 5: $0.17 \text{ kw}/1.89 \times 10^{-3} = 90$, so that any product of flow and heat that equals 90 would result in a power output of .17 kw.

*Heat Energy Needs* (supplied with solar energy):

SOUTHWEST:

| | kw-hrs/yr |
|---|---|
| Space heating | 108,000 |
| Water heating | 26,900 |
| Cooking | 1,200 |
| Total heat energy load | 136,100 kw-hrs |

For the flat-plate collector to supply the total energy load the total area must be sufficient to collect the energy for the coldest month of January. The house heat load in Albuquerque for that month is 274 therms and the water heat load is 30 therms,[51] or a total of 304 therms/month ($10^6$ BTU/day). The available daily solar insolation for Albuquerque in January is 1,113 BTU/ft$^2$/day of horizontal surface (Table I), or 1,303 BTU/ft$^2$/day for a correctly tilted collector. Therefore, the theoretical area needed would be about 767 ft.$^2$; with two glass plates the efficiency of collection might be around 60 percent (Figure 2) so that the actual collector should be about 1,280 ft$^2$. The collector could be distributed over two or three dwellings, or if it is considered too large and expensive (collectors might cost $1.00–$2.50/ft$^2$), a small collector could be built and a conventional heating system installed to use wood, alcohol, or methane. This sometimes proves to be cheaper since much of the collector is wasted during the warmer months of the year.

MIDWEST:

| | kw-hrs/yr |
|---|---|
| Space heating | 47,000 |
| Water heating | 9,400 |
| Cooking | 400 |
| Total heat energy load | 56,800 kw-hrs |

In January, the house heat load is 307 therms and the water heat load is 31 therms, or about 338 therms ($11 \times 10^5$ BTU/day). The available solar energy for Madison in January is 539 BTU/ft$^2$/day (Table I), or with a properly tilted solar collector around 620 BTU/ft$^2$/day. The theoretical collector area needed to supply heat for a winter day would be about 1,800 ft$^2$, or, with 60 percent efficiency, about 2,350 ft$^2$. This collector would be much larger than the floor area of the house and would be quite expensive. In October the heat load for the same building is one-third that of January, so two-thirds of the collector would be wasted in that month. May, June, July, August, and September would use less of the collector than October. It can be seen now how an auxiliary furnace could be very practical. If the local supply of organic fuels is sufficient, the use of a conventional furnace to supply all the heat needs over and above a collector half the maximum size (say 1,200 ft²) would be a much more efficient setup in terms of material and energy savings. In fact, it might be wiser simply to increase the size of the wind plant, taking advantage of the higher midwestern winds and using this energy for water or space heating purposes.

*Chemical Energy Needs* (methane, alcohol, and wood):

### SOUTHWEST:

|  | kw-hrs/yr |
|---|---|
| Cooking | 1,066 |
| Cultivation | 5,600 |
| Transportation | 15,000 |
| Total chemical load | 21,666 kw-hrs |

The net available energy from organic wastes in the form of methane gas is 14,800 kw-hrs or about 70 percent of the total chemical energy loads. If an acre of sugar beets can be fermented and distilled into alcohol, it would provide about 6,600 kw-hrs for various uses. Wood is probably too valuable in the Southwest for use as a fuel. Another possibility is to use the two horses to take over a larger percentage of the transportation and cultivation.

### MIDWEST:

|  | kw-hrs/yr |
|---|---|
| Cooking | 367 |
| Cultivation | 596 |
| Transportation | 7,500 |
| Total chemical load | 8,463 kw-hrs |

From the section Available Energy, the net available energy from methane is 3,900 kw-hrs or about 46 percent of the chemical load. Wood could supply a substantial amount of chemical energy.

Table IX presents a summary of the energy needs and resources of the two settlements.

STORAGE

The amount of storage in any energy system must be selected with care since this aspect of an energy supply adds quite a bit of cost to the system. Again, the various forms of incoming energy are most suited to different types of storage.

*Mechanical Energy Storage* (includes electricity storage in batteries, hydrogen storage, and compressed air storage):

> *Southwest*: The mechanical energy load that needs storage for use at specific times of the day falls under the headings Lighting & Small Power and Miscellaneous Small Power (Table IX), which total 2,850 kw. If it is assumed that these "specific time requirement" loads are used equally throughout the year, about 8 kw-hrs/day would be needed. If wind power is supplying the mechanical energy it would be wise to allow for three windless days storage, or 24 kw-hrs. The total capacity of a 120-volt battery system would have to be at least 200 ampere-hours $\left( \dfrac{24\text{-kw-hrs}}{120 \text{ volts}} \right.$ ; see the section, Batteries). This system could consist of twenty sets of heavy-duty 6-volt batteries (or ten sets of 12-volt batteries) in series with each set consisting of two 100-amp-hr batteries in parallel.

An alternative setup could consist of four heavy-duty 450-amp-hr 12-volt batteries of about 5.4 kw-hrs each. These large industrial batteries cost about $350 apiece and will last at least ten years. Using a 120-volt DC system is a very practical method if the windplant is not too far from the energy loads, since DC power transmission losses are quite high. Lights and heating appliances can run off DC current, and many radios, stereos, small appliances, and motors can be converted to run off DC. Where AC current is needed, a small ½ kw DC-to-AC inverter can be used.

> *Midwest*: The mechanical storage is assumed to be compressed air. The mechanical energy loads requiring storage—Lighting & Small Power and Miscellaneous Small Power—total 950 kw-hrs/yr or 2.6 kw/day. For three days a total

TABLE IX
ENERGY NEEDS AND RESOURCES OF TWO HYPOTHETICAL SETTLEMENTS

| Energy Use | Energy Need Total/Yr (kw-hrs) | | Source | | | |
| | | | Primary | | Secondary | |
| | SW | MW | SW | MW | SW | MW |
|---|---|---|---|---|---|---|
| Refrigeration | 600 | 200 | Wind | Wind | Sol | Org |
| Irrigation | 13,050 | 0 | Wind | — | Org | — |
| Grinding, Mixing | 231 | 116 | Wind | Wind | Org | Org |
| Domestic Water Pump | 250 | 50 | Wind | Wind | Sol | Org |
| Lighting & Small Power | 2,550 | 850 | Wind | Wind | Org | Org |
| Misc. Small Power | 300 | 100 | Wind | Wind | Org | Org |
| Space Heat | 108,000 | 47,000 | Sol | Sol | Org | Org |
| Water Heat | 26,900 | 9,400 | Sol | Sol | Org | Org |
| Cooking | 1,200 | 400 | Org | Org | Sol | Sol |
| Cultivation | 5,600 | 596 | Org | Org | Wind | Wind |
| Transportation | 15,000 | 7,500 | Org | Org | Wind | Wind |

NOTE: SW = Southwest Rural Community; MW = Midwest Homestead
See text for full explanation.

of 7.8 kw-hr needs to be stored. From the section, Compressed Air, 5,400 ft³ of air stored at 160 psi would be enough to store 2 hp. hrs. (1.5 kw-hrs). Therefore, to store 7.8 kw-hrs would require $7.8/1.5 \times 5400$ ft³ $= 28,080$ ft³ stored at 160 psi. The tank needed for this would be about 2,650 ft³ $(28,080 \times 14.7/160)$, or a cube tank 14 feet on a side.

*Heat Energy Storage*: Assume that one should store enough heat energy to take care of three cloudless days in the coldest month (January).

*Southwest*: The space heat and water heat energy load per day in January for Albuquerque is $10^6$ BTUs (see the section, Physical Plant). For three days the total to be stored is $3 \times 10^6$ BTUs.

*Midwest*: Using the same assumptions as above, the average heat load for three days for the coldest month (January) in Madison is $33 \times 10^5$ BTUs. Storage reserves for both settlements are indicated in Table X from "20 degree" data in Table V.

*Chemical Storage*: The storage of methane, alcohol, or wood is much easier than for the other forms of energy. Methane can be compressed into cylinders or stored in inner tubes, floating water tanks, or, porous materials like zeolites. Alcohol can be stored in tanks and wood can be piled in a shed.

TABLE X

STORAGE RESERVES FOR 3 DAYS OF SPACE AND WATER HEAT FOR SOUTHWEST (SW) AND MIDWEST (MW) SETTLEMENTS

| Material | Ft³ Necessary | Tank Size (Ft. on a Side) |
|---|---|---|
| Water: | | |
| SW | 2,400 | 13.4 |
| MW | 2,640 | 14 |
| Rocks: | | |
| SW | 4,200 | 16 |
| MW | 4,600 | 17 |
| Salts: | | |
| SW | 280 | 6.5 |
| MW | 307 | 7 |

The actual amount of storage needed for each form of energy is very difficult to estimate. A great deal depends on the maximum number of windless or sunless days that can be expected. Sometimes the Weather Bureau will have this information or at least records of wind speeds and cloud cover from which the greatest probable number of windless and sunless days can be calculated. Other sources would be local airports and local newspapers. In most situations, the cost of storage has to be weighed against the cost of conventional standby generators and heaters run on locally produced fuels.

### LOAD SHARING AND INTERCONNECTION

Figure 7 presents most of the possibilities of load sharing. An important point to remember is that no machine or device is 100 percent efficient in converting one form of energy into another. Figures 8 and 9 show some very approximate efficiencies of various energy-related devices.

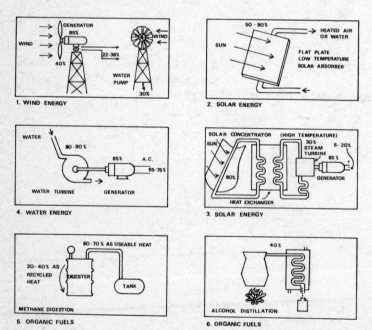

FIGURE 8. Some approximate efficiencies for alternative energy conversion processes.

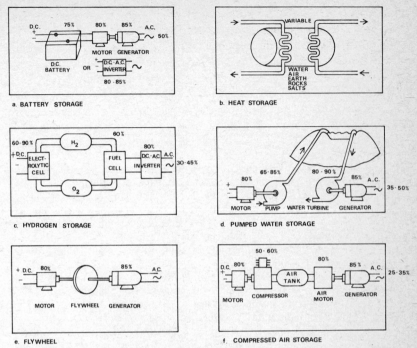

FIGURE 9. Some approximate efficiencies for alternative energy storage processes.

It is obvious from these examples that the local energy resources of separate locales may be very different. The Southwest community has an abundance of solar energy and it should be used to a greater extent to compensate for the low wind-speed and the inability to raise many animals or to grow sufficient wood on the semiarid land. In this example, experimentation with solar gas-absorption refrigeration devices, Stirling hot-air engines, solar steam generators, photovoltaic cells, thermoelectric devices, and Savory solar water pumps would be very valuable.

In the Midwest the solar insolation is low but this part of the country has high average wind speeds. A larger electric wind plant than the one designed could be used for water heating, space heating, or charging electric cars and tractors. In other words, the potential of a geographical area in the different energy forms should be utilized in the proportions to which they are available. The demands of a region are as variable as the supply, and the matching of the two is the art involved in planning and building an alternative power system.

## Notes

1. Farrington Daniels, *Direct Use of the Sun's Energy* (New Haven: Yale Univ. Press, 1955).

2. A British thermal unit (BTU) is the heat required to raise the temperature of one pound of water one degree Fahrenheit. Energy (the ability to do work) equivalents are: 1 kw-hr = 3,413 BTU = 1.341 hp-hr; 1 BTU = $2.93 \times 10^{-4}$ kw-hr; and 1 therm = $10^5$ BTU. Power (the rate of work) equivalents are: 1 kw = .948 BTU/sec = 1.341 hp.

3. Heating & Ventilating Reference Data, April 1954.

4. E. Speyer, "Optimum Storage of Heat with a Solar House," *Solar Energy* (1959): 3–4.

5. H. C. Hottel, "Solar Energy for Heating," in *Mechanical Engineers' Handbook*, L. S. Marks, ed. (New York: McGraw-Hill, 1941).

6. Speyer, "Optimum Storage of Heat."

7. Hottel, *Solar Energy for Heating.*

8. C. D. Davis and R. I. Lipper, "Solar Energy Utilization for Crop Drying," in *New Sources of Energy*, vol. 5, *Solar Energy II*, Proceedings U.N. Conference, 1961; F. H. Buelow, "Drying Crops with Solar Heated Air," in above; G. O. Löf, "Solar Energy for the Drying of Solids," *Solar Energy* 6 (1962): 122–128.

9. E. N. Fales, "Windmills" in *Mechanical Engineers' Handbook*, L. S. Marks, ed. (New York: McGraw-Hill, 1941).

10. Ulrich Hutter, "The Aerodynamic Layout of Wind Blades of Wind-Turbines with High Tip-Speed Ratio," in *New Sources of Energy*, vol. 7, *Wind Power*, Proceedings U.N. Conference, 1961.

11. Hans Meyer, "Wind Generators," *Popular Science*, November 1972.

12. M. Jacobs, "Experience with Jacobs Wind-Driven Electric Generating Plant, 1931–1957," in *New Sources of Energy*, vol. 7, *Wind Power*, Proceedings U.N. Conference, 1961.

13. E. W. Golding, *The Generation of Electricity by Wind Power* (New York: Philosophical Library, 1955); Brace Research Institute, *How to Construct a Cheap Wind Machine for Pumping Water* (McGill University, Ste. Anne de Bellevue 1800. Quebec, Canada); M. A. Hackleman, *Wind and Windspinners* (Saugus, Calif.: Earthmind, 1974).

14. R. Ramanthan, "The Economics of Generating Electricity from Wind Power in India," *Indian Journal of Economics* 50 (1970): 198.

15. S. Visher, *Climatic Atlas of the United States* (Cambridge, Mass.: Harvard Univ. Press, 1954).

16. U.S. Dept. of the Interior, Bureau of Reclamation, *Design of Small Dams* (Washington, D.C.: Government Printing Office, 1960).

17. Hans W. Hamm, *Low-Cost Development of Small Water-Power Sites*, (Volunteers in Technical Assistance, 3706 Rhode Island Ave., Mt. Rainier, Md. 20822); C. D. Bassett, "Your Own Water-Power Plant," parts 1–4, *The Mother Earth News* (P.O. Box 38, Madison, Ohio 44057), January–March 1972; Donald Marier, "Measuring Water Flow," *Alternative Sources of Energy* 1 (July 1971).

18. Harris and Rice, *Power Development of Small Streams* (Orange, Mass.: Rodney Hund Machine Co., 1920).

19. Bassett, "Your Own Water-Power Plant."

20. Hamm, "Low-Cost Development."

21. R. Saunders, "Harnessing the Power of Water," in *Energy Primer* (Menlo Park, California: Portola Institute).

22. K. Imhoff and M. Gordon, *Sewage Treatment* (New York: John Wiley & Sons, 1940).

23. L. John Fry, Richard Merrill, and Yedida Merrill, *Methane Digesters for Fuel Gas and Fertilizer*, Newsletter #3 (Pescadero Calif.: New Alchemy Institute, 1973).

24. Hans Thirring, *Energy for Man* (New York: Greenwood Press, 1968).

25. Phil Carabateas, "More on Alcohol and Wood Gas," *Alternative Sources of Energy* 10 (March 1973).

26. Egon Glesinger, *The Coming Age of Wood* (New York: Simon and Schuster, 1949).

27. George Smith, *Storage Batteries*, rev. ed. (New Rochelle, N.Y.: Soccer Assocs., 1971); C. L. Mantell, *Batteries and Energy Systems* (New York: McGraw-Hill, 1970).

28. "Hydrogen: Likely Fuel of the Future," *Chemical and Engineering News* (26 June 1972).

29. Ibid.

30. Farrington Daniels, "Energy Storage Problems," in *New Sources of Energy*, vol. 1, *General Sessions*, Proceedings U.N. Conference, 1961.

31. F. Bacon, "Energy Storage Based on Electrolyzers and Hydrogen-Oxygen Fuel-Cells," in above.

32. Ramakumar et al., *A Wind Energy Storage and Conversion System for Use in Underdeveloped Countries*, in Fourth Intersociety Energy Conversion Conference, American Institute of Chemical Engineers, 1969.

33. Compressed Air and Gas Institute, *Compressed Air Handbook* (New York: C.A.G.I., 1947); L. S. Marks, ed., *Mechanical Engineers' Handbook* (New York: McGraw-Hill, 1916).

34. Ibid.

35. D. Scott, "Compressed Air 'Stores' Electricity," *Popular Science*, November 1972.

36. H. C. Hottel, *Residential Uses of Solar Energy*, in Proceedings of the World Symposium on Applied Solar Energy, Phoenix, Ariz. (Menlo Park, Calif.: Stanford Research Institute, 1955).

37. Speyer, "Optimum Storage of Heat."

38. Ibid.

39. Hohenemser and McCaull, "The Windup Car," *Environment* 12, no. 5 (June 1970).

40. Golding and Thacker, *The Utilization of Wind, Solar Radiation and other Local Energy Resources for the Development of a Community in an Arid or Semi-Arid Area*, in New Delhi Symposium on the Use of Wind Power and Solar Energy in Arid Areas, 1954.

41. Golding, *Generation of Electricity*.

42. U. Hutter, *Planning and Balancing of Energy of a Small-Output Wind Power Plant*, in New Delhi Symposium on the Use of Wind Power and Solar Energy in Arid Areas, 1954.

43. Ibid.

44. Stephen Willey, "A Practical Self-Contained Power System for Bus or Cabin," *Lifestyle* 2 (December 1972).

45. R. C. Schlichtig and J. A. Morris, Jr., "Thermoelectric and Mechanical Conversion of Solar Power," *Solar Energy* 3, no. 2 (April 1959).

46. Golding and Thacker, *Utilization of Wind*.

47. M. Perelman, "Farming with Petroleum," *Environment* 14, no. 8 (October 1972): 8–13.

48. Fry et al., "Methane Digesters."

49. Ibid.

50. T. G. Hicks, *Pump Selection and Application* (New York: McGraw-Hill, 1957).

51. Speyer, "Optimum Storage of Heat."

52. Willey, "Practical Self-Contained Power System."

## References

*Alternative Sources of Eneregy*, 8 (January 1973). Special issue on wind power. A.S.E., Rt. 1, Box 36B, Minong, Wisc. 54859.

Bossel, Hartmut. *Low-Cost Windmill for Developing Nations.* Volunteers For International Technical Assistance (VITA), 3706 Rhode Island Ave. Mt. Rainier, Md. 20822.

Clegg, D. *New Low-Cost Sources of Energy for the Home.* Charlotte, Vt.: Garden Way Publishing Co.

Golding, E. W. "Power from Local Energy Sources," in *New Sources of Energy*, vol. 1, *General Sessions.* Proceedings U.N. Conference, 1961.

Leckie, J. et al. *Other Homes and Garbage: Designs for Self-Sufficient Living.* San Francisco, California: Sierra Book Club, 1975.

Leo, B. S., and Hsu, S. T. "A Simple Reaction Turbine as a Solar Engine." *Solar Energy* 4, no. 2 (April 1960).

McColly, H. F., and Martin, J. W. *Introduction to Agricultural Engineering.* New York: McGraw-Hill, 1955.

Meyer, Hans. "Wind Energy," in *Domebook II*. Pacific Domes, Box 279, Bolinas, Calif. 94924.

Merrill, R. et al. *Energy Primer: Solar, Water and Wind and Biofuels.* Menlo Park, Calif.: Portola Institute, 1975.

Putnam, P. C. *Power from the Wind.* New York: Van Nostrand, 1948.

Schurr and Netschert. *Energy in the American Economy, 1850–1975.* Resources for the Future, Inc. Baltimore: Johns Hopkins Press, 1960.

Steadman, D. *Energy, Environment and Building.* New York: Cambridge University Press.

# Index